The Health of Aging Hispanics
The Mexican-Origin Population

The Health of Aging Hispanics
The Mexican-Origin Population

Jacqueline L. Angel
The University of Texas at Austin
Austin, Texas

Keith E. Whitfield
Duke University
Durham, North Carolina

 Springer

Jacqueline L. Angel
LBJ School of Public Affairs and
Department of Sociology
University of Texas at Austin
Austin, TX 78713-8925
USA
jangel@mail.utexas.edu

Keith E. Whitfield
Department of Psychology and
    Neuroscience
Duke University
Durham, NC 27708-0086
USA
kwhit1@duke.edu

Cover illustration: From the project "Retrato de Familia" © Lourdes Almeida.

Library of Congress Control Number: 2006938805

ISBN-10: 0-387-47206-1   eISBN-13: 978-0-387-47208-9
ISBN-13: 978-0-387-47206-5   e-ISBN-10: 0-387-47208-8

Printed on acid-free paper.

Printed in the United States of America.

9 8 7 6 5 4 3 2 1

springer.com

Contents

Preface.. viii

Acknowledgments.. xi

List of Contributors xii

1. Setting the Stage: Hispanic Health and Aging in the Americas.......... 1
   Jacqueline L. Angel and Keith E. Whitfield

Section 1. Exploring the Health Consequences of Hispanic Mortality

2. Health Status of Elderly Hispanics in the United States................... 17
   Alberto Palloni

3. Census Disability Rates Among Older People by Race/Ethnicity and
   Type of Hispanic Origin 26
   Kyriakos S. Markides, Karl Eschbach, Laura A. Ray, and
   M. Kristen Peek

4. Disability and Active Life Expectancy of Older U.S.-and Foreign-Born
   Mexican Americans.. 40
   Karl Eschbach, Soham Al-Snih, Kyriakos S. Markides, and
   James S. Goodwin

5. Predictors of Decline in Cognitive Status, Incidence of
   Dementia/CIND and All-Cause Mortality in Older Latinos: The Role
   of Nativity and Cultural Orientation in the Sacramento Area Latino
   Study on Aging.. 50
   Mary N. Haan, Vivian Colon Lopez, Kari M. Moore, and Hector
   M Gonzalez, Kala Mehta, and Ladson Hinton

6. Education and Mortality Risk Among Hispanic Adults in the
United States 65
Sarah A. McKinnon and Robert A. Hummer

7. Does Longer Life Mean Better Health? Not for Native-Born Mexican
Americans in the Health and Retirement Survey 85
Mark D. Hayward, David F. Warner, and Eileen M. Crimmins

Section 2. Contextualizing Support and Mexican-Origin Health in Old Age:
Issues of Family, Migration, and Income

8. Dynamics of Intergenerational Assistance in Middle- and
Old-Age in Mexico 99
Rebeca Wong and Monica Higgins

9. Aging, Health and Migration: The Voices of the Elderly
Poor in Mexico 121
V. Nelly Salgado de Snyder

10. Aging and Health Interrelations at the United States-Mexico
Border 134
Roberto Ham-Chande

11. *Colonias*, Informal Homestead Subdivisions, and Self-Help
Care for the Elderly Among Mexican Populations in the
United States 141
Peter M. Ward

Section 3. Access to Health Care Services Among Elderly Hispanics
with Special Reference to Mexican Americans

12. Disparities and Access Barriers to Health Care Among Mexican
American Elders 165
Fernando Treviño and Alberto Coustasse

13. Lack of Health Insurance Coverage and Mortality Among Latino
Elderly in the United States 181
Rogelio Saenz and Mercedes Rubio

14. Access Issues in the Care of Mexican-Origin Elders:
A Clinical Perspective 195
David V. Espino and Liliana Oakes

15. Cross Border Health Insurance and Aging Mexicans and
Mexican Americans 202
David C. Warner

16. Cultural Myths and Other Fables About Promoting Health in
    Mexican Americans: Lessons Learned from Starr County
    Border Health Intervention Research.. 211
    Sharon A. Brown

17. Health Insurance Coverage and Health Care Utilization
    along the U.S.—Mexico Border: Evidence from the Border
    Epidemiologic Study of Aging... 222
    Elena Bastida, H. Shelton Brown, and José A. Pagán

Section 4. Options for Health Care Reform

18. Impoverishing and Catastrophic Household Health Spending
    Among Families with Older Adults in Mexico:
    A Health Reform Priority.. 237
    *Felicia Marie Knaul, Héctor Arreola-Ornelas, Oscar
    Méndez-Carniado, and Ana Cristina Torres*

19. The Health Care Safety Net for Hispanics.................................. 263
    Ronald J. Angel and Jacqueline L. Angel

Afterword   Globalization and Health: Risks and Opportunities for the
Mexico—U.S. Border... 280

    Julio Frenk and Octavio Gómez-Dantés

Index 289

About the Authors 293

Preface

The emergence of the Hispanic population in the United States and the increasing profile of Latin America on the North and South American continents have raised the profile of "Latino" issues in the American consciousness. Much of this attention, however, has revolved around trends in culture, art, food and music or pathological concerns about poverty, crime, drugs, and immigration. At best, the scholarly and policy assessments of the implications of the impact of Latinos in the United States revolve around Mexican-Americans, Puerto Ricans, and Central Americans (as well as the children of an older Cuban population) and their struggles to achieve a measure of education, health care, employment, and civil rights. What have not been adequately addressed are the demographic trends affecting Latinos in the United States, Mexico, and Latin America, particularly the aging of the Hispanic population and its resultant health and long-term care needs. That Latinos are living longer and shifting from a youthful population to an increasingly older cohort is little known, except among the small cadre of Hispanic gerontologists.

Ironically, this invisibility of aging Hispanics is juxtaposed to the almost panicked response of the U.S. media and public toward the aging of the general U.S. population and the fiscal and political controversies over Social Security, Medicare, and health care costs. The policy debates of an aging U.S. population as a whole arise from substantial research and policy analysis—a level of attention not seen with regard to the aging of the Hispanic population. This book provides an invaluable and substantive contribution to our recognition that Latinos as a subgroup are also aging and that their health needs present serious challenges as well as opportunities. Perhaps for the first time we have a foundation of qualitative, empirical, and analytical assessment of the myriad health concerns of Latino elders and aging Hispanic populations, not just in the United States but in Mexico and Latin America as well. Although there exists a body of literature on Hispanic elders in the United States, few of these studies or publications offer the breadth and depth contained in this book.

The book's authors and contributors present a sophisticated examination of the complex and multifaceted health concerns facing Latino elders and their families and caregivers with a bi-national treatment. This potential paradigm shift in how we view the aging of Hispanics alongside the aging of all groups in North, Central, and South America will constitute the basis for greater attention and study in this crucial arena of social inquiry.

The coming nexus of aging and diversity faced by the United States will require a greater level of scrutiny and analysis if we are to provide policy solutions to aging and health care in the United States and, by example, in Mexico and throughout Latin America. The problems and challenges are daunting. This book addresses many of them within the context of Hispanic health and Hispanic health care policy.

These issues cover a gamut of concerns—disability, health consequences of mortality risks, intergenerational relations, migration, long term care, health disparities, insurance coverage, and health care policy—and examine the issues facing Latino elders in this country, as well as along the U.S.–Mexico border, and in Mexico and other Latin American countries. The complexity of this in-depth overview and recognition of the need to create a new paradigm for such research and policy analysis is reflected in the comprehensive and hopeful views of the book's contributors. While nearly all research on Hispanic aging looks at the problematic status of more Latinos in the United States and more Hispanics becoming older in this country and throughout the hemispheres, this book addresses the advantages and opportunities that the nexus of aging and Latinization presents. Latino elders today, for example, represent the traditions, values, and ethics that have allowed generations of immigrants to the United States to persevere and become successful members of U.S. civil society. First-generation immigrants to the United States generally have better health and mental health status than native-born Mexican-Americans and Whites (although acculturation to U.S. lifestyles erodes those positive health indicators). The relative youthfulness of Latinos in the United States, as well as immigrants from Mexico and Latin America, enables this country to remain relatively young and energetic compared with other developed nations, such as Japan, Germany, France, and Korea.

Yet, despite the strengths that Latinos bring to the table and the absolute need to investigate, understand, and promote those traditions, values, and behaviors, we cannot ignore the serious health and social concerns that create obstacles to the full civic participation of Latinos in the United States and a healthy aging for Latinos in Mexico and Latin America. Nevertheless, the United States will, in due time, find that Latinos are both its largest minority group and a major source of manpower and economic vitality. The growing Latino workforce will be a crucial foundation of productivity and taxes for an aging White population. The extent to which we invest in their health care, education and employment opportunities will determine the extent to which this diverse workforce will shoulder those expectations and responsibilities. Mexico and other Latin American countries are finding that, while they too benefit from increased longevity, they are not well positioned to address the needs of their elderly populations, much less the needs of their youth. These countries face the irony of much of the developing world: They are becoming old before they become rich while the United States and Europe became old after they became wealthy. These multiple demographic, social and economic challenges will tax the political and intellectual ability of Mexico and Latin America to cope with the aging of their populations.

The Health of Aging: The Mexican Origin Population Hispanics provides a prism and the initial elements of an enlightened conceptual framework that can guide researchers, policy makers, and the citizens of the Americas to chart their own demographic journeys and experience the revolutionary opportunity to live long and well. How we respond to the nexus of aging and diversity and how we draw on the global avenues of trade, communications, and information to learn from each other and develop partnerships with North, South, and Central America can determine the state of the future aging and retirement of us all. We are at the doorstep of dramatic demographic changes, and the next 20 to 30 years will reveal how well we handle the health needs of Hispanics and how effectively we create a more compassionate and productive aging for all individuals, regardless of race, country, and economic circumstances.

Fernando Torres-Gil
Director
UCLA Center for Policy Research on Aging

Acknowledgments

This volume would not have been possible without the incredible support of many people. We thank the following individuals for their assistance with reviews of the papers appearing in this volume: Terrence Hill, Michael Gerardo, Dan Hamermesh, and Roland Thorpe. Ms. Debra Moore provided administrative assistance throughout the project. In addition, we acknowledge the following sponsors for providing generous support and in-kind contributions for the Second Conference on Aging in the Americas (SCAIA): Key Issues in Hispanic Health and Health Care Policy Research, which was held on September 21–22, 2005, at the LBJ School of Public Affairs, Bass Lecture Hall, located on The University of Texas at Austin campus:

Center for Population Health and Health Disparities,
The University of Texas Medical Branch at Galveston

Janet F. Harte Lectureship in Population Issues, LBJ School of Public Affairs

Population Research Center, The University of Texas at Austin

Center on Population Health and Aging, The Pennsylvania State University

National Institute on Aging

National Alliance for Hispanic Health

Foundation for Insurance Regulatory Studies in Texas

Bridging Disciplines Program, The University of Texas at Austin

AARP Global Aging Program

Teresa Lozano Long Institute of Latin American Studies, Mexican Center,
The University of Texas at Austin

Task Force on Minority Issues in Gerontology, The Gerontological
Society of America

Center for Health and Social Policy, The University of Texas at Austin

Latin American Initiative, Office of the Executive Vice President and Provost,
The University of Texas at Austin

Lyndon Baines Johnson Library and Museum

List of Contributors

Soham Al-Snih, M.D., Sealy Center on Aging and Internal Medicine, University of Texas Medical Branch at Galveston, Galveston, Texas

Jacqueline L. Angel, Ph.D., LBJ School of Public Affairs and Population Research Center, The University of Texas at Austin, Austin, Texas

Ronald J. Angel, Ph.D., Department of Sociology and Population Research Center, The University of Texas at Austin, Austin Texas

Héctor Arreola-Ornelas, Fundación Mexicana para la Salud, Mexico City, Mexico

Elena Bastida, Ph.D., Department of Sociology and Center on Aging and Health, College of Social and Behavioral Sciences, The University of Texas–Pan American, Edinburg, Texas

Sharon A. Brown, R.N., Ph.D., Vice-President for Research, The University of Texas at Austin, Austin, Texas

H. Shelton Brown, III, Ph.D., School of Public Health University of Texas Health Sciences Center at Houston, Management and Policy Sciences Brownsville, Texas

Vivian Colón López, M.P.H., Department of Epidemiology, School of Public Health, University of Michigan, Ann Arbor, Michigan

Alberto Coustasse, M.D., M.B.A., Dr.P.H., School of Public Health, University of North Texas Health Science Center at Fort Worth, Fort Worth, Texas

Eileen M. Crimmins, Ph.D., USC/UCLA Center on Biodemography and Population Health, University of Southern California, Los Angeles, California

Karl Eschbach, Ph.D., Division of Geriatrics, Department of Internal Medicine, University of Texas Medical Branch at Galveston, Galveston, Texas

David V. Espino, Ph.D., Division of Geriatrics, University of Texas Health Science Center at San Antonio, San Antonio, Texas

Julio Frenk, M.D., Ph.D., Oficinas del C. Secretario (Office of the Secretary), Ministry of Health, Mexico

Octavio Gómez-Dantés, M.D., M.P.H., Office of Director General for Performance Evaluation, Ministry of Health, Mexico

Hector M. González, Ph.D., Department of Family Medicine, School of Medicine, and Institute of Gerontology, Wayne State University, Detroit, Michigan

Tonatiuh Gonzalez-Vasquez, Instituto Nacionál de Salud Pública (National Institute of Public Health, Mexico), Cuernavaca, Mexico

James Goodwin, M.D., Internal Medicine/Geriatrics, University of Texas Medical Branch at Galveston, Galveston, Texas

Mary Haan, Dr.P.H., M.P.H., Department of Epidemiology, School of Public Health, University of Michigan, Ann Arbor, Michigan

Roberto Ham-Chande, Ph.D., El Colegio de la Frontera Norte, Tijuana, Mexico

Mark Hayward, Ph.D., Population Research Center, The University of Texas at Austin, Austin, Texas

Mira M. Hidajat, M.A., Department of Sociology and Crime, Law and Justice, The Pennsylvania State University, University Park, Pennsylvania

Monica Higgins, M.A., Office of Population Research, Princeton University, Princeton, New Jersey

Ladson Hinton, M.D., Department of Psychiatry and Behavioral Sciences, University of California, Davis, Sacramento, California

Robert A. Hummer, Ph.D., Population Research Center, The University of Texas at Austin, Austin, Texas

Berenice Jáuregui-Ortiz, Instituto Nacionál de Salud Pública (National Institute of Public Health, Mexico), Cuernavaca, Mexico

Felicia Knaul, Ph.D., Mexico Ministry of Public Education and Mexican Health Foundation, Fundación Mexicana para la Salud (Mexican Foundation for Health), Mexico City, Mexico

Kyriakos Markides, Ph.D., Division of Sociomedical Sciences, Preventive Medicine and Community Health, University of Texas Medical Branch at Galveston, Galveston, Texas

Sarah A. McKinnon, M.P.H., Population Research Center, The University of Texas at Austin, Austin, Texas

Kala Mehta, D.Sc., School of Medicine, Division of Geriatrics, University of California San Francisco, San Francisco, California

Oscar Méndez-Carniado, Fundación Mexicana para la Salud (Mexican Foundation for Health), Mexico City, Mexico

Kari M. Moore, M.S., Department of Epidemiology, School of Public Health, University of Michigan, Ann Arbor, Michigan

Liliana Oakes, MD., Department of Family and Community Medicine, University of Texas Health Science Center at San Antonio, San Antonio, Texas

José A. Pagán, Ph.D., Department of Economics and Finance, The University of Texas-Pan American, Edinburg, Texas

Alberto Palloni, Ph.D., Center for Demography and Ecology and Department of Sociology, University of Wisconsin-Madison, Madison, Wisconsin

Kristen Peek, Ph.D., Division of Sociomedical Sciences, Preventive Medicine and Community Health, University of Texas Medical Branch at Galveston, Galveston, Texas

Laura Ray, M.P.A., Division of Sociomedical Sciences, Preventive Medicine and Community Health, University of Texas Medical Branch at Galveston, Galveston, Texas

Mercedes Rubio, Ph.D., Division of Adult Translational Research and Treatment Development, National Institute of Mental Health, Bethesda, Maryland

Rogelio Saenz, Ph.D., Department of Sociology, Texas A&M University, College Station, Texas

V. Nelly Salgado de Snyder, Ph.D., Salud Comunitaria y Bienestar Social Instituto Nacionál de Salud Pública (National Institute of Public Health, Mexico), Cuernavaca, Mexico

Ana Cristina Torres, M.P.P., Gender and Development Group, PRMGE, The World Bank, Washington, DC

Fernando Torres-Gil, Ph.D., School of Public Policy and Social Research, University of California-Los Angeles, Los Angeles, California

Fernando Trevino, Ph.D., M.P.H., School of Public Health, University of North Texas Health Science Center at Fort Worth, Fort Worth, Texas

Peter M. Ward, Ph.D., LBJ School of Public Affairs and Department of Sociology, The University of Texas at Austin, Austin, Texas

David C. Warner, Ph.D., LBJ School of Public Affairs, The University of Texas at Austin, Austin, Texas

David F. Warner, Ph.D., Carolina Population Center, The University of North Carolina at Chapel Hill, Chapel Hill, North Carolina

Keith E. Whitfield, Ph.D., Department of Psychology and Neuroscience Sciences, Duke University, Durham, North Carolina

Rebeca Wong, Ph.D., Maryland Population Research Center, University of Maryland, College Park, Maryland

1
Setting the Stage: Hispanic Health and Aging in the Americas

Jacqueline L. Angel and Keith E. Whitfield

1.1. Introduction

Hispanics are redefining the character and look of United States of America. The rate of high immigration from Mexico and other Latin American countries and the higher fertility by native-born Americans has made Hispanic Americans the fastest growing demographic group in the country. Hispanics now outnumber African Americans as the nation's largest minority category. From 2000 through the end of 2001, the Hispanic population grew to 37 million, increasing 4.7 percent, while the black population increased by only 2 percent during the same period, to 36.1 million (U.S. Census Bureau, 2003). Hispanics account for nearly 13% of the U.S. population, which now numbers 284.8 million (U.S. Census Bureau, 2003).

The population explosion of Hispanics in this country is even more complex when one considers individual age. As in a population at large, the elderly are the fastest growing segment of the Hispanic-American population. The number of elderly Hispanics age 65 and older is projected to surpass the number of elderly African Americans by nearly one million by 2030, and by 2050 the number of Latinos in this age bracket will almost double to over 12 million (Angel and Hogan, 2004). Many of these individuals suffer from mental as well as physical problems. The oldest elderly (or "old") Hispanics, those 85 and older, will number 2.6 million by 2050. Providing cost-effective and appropriate medical care to this population will require a clear understanding of their unique needs.

Similar to many of today's older Hispanics, a large proportion of the next decades' Hispanic elders will be poorly educated, with limited economic resources. Among the major race-ethnic groups, older Hispanics, especially persons of Mexican ancestry, who constitute nearly 50 percent of the Hispanic-American older population, have particularly low levels of education, and the highest rates of poverty (Angel and Angel, 1998; Angel, Angel, Lee, and Markides, 1999). As a result, many of tomorrow's older Hispanics' health and long-term care medical needs will require critical attention by public policy makers.

The situation is particularly salient for individuals of Mexican descent approaching retirement, many of whom have spent the majority of their lives outside the United States, toiling in harsh, dangerous conditions for very low pay (Jasso and

Rosenzweig, 1990). Some of these workers have been exposed to environmental health hazards caused by water and air pollution from manufacturing industries, such as maquiladoras in the U.S.–Mexico borderlands have produced (Williams, 1996). The mixture of such physically demanding work, unsafe health, poor remuneration, and safety hazards in factory plants pose serious health risks, resulting in pervasive and chronic public health problems along the U.S.–Mexico border (Becerril, Harlow, Sánchez, and Sánchez Monroy, 1997; Homedes and Ugalde, 2003). Some scholars attribute the threats to public health and safety to relaxed rules and regulations under U.S. international trade agreements like the North American Free Trade Agreement (NAFTA) and the WTO's General Agreement on Trade and Services (GATS). For instance, these policies have led to adverse environmental conditions in Mexican communities, putting workers at risk for serious health problems, including a high incidence of multiple chronic diseases and related disabling chronic conditions requiring long-term care services and supports (Shaffer et al., 2005).

These health conditions, which appear in midlife and are seen less in other populations, highlight health disparities between non-Hispanic whites and Hispanics in late life. Mexican Americans, for example, are at much higher risk for developing type-II diabetes than non-Hispanic whites (Haffner, Hazuda, Mitchell, Patterson, and Stern, 1991). In some Mexican American communities, diabetes affects one out of every two adults (Haffner, Hazuda, Mitchell, Patterson, and Stern, 1991).

These health disparities associated with Mexican American ethnicity are also linked to obesity (Angel and Angel, 2006). Other research suggests that adoption of Westernized lifestyles often associated with high-fat diets, unwanted daily stressors, and insufficient exercise routines are responsible for the higher rates of complications from the disease across the life course.

Arguably, a lifetime of low pay fails to build the financial resources such as those necessary to support Hispanic healthcare needs in the United States. Specifically, this observation: Elderly Mexican immigrants and individuals who never participated in the Social Security program (often related to citizenship status), or those who immigrated to the United States in middle age, face particular health problems but may lack the time needed to contribute to the system so that coverage for their treatment are available (Angel, Angel, Lee, and Markides, 1999). Analyses of the U.S. Census 1990 Public Use Microdata Sample reveal that foreign-born Mexican Americans aged 65 and over depend heavily on Supplemental Security Income or rely on family for support (Angel and Angel, 1997, p. 62). Conversely, a lower proportion of immigrants report that they participate in other social security programs, such as Old Age Survivors, Disability, and Health Insurance (OASDIHI).

The difficulties of elderly legal residents are exacerbated because they are much more likely to live in poverty than their citizen counterparts (Angel and Angel, 2006). Elderly immigrants typically retire on less than two-thirds of what all aged U.S. citizens receive—$950 and $1,438 per month, respectively (Friedland and Pankaj, 1997). In 1995, economically disadvantaged elderly immigrants had an annual income of $5,958 (roughly $497 per month), almost 95 percent of which was directly from SSI, the federal cash assistance program for the aged, blind, and disabled. In concert, these conditions help explain the large amount of federal public assistance necessary to cover the elderly population in the period preceding the Personal Responsibility and Work Opportunity Reconciliation Act (PRWORA) period and underscore the unmet need created by the consequences of welfare reform (Angel, 2003).

The exponential growth of the older population will require far-reaching economic and social adjustments in many countries in the Americas, where at least one-fifth of the population is now likely to be age 60 and over (Chamie, 2005). Although the populations of developed countries are aging more rapidly than those of developing nations, the latter, countries like Mexico (which, because of high fertility have young populations), find they too are facing the growing fiscal burden of caring for a growing number of older citizens (Frenk, 2004). The number of older Mexicans will increase by 227 percent over the next 30 years and will exacerbate the elderly dependency ratio. The current number of workers available to support current Mexican retirees is expected to fall to four working-age persons for each person 65 years or older early into the middle of the century (Ham-Chande, 1990). In the absence of developed Social Security systems, and the migration of younger people out of rural areas and even out of the country, traditional support systems are less able to perform their functions.

Mexico presents a unique and highly salient example. Nowhere in the world does a developing nation border directly on such a wealthy developed nation as the United States. While jobs and economic opportunities are limited in Mexico, they are more abundant in the United States, and the attraction for the young who seek opportunities is irresistible. This geographic and historical phenomenon has important implications for both nations. Although Mexico loses a large fraction of its potential labor force to the U.S., those workers represent an important source of income in the form of remittances. For the U.S., a high standard of living among citizens is buttressed by a cheap and docile labor pool. For those who study aging the process has another consequence: In Mexico the loss of the young leave many rural towns occupied by the elderly alone. For the United States much of the care of the elderly is provided by those same young workers who left Mexico the seek opportunities unavailable at home.

The effect of a graying population in Mexico is particularly significant for women (Palloni and Peláez, 2005). In 2000 the number of women exceeded men aged 60 or older by roughly 1.63:1. Because female life expectancy is higher than that for men, there are two to five times as many women as men at age 85 and, they are particularly vulnerable to financial hardship and the harmful health effects of aging (Wong, 2003). As the numbers of the oldest-old grow, Mexican women will increasingly rely on informal support such as they receive from family caregivers as health declines with increasing age. In addition, the potentially declining support ratios will also influence the capacity of formal support and the health care infrastructure in a context of dwindling resources, widening social inequalities, and constraints on macroeconomic conditions, including high unemployment. In addition, evidence shows that out-of-pocket expenditures are greater in family households with at least one elderly person and there is a more concentrated demand for health care services in such households (Frenk, 2004).

With an increasing proportion of the population falling into the adult and elderly age groups, the epidemiological profile of Mexico increasingly reflects the diseases and health problems of adults rather than of children (Gribble and Preston, 1993). Improved nutrition and living conditions in early life have reduced mortality rates among younger Mexicans, but longevity brings its own problems. For example, the oldest-old population is growing swiftly, and, with this growth, chronic and degenerative diseases such as cancer are becoming leading causes of death in Mexico (Frenk, 2004). In other words, Mexico is suffering from the "New World Syndrome" whereby individuals no longer die from poverty but from diseases of affluence.

1.2. Hispanic Aging and Immigrant Health: Selection or Socioeconomic Incorporation?

The exponential growth of the Mexican American population in the United States has been accompanied by a growing need to improve and protect the health of its members. Despite these demographic facts there is a significant dearth of information about the notable strengths and characteristics of aging Hispanic groups. After years of research, investigators have a grasp of some of the complex health-related issues and social and behavioral patterns that impact older Hispanic people. As Jasso (2004) and her colleagues note:

Despite overall improvements in health ... ethnic health disparities are inherently linked to immigration because ethnic identities are traced to the country of origin of an immigrant or his or her ancestors. The average healthiness of the original immigrants, the diversity in health status among immigrants, and the subsequent health trajectories following immigration both over the immigrants' lifetime and that of their descendants all combine to produce the ethnic health disparities we observe at any point in time. Identifying the determinants of the original health selection of migrants and the forces that shape health paths following immigration is critical to understanding ethnic health differences (p. 227).

Even though Mexican Americans have a generally poor socioeconomic profile, their mortality rates are lower than non-Hispanic whites in almost every age bracket (Markides and Coreil, 1986; Rosenwaike, 1987). Several studies document a potential Hispanic mortality advantage, one that focuses on social, cultural, genetic, and health care system factors (Jasso, Massey, Rosenzweig, and Smith, 2004). Research has animated the positive health profiles of Hispanics, supporting the idea of positive migrant selection (Hummer, Rogers, Nam, and LeClere, 1999; Vega, Kolody, Aguilar-Gaxiola, Alderete, Catalano, and Caraveo-Anduaga, 1998). Immigrants may benefit from positive initial selection or suffer from the challenges of cultural assimilation and socioeconomic incorporation, but their overall influences on patterns of Hispanic health in the U.S. are as yet unknown (Palloni and Arias, 2004).

Alternatively, sociologists have considered a set of interrelated propositions to explain how age at migration may affect Hispanic illness profiles, notably (1) an *assimilation hypothesis*, and (2) a *cumulative disadvantage hypothesis*. Several studies document a positive correlation between length of residence and socioeconomic incorporation as evidence to support the assimilation hypothesis (Angel, Buckley, and Sakamoto, 2001). The assimilation process begins with uprooting from the culture of origin, which leads to geographical displacement, and ends with the protracted and often difficult insertion of the migrant into a foreign culture. To the extent that migration severs supportive ties and increases an individual's isolation, it places the migrant at high risk of ill health (Lara, Gamboa, Kahramanian, Morales, and Bautista, 2005).

The cumulative disadvantage hypothesis provides an alternative explanation to the process of social integration (Dannefer, 2003). This perspective focuses on the persistence of barriers to immigrant adaptation. Rather than focusing on challenges easily overcome with time, scholars argue that immigrants are unlikely to lose their disadvantaged social or economic status. Duration of residence correlates not with increased integration, but with exposure to the detrimental effects of social and economic disadvantage. Although the cumulative disadvantage and cultural assimilation models operate on completely different assumptions, both approaches emphasize economic factors, such as income, education, and occupation. These relationships are complex, and a multifaceted array of variables often associated with lower-class membership increase the risk of frailty and poor health among aging Mexican Americans.

Fresh data shed light on the magnitude of health care disparities and the unique factors that affect Hispanic health in the United States and Mexico (Wong and Espinoza, 2004). Marked differences in health and economic profiles distinguish those who live in Mexico from those who live in the United States. For example, there are special risk factors distinct in the Hispanic populations, such as immigration occurring at the U.S.–Mexico border and other job related occupational hazards (Becerril et al., 1997). It is also widely recognized that migration, defined in terms of moving from one's culture of origin in middle and later life entails significant life events and numerous chronic strains that can undermine an older Mexican immigrant's general health and psychological well-being (e.g., Angel and Angel, 1992). Research examining the relationship between chronic stressors and poor health support this linkage among Hispanics on the cusp of retirement and beyond (Angel, Buckley, and Sakamoto, 2001; Santiago and Muschkin, 1996). Many foreign-born Mexican-origin elderly have poor health when they arrive in the United States and thus co-reside with adult children (Angel and Angel, 1997). But, as of yet, there is no consensus on the specific relationship between nativity and Hispanic ethnic identity, how it varies by group, or what role nativity and the timing of immigration play in generating racial/ethnic differences in health.

Limited research has also shown that the rapid growth of the older Hispanic population has serious implications for family elder care (Roberts and Latapi, 1997). Younger generations are facing the burden of caring for older generations. This also includes financial responsibilities for the young and old, because in both Mexico and the United States most elderly people live outside institutions designed specifically for their care (Palloni, Soldo, and Wong, 2002; Wallace, Levy-Storms, and Ferguson, 1995). Informal sources of emotional as well as material support, which include relatives, friends, and neighbors, are vital supplements to the available formal support systems, especially in situations in which the formal support system is underdeveloped. For this reason, innovative methods for overcoming personal and institutional obstacles in caring for Hispanic elderly parents will need to be developed.

1.3. Policy Considerations: Access to Health Services

The United States is alone among highly developed nations in how it does not provide free or low cost medical care to all citizens. Rather than relying on public funding for health care for all but the poor, the health care financing system relies on private health insurance provided primarily through employer-sponsored group plans. Put simply, in America health insurance is an employment benefit. As a consequence, those individuals who are disadvantaged in the labor market are also disadvantaged in terms of health insurance coverage. Statistics show that there are currently more than 45 million uninsured Americans, and this number has been increasing (Angel, Angel, and Lein, forthcoming). Individuals of Mexican origin are disproportionately represented among those who suffer labor market disadvantages, a fact that has serious negative consequences for their material well-being and health care access throughout the life course. Mexican American children and adults have the lowest rates of health insurance coverage of any group in the nation (Angel, Angel, and Lein, forthcoming).

Even after the age of 65 when they become eligible for Medicare, about 5 percent of elderly Mexican Americans do not participate in the Medicare program, and those who do are less likely than other groups to own supplemental Medigap plans to cover the costs of what Medicare will not pay (Angel, Angel, and Lein, forthcoming). The data reveal a distinct disadvantage in terms of health care coverage associated with immigrant status among Mexican-origin elderly people: Foreign-born Hispanics have lower levels of private employer-sponsored coverage than either born Hispanics of Mexican ancestry or non-Hispanics. Those who immigrated in adulthood are at particularly elevated risk of having inadequate insurance coverage (Angel and Angel, 2005).

Although researchers do not yet fully understand other reasons for the low rates of health insurance coverage among Mexican Americans, some aspects of the problem are clear. Such factors relate to regional concentration and labor market differences, poverty, transportation, immigration status, language difficulties, and other social and cultural barriers that increase the risk of inadequate coverage (Angel and Angel, 1996). Poverty is a key determinant of access to health care, although structural and cultural factors are important barriers to seeking medical treatment. Low wage jobs in the restaurant, hotel, cleaning, and other service sectors do not offer health insurance, or, even if they do, the premiums an employee must pay for family coverage make it an unrealistic luxury. Once again, the Mexican-origin population is overrepresented among those who cook our food, clean our offices and homes, and care for our children and gardens. They receive minimum pay and no benefits. Because of life-long labor force disadvantages, retirement-age Mexican Americans have far less wealth with which to buy health care services or long-term care than do non-Hispanic whites.

A specific example of Hispanic health care disparities is the underserved Mexican-origin elderly population in the Southwest. Young Mexican-origin individuals living in Texas face rather unique and serious risks for inadequate coverage (Angel, Angel, and Lein, forthcoming). Children of Mexican immigrants in Texas suffer significantly higher levels of hardship in the areas of food, health care, and housing relative to those in other states. One in every three Texas immigrant children lives in poverty compared to less than one in every four immigrant children nationwide. In South Texas the situation is even more serious. This area is characteristic of the indigent community presently residing at the junction of the U.S.–Mexico border. For the poorest Mexican-origin families and especially those headed by a single mother, Medicaid and the new State Children's Health Insurance Program (SCHIP) have served as the health safety net. However, for the working poor and for poor adults in general, a health care safety net simply does not exist.

1.4. A Summary of What's To Come

In the not too distant past the major health problems that developing countries like Mexico faced were those of young populations, such as infectious disease and high infant mortality. Developed countries like the United States, on the other hand, that were past the demographic transition experienced falling birth rates and declines in illness and deaths from these causes. Along with aging populations they experienced increasing morbidity and mortality from the diseases of aging, including heart disease and cancer. Because of the fact that the number of older individuals is growing rapidly, Mexico and other countries that still have relatively young populations are now finding that they must deal with the diseases of aging at the same time that they still have serious problems of infectious disease and other diseases related to poverty. As a consequence, policy makers throughout the Americas are having to deal with growing population health care needs, which are propelled by forces at both ends of the life course. The management of chronic disease in ongoing and expensive and aging populations require more resources be devoted to the care of the elderly, even as the young need vaccinations and treatment for acute illnesses. The collection of papers that follow provide new and useful information on the numerous factors affecting the health security of Mexican-origin families and the new challenges that communities and states face in providing care for children and the elderly simultaneously.

It is clear that both the United States and Mexico must confront the problems of growing numbers of older Mexican-origin individuals with all of their health and social support needs. The studies reported in this book offer information and analysis that will help both countries to better understand these problems and to begin formulating policy solutions. The research presented here may even identify problems or such strategies that will be fielded in cooperation between the U.S. and Mexico, given their interdependency.

In the first section of this book, population health scholars examine the health consequences of the apparent Hispanic mortality advantage across the life span (Markides and Eschbach, 2005). Several papers look at the positive and negative health effects of migration. Alberto Palloni probes the apparent Hispanic epidemiological paradox by examining competing explanations of migratory attributes linked to leading causes of morbidity and chronic illnesses. He seeks to unravel the mystery surrounding whether poor health and well-being is due to selective migration or to the health consequences associated with migration (i.e., socioeconomic incorporation and cultural assimilation). Sarah McKinnon and Robert Hummer augment that discussion with empirical analyses of the National Health Interview Survey (NHIS), linked with the National Death Index (NDI) to document a *true* health paradox in the Mexican American population. Reaching a similar conclusion, Mary Haan and the investigative team from the Sacramento Area Latino Study on Aging provide compelling evidence of a high risk of death and dementia due to type-II diabetes among native-born as opposed to foreign-born Mexican-origin elders. They note, however, that lifetime exposure to the environment associated with low socioeconomic status may increase the risk for both groups.

Two studies present their current analytic work on the prevalence of physical chronic conditions and disablement among older Mexican Americans. Employing census data, Kyriakos Markides and colleagues document that older Hispanics, as a group, have greater disability rates than older non-Hispanic whites and that rates are also slightly lower than for African Americans. Karl Eschbach and his associates identify the exact causes of the variations in active life expectancy for older Mexican American immigrants and natives, as well as social risk factors, such as education, acculturation, and neighborhood characteristics.

Mark Hayward and his biodemography collaborators employ data from the Health and Retirement Study (HRS) and find scant evidence of an Hispanic paradox with regard to disability. In fact, they find that middle-aged Mexican Americans are more likely than their non-Hispanic white peers to be disabled and to be disabled for longer periods of time compared to whites. Nelly Salgado de Snyder and colleagues examine the observed variability in long-term mental and physical health consequences of Mexican labor migration to the United States. Although the relationship between migration and health is mediated by financial help they receive from the U.S., subjective factors such as the value men place on their culture of origin (e.g., values toward family, freedom, land ownership) are also meaningful in successful incorporation.

In the second part of the book, researchers focus on several issues related to health and family welfare, including the extent to which the timing of immigration from Mexico affects how Hispanics grow old and their propensity to use informal and formal supports, and how these outcomes are associated with indicators of cultural and structural incorporation. In both Mexico and the United States most informed elderly people live outside institutions designed specifically for their care. As a result, they must rely on their families when they can no longer care for themselves. Comparisons of how institutional and social factors influence the situation of elderly Mexican Americans in the United States to elderly Mexican nationals in Mexico will provide new insights in areas where health and social service coordination is needed for Hispanic families.

Peter Ward examines how low-income Mexican households now living in the United States are going to care for their aged parents in the context of dramatic demographic change. Although remittances to Mexico to those siblings still living in Mexico already exist, he argues that the dynamics of transnational families are changing, giving rise to (1) a greater proportion of permanency in the United States; (2) dual nationality and large number of native-born Mexican origin citizens; (3) declining remittances as Mexicans focus on their U.S. home and family commitments; and (4) growing pressures for family reunification, namely, elderly Mexicans coming to live with their U.S. resident children and grandchildren. He offers new and valuable insights into how *colonias* and informal homestead subdivisions may be pathways to providing a source of housing supports for family elder caregivers.

It is vitally critical for researchers to understand the demography of the bi-national region constituting the U.S.–Mexico border and the existence of distinctions in health and illness behaviors so they may identify specific policy interventions. Roberto Ham-Chande examines transnational challenges like border health threats. He examines the constellation of trans-border issues on elderly health and the bi-national actions for the study and prevention of diseases, and the transfers of resources from the United States to Mexican households aimed at health care from kinship on both sides of the border.

Rebeca Wong and Monica Higgins present new evidence of the ways in which population aging affects intergenerational relationships, including the probabilities of financial contributions to parents as well as the effects of parents' aging on health status and medical care use. Their analysis of the Mexican Health and Aging Study reveals that the path of assistance seems either non-persistent or responsive to deterioration and improvement in health, and is driven by more than just cultural norms or older age. These results critically contribute to the literature demonstrating that children whose parents invest more in education, health, and other social capital in their upbringing will be more likely to make financial transfers to their parents, while children who received fewer opportunities will be more likely to make time transfers to their parents. Parents with low levels of wealth or health will receive more cash or in-kind assistance than parents with greater resources. Because established networks on both sides of the U.S.–Mexico border have made the migration process easier, families will tend to be less selective in choosing which family members should migrate and, because of a lack of selectivity, current levels of health and education of Mexican-born individuals in the United States will decrease compared to earlier migrants.

The third section includes papers addressing touchstones of the health care crisis as they pertain to the unique health care needs of the Hispanic population across the life course. Even though Medicare, the publicly funded health insurance program for elderly Americans, has eliminated many of the health disparities associated with old age, Hispanics are more likely than any other subgroup of the elderly population to be at risk for inadequate health insurance coverage. Mexican Americans are the most vulnerable group of Hispanic elderly who lack health insurance.

In their paper, investigators Rogelio Saenz and Mercedes Rubio document the constellation of factors associated with health insurance disparities for Mexican Americans in the Hispanic Established Populations for Epidemiologic Studies of the Elderly (H-EPESE). What they discover is a complex relationship that has a multifaceted array of factors giving rise to the fragility and health shocks faced by aging Mexican Americans. They find that the uninsured tend to be younger elderly, recent immigrants, those with greater economic difficulties, and those who are employed (perhaps working out of necessity). In-depth analyses reveal the serious consequences of these life-long disadvantages in health care coverage, which ultimately place older Mexican Americans at elevated risk of preventable health problems, such as the complications or sequelae of diabetes mellitus (e.g., amputation, blindness, hypertension, etc.) and a diminished quality of life.

Elena Bastida, Shelton Brown and José Pagán provide new evidence of the serious health consequences of such inadequate health care coverage based on a cohort study of older Mexican-origin adults residing along the Lower Rio Grande Valley. Analyses of illness behaviors point to a pattern of increased and expanded chronic diseases and socioeconomic problems as these cohorts advance in age. Sharon Brown's paper disproves the conception that Mexican Americans are genetically predisposed to developing type-I diabetes. Employing data collected from adult participants in Starr County, the poorest county in Texas and amongst the poorest in the United States, she proves the costs versus benefits of health education interventions for Mexican American family members reporting type-II diabetes. These findings have important policy implications that underline the need for a border health policy directed toward the unique health care needs of poor elderly Mexican Americans residing along Texas–Mexico border counties.

The issue is also examined from the position of a geriatrician, David Espino, who believes a greater understanding of the cultural differences coupled with increased cultural awareness would aid in the delivery of more efficient and appropriate care for Mexican-origin elders. Fernando Trevino and Alberto Coustasse continue the discussion of cultural issues in health care delivery to suggest that Hispanic health leadership development, empowerment and advocacy are crucial in developing a coherent course of action designed to 1) promote healthy behaviors and 2) reduce and ultimately eliminate barriers that limit the access to adequate health care for elderly Mexican Americans.

Finally, David Warner looks at policy alternatives in the absence of comprehensive national health insurance and identifies potential solutions to eliminate health and health care inequalities. He recommends that cross-border health insurance is just one of many alternatives that are under development by state government to care for retirees in Mexico.

In section 4, innovative strategies of eliminating health disparities and concrete solutions designed to protect and improve the health and social welfare of the Hispanic population are explored in great detail. The papers address the roles of states, markets, and non-governmental organizations in dealing with the health and welfare problems associated with Mexican ethnicity. Ronald Angel and Jacqueline Angel provide an informative summary capturing policy options of what is to be done about the growing number of uninsured and underinsured Hispanics with special emphasis on Mexican Americans. Toward that end, Felicia Knaul and associates provide a substantive and insightful overview of effects of the epidemiologic transition on the economic well-being of Mexican households. She presents analytic work documenting the high prevalence of out-of-pocket expenditures for catastrophic health care coverage borne by Mexicans. The findings are discussed in the context of recent health care reform in Mexico and the aging population. Finally, the discussion by Julio Frenk and Octavio Gómez-Dantés underscore many important issues related to the intense political debate on contemporary immigration and health care reform in the United States and in Mexico.

1.5. Conclusion

This collection of papers provides a unique, provocative, and stimulating discussion about the unique health care needs of older Hispanics, and it provides particularly useful insights into the situations of those residing along the Texas–Mexico border region. The overarching conclusions that can be made from the results of these studies suggest that while Mexicans and Mexican Americans are living longer lives than in the past, they may not be living healthier lives. Chronic mental illnesses and physical decline accompany aging for many Hispanics. Many suffer from disabling conditions like diabetes that take their toll over a lifetime and compromise health and well-being in later life. The evidence clearly shows that cultural, economic, and political factors associated with Mexican American ethnicity often restrict access to high quality health and chronic care services, in many cases interfering with access to effective prevention and management of disease.

The research reported in the following chapters also makes it clear that economic need and cultural factors affect the role of Mexican American adult children in providing help to their elderly parents. The data indicate that as a group Mexican-origin individuals who come to the United States when they are older remain heavily dependent on family members, often as a result of poor health and disability. Older immigrants possess few job skills and because of recent changes in immigration policy they do not qualify for social services for at least five years. During periods of low unemployment and relative economic prosperity some families with dependent elderly members may find it possible to bear the burden. In more difficult times these same families may be placed under greater financial and emotional strain and the needs of older family members may compete for limited resources with those of younger family members. Among groups that are struggling to survive and to accumulate assets, such a burden may hinder the social mobility of the entire family. For these reasons it is essential that provisions be made for building new partnerships inside and outside the Mexican and U.S. governments and non-governmental organizations in a bi-national setting with an eye toward the relevant health policy initiatives, such as transnational health care, and other issues affecting the U.S.–Mexico border. The knowledge will be of particular interest to those who wish to learn more about social policies and new institutional arrangements in an aging Hispanic population.

References

Angel, J.L. (2003). Devolution and the social welfare of elderly immigrants: Who will bear the burden? *Public Administration Review, 63,* 79–89.

Angel, J.L. (2001, February). Proceedings of *Policy Roundtable Aging in the Americas: Critical Social Policy Issue.* LBJ School of Public Affairs, The University of Texas at Austin, Austin, TX [Internet] http://www.utexas.edu/lbj/faculty/angel/latinfinal2.pdf

Angel, J.L., and Angel, R.J. (1992). Age at migration, social connections, and well- being among elderly Hispanics. *Journal of Aging and Health, 4,* 480–499.

Angel, J.L., and Hogan, D.P. (2004). Population aging and diversity in a new era. In K.E. Whitfield (Ed.),*Closing the gap: Improving the health of minority elders in the new millennium* (pp. 1–12). Washington, DC: The Gerontological Society of America.

Angel, J.L, Buckley, C.J., and Sakamoto, A. (2001). Duration or disadvantage? Exploring nativity, ethnicity, and health in midlife. *Journal of Gerontology: Social Sciences, 56,* S275–284.

Angel, R.J., Angel, J.L., and Lein, L. (forthcoming). The health care safety net for Mexican American families. In R. Crane and E.S. Marshall (Eds.), *Handbook of families and poverty: interdisciplinary perspectives.* Thousand Oaks, CA: Sage Publications.

Angel, R., and Angel, J.L. (2006). Diversity and Aging. In R.H. Binstock, and L.K. George (Eds.), *Handbook of Aging and the Social Sciences* (pp. 94–110). San Diego, CA: Academic.

Angel, R.J., Angel, J.L., Lee, G.Y., and Markides, K.S. (1999). Age at migration and family dependency among older Mexican immigrants: Recent evidence from the Mexican American EPESE. *The Gerontologist, 39,* 59–65.

Angel, R.J., and Angel, J.L. (1997). *Who will care for us? Aging and long-term care in multicultural America.* New York: New York University Press.

Angel, R.J., and Angel, J.L. (1996). The extent of private and public health insurance coverage among adult Hispanics. *The Gerontologist, 36,* 332–340.

Becerril, L.A., Harlow, S.D. Sánchez, R.A., and Sánchez Monroy, D. (1997). Establishing priorities for occupational health research among women working in the maquiladora industry. *International Journal of Occupational and Environmental Health, 3,* 221–230.

Chamie, J. (2005). *World population ageing: 1950–2050.* Population Division, Department of Economic and Social Affairs, United Nations Secretariat, New York, NY.

Dannefer, D. (2003). Cumulative advantage/disadvantage and the life course: Cross-fertilizing age and social science theory. *Journal of Gerontology,* 58b, S327–S337.

Frenk, J. (2004, May). Ageing and health. Paper presented at the OCDE Forum 2004 http://www.oecd.org/dataoecd/26/48/31860049.pdf#search='frenk%20out%20of%20 pocket%20expenditures%20elderly'

Friedland, R.B., and Pankaj, V. (1997). *Welfare reform and elderly legal immigrants.* Washington, DC.: Henry J. Kaiser Family Foundation.

Gribble, J.N., and Preston, S.H., (Eds.). (1993). *The epidemiological transition: Policy and planning implications for developing countries.* , Washington, DC: National Research Council, National Academy Press.

Haffner S., Hazuda H., Mitchell B., Patterson J., and Stern M. (1991). Increased incidence of type II diabetes mellitus in Mexican Americans. *Diabetes Care, 14,* 102–108.

Ham-Chande, R. (1990). The young and the elderly at the U.S. –Mex Border. In J.J. Kattakayam (Ed.), *Contemporary Social Issues* (pp. 81–89). Trivandrum: University of Kerala, Karyavattom.

Hummer, R., Rogers, R., Nam, C., and LeClere, F. (1999). Race/ethnicity, nativity, and U.S. adult mortality. *Social Science Quarterly, 80,* 136–153.

Jasso G, Massey D.S., Rosenzweig M.R., and Smith J.P. (2004). Immigrant health- selectivity and acculturation. In N.A. Anderson R.A. Bulatao and B. Cohen (Eds.), National Research Council, *Critical perspectives on racial and ethnic differences in health in later life* (pp. 227–266). Committee on Population, Division of Behavioral and Social Sciences and Education. Washington, DC: National Academy Press.

Lara, M., Gamboa, C., Kahramanian, M.I., Morales, L.S., and Hayes Bautista, D.E. (2005). Acculturation and Latino health in the United Status: A review of the literature and its sociopolitical context. *Annual Review of Public Health, 26,* 367–97.

Markides, K.S., and Eschbach, K. (2005). Aging, migration, and mortality: current status of research on the Hispanic paradox. *Journal of Gerontology: Social Sciences, 60,* S68–75.

Palloni, A., and Arias, E. (2004). Paradox lost: Explaining the Hispanic adult mortality advantage. *Demography, 41,* 385–415.

Palloni, A., and Peláez, M. (2005). Final report: Survey on health and well-being of elders in Latin America. http://www.ssc.wisc.edu/sabe/docs/ informeFinal%20Ingles%20 noviembre%202004.pdf

Palloni, A., Soldo, B.J., and Wong, R. (2002). Health status and functional limitations in a national sample of elderly Mexicans (Paper presented at the Population Association of America Conference, Atlanta, May).

Roberts, B.R., and Latapi, A.E. (1997). Mexican social and economic policy and emigration. In F.D., Bean, R. de la Garza, B.R. Roberts, and S. Weintraub (Eds.), *At the Crossroads: Mexican migration and U.S. policy* (pp. 47–78). New York: Rowman and Littlefield.

Shaffer, E., Waitzkin, H., Brenner, J., and Jasso-Aguilar, R. (2005). Global trade and public health. *American Journal of Public Health, 95,* 23–34.

U.S. Census Bureau. (2003). Census bureau releases population estimates by age, sex, race and hispanic origin [internet] http://www.census.gov/Press-Release/www/2003/ cb03-16.html

U.S. Census Bureau. (2004). *Global population at a glance: 2002 and beyond.* Demographic Programs, International Population Reports, Washington, DC. http://www.census.gov/ ipc/prod/wp02/wp02-1.pdf.

Vega, W.A., Kolody, B., Aguilar-Gaxiola, S., Alderete, E., Catalano, R., and Caraveo-Anduaga, J. (1998). Lifetime prevalence of DSM-III-R psychiatric disorders among urban and rural Mexican Americans in California. *Archives of General Psychiatry, 55,* 771–778.

Wallace, S.P., Levy-Storms, L., Ferguson, L.R. (1995). Access to paid in-home assistance among disabled elderly: Do Latinos differ from non-Latino whites? *American Journal of Public Health, 85,* 970–975.

Williams, E.J. (1996). The maquiladora industry and environmental degradation in the United States–Mexico borderlands. *St. Mary's Journal of Law, 27,* 765–815. http://americas.irc-online.org/borderlines/1998/bl47/bl47tab1.html

Wong, R. (2003). La relación entre condiciones socioeconómicas y salud en edad media y avanzada en México: diferencias por género, In N. Salgado and R. Wong (Eds.), *Calidad de vida y vejez en la pobreza: Una perspectiva de género* (pp. 97–1122). Instituto Nacional de Salud Publica, Mexico.

Wong, R., Díaz, J.J., and Espinoza, M. (2004). Health care use among elderly Mexicans in the U.S. and in Mexico: The role of health insurance. Paper presented at the Conference on "Changing Demographics, Stagnant Social Policies," Syracuse University Gerontology Center, May 2004; and at the Population Association of America Meetings, Philadelphia, April 2005.

Section 1
Exploring the Health Consequences
of Hispanic Mortality

2
Health Status of Elderly Hispanics in the United States

Alberto Palloni*

2.1. Introduction

Because the theme of this conference is Hispanic health, it is inevitable that the issue of the "Hispanic paradox" will again make an appearance. In this short intervention I offer a few reflections with two goals in mind. The first is to try to convince the reader that the *scientific* problem associated with the Hispanic paradox is unsolvable, at least by the means available. The best we can do is to put forward a set of plausible conjectures, none of which can be completely falsified by empirical evidence of the sort we can realistically collect. The most immediate conclusion from this appraisal is that it is too optimistic and/or premature to hope that by insisting on studying this problem we will either move the field forward or learn anything useful to illuminate the determinants of healthy aging. We may learn something along the way about the benefits of using some data sets over others, the role of some set of determinants, and the nature of some migration processes, but we will not answer the questions with which we started (this idea is posed more precisely below).

The second goal is to argue that a narrow focus on the epidemiological paradox is blindsiding and cornering us into an area that does not contain, and indeed may even conceal, the most pressing health problems that both young and adult members of the Hispanic population in the U.S. will face in the next 20 years or so. These problems are related to increasing health gradients, a turn for the worse in a number of childhood and adult health conditions, and, worse yet, the creation of preconditions that will reproduce inequalities and health problems among higher order generations.

If I accomplish these two goals I will have addressed at least two of the concerns of this session, namely, what factors make Mexican-origin adults in the U.S. vulnerable to leading causes of morbidity and chronic conditions; and what lessons can we learn from the epidemiological paradox.

---

* The "Hispanic paradox" (or epidemiological paradox) refers to the fact that Hispanic in the United States appear to do better than the rest of the population on a number of health outcomes.

2.2. Epidemiologic Paradox: An Unsolvable Conundrum

Below I consider the following issues: (a) What is the origin of the epidemiological paradox? (b) To whom does it apply? (c) What outcomes are relevant? and (d) How significant is it?

2.2.1. Origins

The idea that migrants from a given region tend to have better health and mortality than those who do not migrate is neither new nor confined to Hispanics in the U.S. There is a long history of epidemiological research on this subject, and most of it is discouragingly inconclusive with regard to the explanatory mechanisms (for a succinct summary see Palloni and Ewbank, 2005). The case for an Hispanic paradox in the U.S. began to unfold with findings that pertained to infant mortality in the area of San Antonio. The term "paradox" was used due to a belief that lower Hispanic (mostly Mexican) infant mortality levels were produced in a population whose composition by *presumed determinants of infant mortality* was unfavorable. We know now that at least some disparities among Hispanics (mostly Mexicans and foreign-born other Hispanics) are maintained even when one adjusts for composition by those factors. That is, that being a person of Mexican or of other Hispanic origin *and* foreign-born somehow confers an advantage over black and white non-Hispanic populations, irrespective of the value gained by those factors. Why this continues to be called a "paradox" is a mystery to me. It is surely an effect or difference to be explained, but not necessarily paradoxical. The advantage just referred to has turned up in a subset of birth outcomes (specifically birth weight), indicators of maternal health, some (not all) indicators of adult health (self-reported health status and chronic conditions), and adult mortality, particularly at older ages.

2.2.2. To Whom Does the Hispanic Advantage Apply?

Although the answer to this question depends on the outcome being examined, Mexicans and foreign-born other Hispanics consistently express a more favorable position relative to both non-Hispanic whites and blacks. Cubans tend to have a slightly better profile in some outcomes but are generally closely similar to non-Hispanic whites. Puerto Ricans, on the other hand, are at the other extreme and tend to have a worse profile, closer to that applicable to the non-Hispanic black population. Finally, outcomes for Mexicans and other Hispanics of the first generation (not foreign-born) converge with those of the non-Hispanic white population and, in some cases, at least (e.g., teenage pregnancy, obesity), are closer to those of the non-Hispanic black populations.

So, here is the first lesson we learned: *The epidemiological paradox is largely circumscribed to a handful of Hispanic groups and, among these, to those who are foreign-born.*

2.2.3. How Pervasive Is It?

Although it is difficult to make unambiguous inferences, I believe I am not mistaken to confine the epidemiological paradox to two and only two health outcomes: those outcomes associated with birth and the survival of infants during the first year of life, and male adult mortality. The evidence we have suggests that in a number of relevant health outcomes just about all Hispanic groups fare worse than their non-Hispanic white—though not non-Hispanic black—counterparts. These are: health status of children, adult and elderly self-reported health and chronic conditions (particularly diabetes, hypertension, obesity), and even mortality, most notably among women. In their paper on cognitive status, Haan and colleagues (Haan et al., 2005) show that crude effects of nativity, which appear to show a better cognitive status among foreign-born Hispanics, disappear altogether after adjustments for indicators of SES. In their paper on disability rates (with all the caveats regarding cultural biases affecting self-reports), Markides and colleagues (Markides et al., 2005) make the case that, from what we can gauge from Census data, older Hispanics fare worse than other groups, as do those who are both Hispanic and foreign-born. However, data from other sources (Arias, 2002) appear to indicate lower rates of morbidity and of functional limitations for *the entire Hispanic population*, suggesting that relative advantages and disadvantages are heavily age-dependent. Thus, for example, Hispanic minors have a slightly lower prevalence of developmental delays, attention deficit disorders, learning disabilities, and asthma than non-Hispanic whites; and Hispanics, including Mexicans, Cubans, and other Hispanics (native and foreign-born) have slightly lower rates of functional limitations than non-Hispanic whites (Ioanotta, 2002).

Finally, in a paper reporting on results using the Hispanic Established Populations for Epidemiologue Studies of the Elderly HEPESE (Patel et al., 2004) the authors conclude that once errors in matching of records are removed the mortality patterns of Hispanic women turn out to be worse than for the non-Hispanic white population, while an advantage is preserved only for males at all ages.

So, here is a second lesson we learned: *A clear-cut epidemiological paradox is largely circumscribed to a few and not necessarily the most important health outcomes, notably birth weight, male adult mortality, and, perhaps, broad health status at younger ages.*

2.2.4. How Significant Is It?

Let us assume for a moment that the advantage for the outcomes and groups previously noted are genuine. Namely, these outcomes are neither the artifact of data defects or measurement procedures or of the processes that may provide the appearance of an advantage when there is none. Are they significant, that is, do they have durable consequences? Consider the example of birth weight. Many researchers have made careers trying to explain why Hispanic mothers (especially Mexican-born mothers) tend to produce lower proportions of LBW babies than do non-Hispanic whites and blacks. Although the effect is statistically significant in most cases, the actual differences are not very large and may not be consequential at all for the future health status of infants. Thus, for example, the proportion of LBW babies among Hispanics is about 7% whereas that of non-Hispanic blacks is a bit over 13%. And Mexican Americans (not foreign-born Mexican migrants) have rates slightly lower than 7 percent. Whether or not these differences are significant hinges on the role played by birth weight in shaping outcomes. The most central is infant mortality but there might be others such as cognitive skills, ability to learn, physical appearance, stamina, creativity, industriousness, and other factors that are highly valuable in labor markets. But since the associated effects of birth weight are small, the differences in proportion-attributable risks in a particular outcome between the groups that appear to have an advantage and those that do not turn out to be minuscule and may be completely offset by the operation of other, more powerful determinants.

There is more to this idea, though. By focusing on LBW we conceal the fact that birth weight is a continuous variable that does have a right tail after all. In particular, it is known that excessively high birth weight may have as many if not more deleterious health connotations than LBW. On this score, Mexican and other Hispanic mothers do not fare well at all (Morenoff, 2002). And yet, this fact has received remarkably little attention.

With respect to adult mortality, we calculate that, adjusting for a number of factors, Mexicans and foreign-born other Hispanics enjoy an advantage of about 6 to 7 years of life in their life expectancy at age 35 (Palloni and Arias, 2004). Now, a 7-year difference is absolutely and relatively speaking quite large. But, if the findings by Markides and colleagues and those from Smith and Kington (1997) with HRS are approximately correct, it simply means that the excess number of years is lived with more disability and illness, at which point the paradox arguably ceases to connote an advantage and instead points to an absolute disadvantage.

So, the third lesson that we can draw from the evidence is this: *Even if genuine, the epidemiological paradox may translate into a cascade or chain of effects ranging from the innocuous to some that are outright deleterious.*

2.2.5. What Factors Explain the Epidemiological Paradox?

By now the reader has probably guessed that perhaps it is not that productive for one to make a career out of studying the Hispanic paradox. But there are some interesting methodological issues raised by it that are worth examining. Since the alternative explanations put forward change slightly according to the outcome, I will, in the interest of brevity, focus only on adult mortality. Here, the explanations for Mexican and foreign-born other Hispanic advantage can be classified into four types. The first is *selection through immigration.* Simply put, this refers to the possibility that individuals who immigrate into the U.S. are blessed with lower than average levels of frailty than individuals in the population of destination. In the particular case of Mexico and the U.S. this also means that immigrants experience mortality conditions that are *better* than what is the average in the country of origin. And this is, in fact, the case (Swallen, 1999). The problem with this explanation is that it cannot be disproved very easily, for two reasons: First, the data we have available are not of a bi-national nature; second, we do not know all the factors that contribute to individual frailty. Even if we did, in fact, conduct a bi-national study we would not necessarily be able to control for all the relevant characteristics, and so doubts would continue to linger. To get around these problems we would need to mount an experiment in which people who migrate to the U.S. and those who do not live their life twice, once as immigrants and again as non-immigrants!

Analytic approaches to the selection problem (selection models, propensity matching, and the like) all produce unconvincing results as they are tied too closely to the formulation of explicit models where assumptions replace the unknowns. The evidence I have examined so far suggests that selection is only part of the story and in some cases may not play a part at all. For example, an important fraction of foreign-born other Hispanic adults living in the U.S. are refugees from Central America, a group that is unlikely to have been subjected to positive selection and could even have been selected negatively (from among the very poor and infirm populations).

The second explanation has to do with data artifacts. Although there are a number of candidates among them, the most important data are generated by return migration of those at high risk of mortality who then die outside the U.S., but who contribute to mortality exposure in the U.S. Death abroad is only one of the mechanisms producing a data artifact. Patel and colleagues (Patel et al., 2005) as well as Arias (2005) argue that imperfect matching of survivors at baseline and death records from the National Death Index may be a second mechanism. While Patel and colleagues make a convincing argument that diminishes the potential role of return migration, at least when using the HEPESE as a baseline, Arias' analyses of national data suggest that imperfect matching is proportionately worse among young adults, a segment of the age span *where there is no paradox at all.*

The remaining two explanations have rounded up a great deal of adherents but they remain, to me at least, largely obscure and somewhat implausible. The argument goes as follows: Mexican and other Hispanic immigrants to the U.S. are blessed with adherence to cultural features and social interaction patterns that exert favorable effects on the outcome of interest. These features have been vaguely identified with well-defined intermediate variables such as behavioral profiles (smoking, alcohol, diet, and exercise) and other variables more vaguely connected to social support and the spillover beneficial effects of membership in tight social networks. There are two problems with these explanations. The first is that the advantage generally remains when one controls for well-measured constructs reflecting life style features (for example, drinking or smoking). The second problem is that it avoids an important issue: Why would these immigrants have better outcomes—as indeed they do—than their non-immigrant counterparts in the country of origin, that is, than those in a population blessed with similar cultural features?

The conclusion I draw from the foregoing discussion is this: *The apparent epidemiological paradox or Hispanic advantage should not be a dominant scientific problem because it applies to only a few (and shifting) groups, to only a few (and*

*shifting) outcomes, has only small ultimate significance (regardless of outcome)
and its explanation cannot be decided except by unfeasible studies or unrealizable
counterfactuals.*

2.3. What's in Store for the Hispanic Elderly in the U.S.?

There are a number of factors that conspire against a bright future and that may
ultimately impede a healthy profile for the Hispanic elderly population living in
the U.S. In my view these factors overshadow the apparently favorable position
of some groups associated with the paradox. Most important, those factors can be
studied with data sets available to us now.

2.3.1. Low Levels of Human Capital

Although we profess ignorance, we know all too well that wallet size is directly
proportional to the length of life and to the quality of that life (Smith, 1999; Palloni
and Spittel, 2004); a well-endowed wallet ultimately is the result of the quantity
and quality of human capital, especially educational level. We also know that
most Hispanic immigrants work at low-paying, low-skill jobs, with little or no
opportunity to embark upon careers of any sort. Each one's progress is impeded
by his or her level of education, by language barriers, by legal status, and by overt
or covert discrimination. And, although this state of affairs may not necessarily
lead to bad health and mortality when incumbents of these positions are young, it
surely will become more visible at older ages.

Thus, if Hispanic immigrants enter the social stratification system from below
(as many immigrants to the U.S. did before them), but are unable to climb the
hierarchy (as others were able to do), we can only expect less than paradoxical
results for health status and mortality.

However, this conclusion may be premature. If the results obtained by McKinnon
and Hummer in their paper hold true (McKinnon and Hummer, 2005), the inference
regarding mortality (and perhaps health status) differentials may not ultimately
apply to Hispanics, as the mortality and health status gradients by education do
not appear to be as relevant as they are for other groups. Before one can make
claims on this score we should wait for more definitive results regarding the health
status and mortality gradients with respect to wealth, not just education.

Finally, although these remarks refer mostly to first generation migrants, they
have implications for subsequent generations as well (see below).

2.3.2. Access to and Use of Health Care

First generation migrants of Hispanic origin, but particularly Mexicans and foreign-
born other Hispanics, are the group least likely to have health insurance. About 35%
of Hispanics ages 0 to 64 are uninsured compared to 11% of non-Hispanic whites
and 20% of African Americans (Brown, 2002; Schur, 2002). This is a result of both

low levels of uptake of employment-based insurance and low rates of employment in jobs that offer any kind of insurance. Although the effects of low levels of coverage may not be visible when incumbents are of working age, they may become more important at later ages as lack of prevention and untreated ailments begin to take a toll. Because health insurance not only protects an incumbent in an occupation but also his or her family, low levels of coverage eventually translate into deleterious consequences for members of the second generation. Their health status can be affected directly (as a conseqence of lack of coverage) and indirectly (as a consequence of lack of care for the main procedure). Furthmore, to the extent that the second generation health status is jeopardized, so are their members' chances to succeed as adults as early health status has a non trivial effects on socioeconomic status.

2.3.3. Increased Westernization

Individual risk profiles are proximate determinants of health status and mortality. And, although elderly Hispanics fare well in some respects, on some other important dimensions they do not. Thus, for example, elderly Hispanic men have high rates of smoking and binge drinking as well as low rates of physical activity. This is quite similar to the patterns observed among elderly in countries of origins, such as Mexico and Puerto Rico. By the same token, obesity, diabetes, and hypertension are prevalent among elderly Hispanics, particularly those of Mexican origin but also among Puerto Ricans (Black and Markides, 1999). Again, this mirrors patterns observed among elderly living in Mexico and Puerto Rico (Palloni et al., 2005).

The adoption of a Western life style, with a diet based on highly processed food, high intakes of sugars and saturated fats, low levels of physical activities, and increased smoking and alcohol consumption is part of a process of assimilation of Hispanic migrants. None of this is likely to translate into health benefits in the near future.

2.3.4. Lingering Effects of Early Child Health

Elsewhere I have argued that cohorts of elderly in Latin America that reached age 60 after the year 2000 experienced very peculiar conditions in childhood: they belong to groups whose chances of survival to older ages were enhanced largely because of the diffusion of medical technology and not because of improvements in standards of living. If Barker-type conjectures are approximately on target, one should then expect that these older cohorts will experience a relatively high burden of disease and disability. In work completed with data on elderly in Mexico and Puerto Rico we shown that markers of early nutritional status exert important effects on the risks of obesity, diabetes, and cardiovascular disease (Palloni et al., 2005). All these considerations apply equally well to elderly Hispanic migrants in the U.S.; they are particularly germane because migrants into the U.S. not only belong to these potentially frail cohorts but they are also exposed to environments created by the process of migration that enhance exposure to risks (stress, changes of diet, shifts in physical activity, uptake of smoking and alcohol consumption).

One last word regarding the second higher order generation of immigrants: Some of the data available indicate that second and third generations of Hispanics, particularly those of Mexican and Puerto Rican origins, do not fare well on a number of dimensions, including early life health status (obesity, patterns of illnesses), cognitive ability, high school completion, college attendance and rates of permanent employment (NRC, no date). Although making extrapolations into the future is always dangerous, I would venture to say that if these regularities reflect at all the conditions of second and third generation young adult Hispanic migrants, the future health status of elderly Hispanics in the U.S. may be compromised for some time to come. If so, we will likely see the reproduction or increase of current gradients on health and mortality across Hispanic and non-Hispanic populations. If so, the relevance of the paradox will recede and become an historical curiosity.

References

Arias, E. (2002). The health status of Hispanics. In J.G. Ioanotta (Ed.) *Emerging issues in Hispanic health: Summary of a workshop*. Washington DC: National Academy Press.

Arias, E. (2005). Evaluation of NHIS–NDI linkage by Hispanic origin. Unpublished draft, NCSH–CDS.

Black, S.A., and Markides, K. (1999). Depressive symptoms and mortality in older Mexican Americans. *Annals of Epidemiology, 9*, 45–52.

Brown, E.R. (2002.) Health insurance coverage of Hispanics. In J.G. Ioanotta (Ed.) *Emerging issues in Hispanic health: Summary of a workshop*. Washington, DC: National Academy Press.

Haan, M.N, Colon Lopez, V., Moore, K.M., Gonzalez, H.M., Mehta, K., and Hinton, L. (2005). Predictors of decline in cognitive status, incidence of dementia/CIND and all cause mortality in older Latinos: The role of nativity and cultural orientation in the Sacramento Area Latino Study on Aging. Paper presented at the 2nd Aging of the Americas Conference, Austin, Texas, September 22nd, 2005.

Ioanotta, J.G. (2002). Introduction. In J.G. Ioanotta (Ed.), *Emerging issues in Hispanic health: Summary of a workshop*. Washington, DC: National Academy Press.

Markides, K., Eschbach, K., Ray, L.A., and Peek, M.K. (2005). Census disability rates among older Hispanics by race/ethnicity and type of Hispanic origin. Paper presented at the 2nd Aging of the Americas Conference, Austin, Texas, September 22nd, 2005.

McKinnon, S., and Hummer, R.A. (2005). Education and mortality risk among Hispanic adults in the United States. Paper presented at the 2nd Aging of the Americas Conference, Austin, Texas, September 22nd, 2005.

Morenoff, J.D. (2002). Unraveling paradoxes of public health: Neighborhood environments and racial/ethnic differences in birth outcomes. Ph.D. Dissertation, University of Chicago.

National Research Council (n.d.). Multiple origins, uncertain destinies: Hispanics and the American future. Unpublished. Washington, DC: National Research Council.

Palloni A., and Arias, E. (2004). Paradox lost: Explaining the adult Hispanic mortality advantage. *Demography, 41(3)*, 385–415.

Palloni A., and Ewbank, D. (2005). Selection processes and heterogeneity: Tools for making inferences about social and economic determinants of health and mortality. In *Race, ethnicity and health in later life,* Washington, DC: National Research Council Press, National Research Council, 2003.

Palloni, A., McEniry, M., Davila, A.L., andand Garcia, A. (2005). Health among elderly Puerto Ricans: Analysis of a data set. Paper presented at the IUSSP Conference, Tours, France, July 18–23, 2005.

Palloni A., and Spittel, M. (2004). Persistent health and mortality differentials in the US. Draft. Center for Demography and Ecology, University of Wisconsin.

Patel, K., Eschbach, K., Ray, L.A., and Markides, K. (2004). Evaluation of mortality data for older Americans: Implications for policy. *American Journal of Epidemiology, 159(7),* 707–715.

Schur, C.L. (2002). A brief overview of employer-sponsored health insurance among Hispanics. In J.G. Ioanotta (Ed.) *Emerging issues in Hispanic health: Summary of a workshop.* Washington, DC: National Academy Press.

Smith, J.P. (1999). Healthy bodies and thick wallets: The dual relation between health and economic status. *Journal of Economics Perspectives, 13(2),* 145–167.

Smith, J.P., and Kington, R.S. (1997). Demographic and economic correlates of health in old age. *Demography, 34(1),* 159–170.

Swallen, K. (1999.) Mortality and nativity: A consideration of race, ethnicity and age at time of immigration. Unpublished manuscript, Center for Demography and Ecology, University of Wisconsin–Madison.

3
Census Disability Rates Among Older People by Race/Ethnicity and Type of Hispanic Origin

Kyriakos S. Markides, Karl Eschbach, Laura A. Ray,
and M. Kristen Peek

3.1. Introduction

Over the last 20 years or so there has been mounting evidence that Hispanics as a group as well as Hispanic populations from individual countries are characterized by relatively favorable mortality profiles despite generally disadvantaged socioeconomic profiles (Markides and Coreil, 1986; Markides, Rudkin, Angel, and Espino, 1977; Markides and Eschbach, 2005; Franzini, Ribble, and Keddie, 2001; Palloni and Morenoff, 2001; Sorlie, Rogot, and Johnson, 1993). The advantage shown in vital statistics is greatest, though such data likely underestimate mortality rates because of misclassification of Hispanic ethnicity on death certificates (Sorlie, Rogot, and Johnson, 1992). A somewhat lower advantage is shown with data from large population surveys linked to the National Death Index where misclassification of ethnicity on death certificates is not an issue, though completeness of record linkage may be (Hummer, Benjamins, and Rogers, 2004). The advantage is further reduced—but remains substantial—among older people when data are taken from the Social Security Administration's Master Beneficiary record and the NUMIDENT file (Elo et al., 2004).

While there is a lively debate about the source of a Hispanic mortality advantage that persists into old age (Abraido-Lanza, Dohrenwend, Ng-Mak, and Turner, 1999; Franzini et al., 2001; Markides and Eschbach, 2005; Palloni and Arias, 2004), it is generally accepted that the advantage is real and is also present at older ages. The data appear to show that the advantage is greatest among immigrants from Mexico, as well as among immigrant members of the residual category that the U.S. statistical system labels "other Hispanics" (Palloni and Arias, 2004). "Other Hispanics" are a diverse group that includes Dominicans, Spaniards; Central and South Americans, some Mexicans, Puerto Ricans, and Cubans who do not disclose a specific national origin; and persons of mixed ancestry.

If indeed Hispanic populations experience low mortality rates—including at older ages—one would expect that they would also enjoy other health advantages; for example, in self-ratings of health. However, what may be most paradoxical about the Hispanic paradox is that this expectation is not what the data show. While older Hispanics appear to live longer than other groups, they appear to do so despite poorer health and a greater prevalence of functional limitations. For example, older Mexican Americans have been repeatedly found to report poorer self-rated health than older non-Hispanic whites (Hummer et al., 2004; Markides et al., 1997). Older Hispanics overall and Mexican Americans in particular are more likely to report activity limitations in national surveys, like the National Health Interview Survey, than do older non-Hispanic whites and older Asian and Pacific Islanders (Hummer et al., 2004).

With respect to specific national origin groups, data from the Hispanic Established Population for Epidemiologic Studies of the Elderly (HEPESE) have shown that older Mexican Americans report more disabilities in activities of daily living (ADL) as well as instrumental activities of daily living (IADL) than do older non-Hispanic whites. Other studies have suggested that older Cubans have similar disability rates (as well as health profiles in general) to those of older non-Hispanic whites (Markides et al., 1997). At the same time, older Puerto Ricans typically report the highest disability rates among the Hispanic groups (Markides et al., 1997; Tucker, Falcon, Bianchi, Cacho, and Bermudez, 2000).

It has been suggested that self-reports of health are subjective and thus do not reflect the true health status of populations. A recent study has found evidence that, unlike the case with other populations, self-ratings of health are not as predictive of mortality among Hispanic immigrants to the United States (Finch, Hummer, Reindl, and Vega, 2002). Older Mexican Americans are also thought to be "health pessimistic" in that they define their health as poorer than it "actually" is. However, it has also been suggested that such self-reports may be "health realistic" because poor health is likely to have greater negative consequences on the lives of people from poor socioeconomic backgrounds (Markides et al., 1997).

It is also possible that older Mexican Americans, as well as older Puerto Ricans and other Hispanics, are living longer but with more disability and in poorer health. High disability rates among older Mexican Americans have been partly attributed to high rates of obesity and diabetes as well as sedentary life styles (Markides et al., 1997; Rudkin et al., 1997; Wu, Haan, and Liang, 2003). The same may be said about older Puerto Ricans (Markides et al., 1997; Tucker et al., 2000). So it is possible that recent increases in life expectancy have been accompanied by increases in disability and poorer health, a situation that was observed in the general population during the 1970s. More recent evidence suggests that disability rates in the general United States population have been declining since the early 1980s (Manton and Wu, 2001) a situation also observed in a number of other Western countries (Crimmins, in press; Satariano, 2007). Data on such trends in disability, unfortunately, are not available for older Mexican Americans and other Hispanic groups.

So where are we with respect to disability rates among older Hispanics in comparison with older people in other groups? While they may live longer than other populations, older Hispanics as a group tend to have high disability rates that are almost as high as those for African Americans. Within the Hispanic populations, disability rates are highest among older Puerto Ricans, lowest among older Cubans, and intermediate among Mexican Americans. Little to nothing is known about disability rates in older Hispanics from Central and South America because very few of them are included in studies of Hispanics or studies of the general population. Also, little is known about groups who report being of "other" Hispanic origins, that is, groups other than Mexican, Puerto Rican, Cuban, Central American, and South American origins.

We also believe that older Asian Americans and Pacific Islanders as a group probably have lower disability rates than older non-Hispanic whites, while older Native Americans probably have considerably higher disability rates (Hummer et al., 2004. See also Markides and Wallace, in press; and John, in press).

A problem with available data on disability among older people from the various Hispanic populations is that they are based on relatively small samples, whether they are from national surveys or small regional studies, thus limiting their usefulness. Thus, disability rates may not provide stable estimates for certain populations such as older people from Central or South American origins. The same may be said for Mexican, Puerto Rican, Cuban, and other Hispanic origins if one is interested in variation by such variables as age, gender, and nativity.

A largely underutilized resource that can answer some questions about the prevalence of disability among detailed Hispanic and racial subgroups comprises the data that come from responses to the redesigned disability module in the 2000 U.S. Census. To investigate intergroup differences in patterns of disability we used the Census's *five percent public-use micro data sample* (PUMS) file, which includes disability reports for a sample of approximately 1.8 million Americans who were age 65 or older in April 2000. We report age-standardized disability rates for each of five Census disability items, as well as for overall disability measured as having any one (or more) of the five disabilities measured.

If our knowledge about ethnic patterns in old age disability from smaller and regional samples is correct, we expect the following hypotheses to be supported:

1. Older Hispanics as a group will have greater disability rates than older non-Hispanic whites, rates that are only slightly lower than those for African Americans.
2. Older Asian and Pacific Islanders will have disability rates that are somewhat lower than those of older non-Hispanic whites.
3. Older Native Americans (American Indians and Alaska Natives) will have disability rates higher than any other population.
4. Among the Hispanic populations older Puerto Ricans will have the highest disability rates, with older Cubans having the lowest and Mexican Americans having intermediate rates.
5. It is not clear what rates for older persons who are Central Americans, South Americans, Dominicans, and Spaniards would be given absence of guidance from existing literature. We might hypothesize that since each of these groups consists overwhelmingly of immigrants, their rates might be somewhat

lower than rates for Mexican Americans if indeed they are selected through migration. Given that men are more likely to be selected than women, men will have lower disability rates. Among older Hispanic populations, Spaniards and South Americans have relatively good socioeconomic profiles (higher average years of schooling and lower poverty rates). Dominicans have the most disadvantaged socioeconomic profile of any Hispanic group. Thus, we might expect higher disability rates for Dominicans than for these other immigrant based groups, lower disability rates for Spaniards and South Americans, and intermediate rates for Central Americans.

6. It is also not clear what the rates for other Hispanics might be like. As we use the term here, "other Hispanics" denotes primarily U.S.-born persons who self-identify as Hispanic but do not report membership in a specific national origin group. A majority of other Hispanics was born in southwestern states whose Hispanic populations were overwhelmingly Mexican American during the decades when these cohorts were born. Given this mix, we might expect them to exhibit average disability rates perhaps similar to those for U.S.-born Mexican Americans, who are probably the most numerous members of this group.

7. Among older persons of Mexican origin, the foreign-born will have lower disability rates than the native-born. This is more likely to be the case for men than for women.

3.2. Methods

Disability items were obtained from the five percent PUMS sample of the 2000 U.S. Census. The Census obtained data on each person in the selected household on six disability items (Stern, 2004). The first question asked whether the respondent or others in the household who were 5 years old or over had: (1) *sensory disability,* defined as blindness, deafness, or a severe vision or hearing impairment; or (2) *physical disability,* defined as a long-lasting condition which substantially limits one or more basic activities. The second question asked whether a physical, mental, or emotional condition lasting six months or longer made certain activities difficult: (3) *mental disability* referred to learning, remembering, or concentrating; (4) *self-care disability* referred to difficulty in dressing and bathing; (5) *going outside the home disability* asked about difficulty going outside the home alone to shop or visit a doctor; (6) *Employment disability* asked about difficulty working at a job or business (persons 16 year old and older).

In the analysis below we use the first five items. Employment disability was dropped because it has little relevance to most people aged 65 and over. In addition, the employment disability item was judged to be of low reliability (Stern, 2004).

The five items we employ are not representative of typical disability items used in the literature such as the standard ADLs and IADLs. Self-care disability is similar to two ADLs and going outside the home disability is similar to an IADL item. As such it may be subject to certain gender and cultural biases typical of some IADLs.

Sensory and mental disabilities are close to what the literature usually refers to as "functional impairments" (Verbrugge and Jette, 1994).

Of the five items, sensory, physical, mental and self-care disabilities were judged to be fairly reliable. The going outside the home item may be somewhat less reliable (Stern, 2004). Another limitation is that many of the responses were via proxy since one person typically filled out the questionnaires. Despite such limitations the Census disability items provide a convenient and easy way to estimate disability prevalence in a large number of people and can be generalized to the U.S. population.

Below we report disability rates for persons 65 and over by race/ethnicity and by type of Hispanic origin. The sample size is approximately 1.8 million subjects. Rates were directly age-standardized to the age structure of the 65+ population of the U.S. all racial/ethnic groups in the year 2000 for ages 65–69, 70–74, 75–79, 80–84, 85–89, and 90+.

We first present comparisons between the major race/ethnic groups: non-Hispanic whites, all Hispanics, blacks (African Americans), Asian Americans and Pacific Islanders, and Native Americans. Comparisons are presented separately by gender. We then present comparisons by type of Hispanic origin: Mexican, Puerto Rican, Cuban, Central American, South American, Dominican, and other Hispanic. Subsequently we present comparisons between foreign-born and U.S.-born persons from the major racial/ethnic groups and for Mexican Americans, the Hispanic group for which nativity differences are relevant. We do not make foreign-born to U.S.-born comparisons for persons of Cuban, Central American, South American, and Dominican origin because these populations are overwhelmingly foreign-born. We also do not distinguish between native and "immigrant" Puerto Ricans because virtually all Puerto Ricans regardless of place of birth are natural-born citizens of the United States. Migration between the U.S. mainland and Puerto Rico is internal migration and follows a different pattern of dynamics compared to migrations between Mexico the U.S. or between Central or South America and the U.S. (Rivera-Batiz and Santiago, 1996).

In preparing these estimates, we recoded non-specific Hispanic origin responses to specific national origin groups based first on responses to the census ancestry item, and then, for the foreign or Puerto Rico-born, on country of birth. Thus, for example, a person who reported Hispanic origin but who did not report a specific national origin was recoded as Mexican if coded responses to the Census ancestry items indicated that the person's Hispanic ancestry was Mexican. Mixed Hispanic ancestry responses were not recoded. Recoding of Latin American born persons among persons not already assigned on the basis of reported Hispanic origin and or ancestry was straightforward. We did not reassign any person to a specific Hispanic origin group based on ancestry or country of birth if they were not identified as Hispanic on the Hispanic origin item.

Findings are discussed in relation to existing disability literature among America's racial and ethnic groups as well as literature for different types of Hispanic origin. Even with such a large sample, by necessity we ended up lumping potentially diverse ethnic groups (e.g., persons of different South American origins;

persons in the "other Hispanic" category). Nevertheless, despite data limitations we feel this is a worthwhile exercise that is likely to yield important new information about ethnic differences in disability among older people, in particular those of Hispanic origin.

3.3. Results

Table 3.1 presents age-standardized disability rates (percentages) for all persons aged 65 and over by major racial and ethnic categories. The table shows that, as expected, Hispanic men, as a group, are considerably more disabled than non-Hispanic white men on each of the five items, as well as for any disability (49.3 vs. 42.4%). The gap is greatest for "mental" disability and "going outside" the home. Black men are slightly more disabled than Hispanic men while Asian American men and "other" men are very similar to non-Hispanic white men. An exception is that Asian American men are more likely to report that they have difficulties going outside the home—an item we feel may be subject to cultural bias in reporting. Native American men are considerably more disabled on most items than men from the other racial/ethnic groups. Their overall ("any disability") rate was by far the highest at 59.7 percent.

A similar picture emerges when examining rates for females. Hispanic and black females are more disabled than non-Hispanic white females on each item while Asian American and "other" females are similar to non-Hispanic white women. Asian American women are considerably more disabled than non-Hispanic white women on two items: mental and going outside the home. As was the case with

TABLE 3.1. Age-standardized census disability rates (percent) by race/ethnicity for persons 65 over (United States Census, 2000)

Race/ethnicity	Sensory	Physical	Mental	Self-care	Go outside	Any disability
Males						
Non-Hispanic white	17.2	27.8	11.3	9.5	18.5	42.4
Hispanic	18.2	30.1	16.4	12.5	27.1	49.3
Black	17.4	35.8	18.1	16.3	28.7	53.1
Asian American	15.7	24.6	15.0	10.3	24.8	43.4
Native American	30.3	41.8	19.2	15.4	25.6	59.7
Other	16.2	26.0	11.8	9.9	22.7	43.5
Females						
Non-Hispanic white	13.2	31.0	12.0	12.6	23.5	42.4
Hispanic	16.5	36.8	19.4	16.4	32.2	53.5
Black	15.1	42.7	19.6	20.2	34.1	57.4
Asian American	14.3	29.9	18.3	13.4	29.1	45.9
Native American	23.4	48.2	19.9	18.8	31.6	61.0
Other	16.4	33.6	15.8	15.7	27.0	46.8

Note: Directly standardized to age 65+ population, U.S., all racial/ethnic groups in 2000, for ages 65–69, 70–74, 75–79, 80–84, 85–89, 90+.

TABLE 3.2. Age-standardized Census disability rates (percent) for persons 65 and over by type of Hispanic origin (United States Census, 2000)

Type of Hispanic origin	Sensory	Physical	Mental	Self-care	Go outside	Any disability
Males						
Non-Hispanic white	17.2	27.8	11.3	9.5	18.5	42.4
All Hispanic	18.2	30.1	16.4	12.5	27.1	49.3
Mexican	20.1	31.4	16.8	12.9	26.7	50.6
Puerto Rican	19.9	36.2	20.3	16.0	31.1	54.8
Cuban	11.8	22.9	13.9	10.2	26.9	43.8
Central American	13.6	25.2	12.5	9.6	24.6	42.4
South American	12.3	22.3	12.4	8.0	24.6	41.8
Dominican	18.0	30.7	17.4	14.0	30.8	50.2
Spaniard	14.2	25.2	11.4	9.3	21.9	40.1
All other Hispanic	21.4	34.4	18.0	14.0	28.1	52.8
Females						
Non-Hispanic white	13.2	31.0	12.0	12.6	23.5	42.4
All Hispanic	16.5	36.8	19.4	16.4	32.2	53.5
Mexican	17.8	37.8	19.1	17.0	32.1	53.9
Puerto Rican	19.3	44.4	22.9	20.0	36.8	61.5
Cuban	12.0	31.7	18.9	14.7	31.8	49.9
Central American	14.3	34.3	19.5	12.9	31.3	51.7
South American	13.4	29.3	18.0	12.2	28.6	46.7
Dominican	16.4	36.8	23.0	16.6	32.5	56.0
Spaniard	13.8	27.2	14.0	12.3	27.0	44.5
All other Hispanic	16.7	38.1	17.4	16.7	30.7	53.5

Notes: Hispanic origin persons were recorded based on ancestry or country of birth if they reported general Hispanic. "All other" includes primarily U.S.-born Hispanics who reported a non-specific Hispanic identity and who live in a Southwestern state. Directly standardized to age 65+ population, U.S., all racial/ethnic groups in 2000, for ages 65–69, 70–74, 75–79, 80–84, 85–89, 90+.

men, Native American women report the highest overall disability rate at 61.0 percent.

Women are only slightly more disabled than men in overall disability. They report somewhat lower sensory disability than men in all ethnic groups except other. In contrast they report greater physical, mental, self-care, and going outside the home disabilities, as might be predicted from the literature.

Table 3.2 presents age-standardized disability rates (percent) by type of Hispanic origin for persons aged 65 and over. To facilitate comparison rates for non-Hispanic whites, all Hispanics are also presented. As expected, the overall disability ("any disability") rate is highest among Puerto Rican men (54.8%) followed by all other Hispanic men (52.8%), Mexicans (50.6%), and Dominicans (50.2%). Overall disability rates are substantially lower for Spaniards (40.1%), South Americans (41.8%), Central Americans (42.4%), and Cubans (43.8%).

Puerto Rican men report more physical, mental, self-care, and going outside the home disability rates than men from other Hispanic origins. South American men report the lowest rates of physical and self-care limitations, and male Spaniards have the lowest rates of mental and going outside the home disability. In fact, South American and Spaniard men report less overall disability and lower rates of sensory, physical, and self-care disability than non-Hispanic white men. These groups report higher rates of going outside the home disability than do non-Hispanic whites, though this may reflect cultural biases in reporting on this relatively low reliability item. Central American and Cuban men are similar to non-Hispanic white men on most comparisons. Mexican, Dominican, and Other Hispanic origin men report disability rates that are somewhat similar, and they are generally somewhat higher than those reported by non-Hispanic whites.

As with men, Puerto Rican women report the highest overall disability rate (61.5%) followed by Dominican women (56.0%), Mexican origin women (53.9%), All Other Hispanic women (53.5%), Central American women (51.7%), Cuban women (49.9%), South American women (46.7%), and Spanish women (44.5%). Puerto Rican women report more disability on four of the five items than women from all other origins. The exception is mental disability, for which Dominican women have a slightly (though non-significantly) higher rate.

A difference from men is that for no group of women are overall disability rates lower than those of non-Hispanic white women, suggesting weaker migration selection effects for women. This pattern is also apparent on most specific disability items for those groups that are predominantly immigrant in composition. South American women reported greater disability than non-Hispanic White women on three of the five items (sensory, mental, and going outside the home) and only slightly less disability on two items (physical and self-care), again suggesting a lower level of migration selection. Central American and Dominican women were clearly disadvantaged in overall disability compared to non-Hispanic white women (51.7 and 56.0% vs. 42.4%). Cuban-origin women also report more overall disability than non-Hispanic white women (49.9 vs. 42.4%). Finally, "all other" Hispanic women report similar disability rates than Mexican-origin women on each item as well as for overall disability (53.5 vs. 53.9%).

Table 3.3 presents disability rates for U.S.-born and foreign-born men for the major racial and ethnic groups. Foreign-born non-Hispanic whites, Hispanics, blacks and other persons report lower overall disability than native-born persons from their respective ethnic group. An exception are Asian Americans who report more disability than U.S.-born Asian Americans on four of the five items, possibly reflecting compositional differences in the foreign-born and native-born populations (more on this later). Foreign-born non-Hispanic white men report the lowest overall disability rate followed by other foreign-born men.

Table 3.4 shows that foreign-born non-Hispanic white women report an overall disability rate that is just below the rate for U.S.-born women (41.5 vs. 42.4%). Hispanic and black foreign-born women report lower disability than U.S.-born Hispanic and black women on each of the five items while other foreign-born women report only slightly lower overall disability than U.S.-born other women (46.2 vs. 47.1%). As was the case with men, foreign-born Asian women report consistently more disability on each item than U.S.-born Asian women, as we discuss in more detail below.

TABLE 3.3. Age-standardized census disability rates (percent) for U.S.-born and foreign-born men 65 and over by race/ethnicity (United States Census, 2000)

Race/ethnicity	Sensory	Physical	Mental	Self-care	Go outside	Any disability
U.S.-born						
Non-Hispanic white	17.4	28.1	11.3	9.5	18.4	42.6
Hispanic	21.4	33.0	17.2	13.9	26.7	51.6
Black	17.6	36.4	18.4	16.5	28.7	53.5
Asian American	16.4	23.5	11.9	9.5	21.9	41.8
Native American						
Other	18.7	27.9	11.6	9.7	20.1	46.4
Foreign-born						
Non-Hispanic white	13.1	22.6	12.1	9.4	21.2	39.4
Hispanic	15.9	28.1	15.8	11.6	27.5	47.7
Black	13.6	25.7	13.7	12.8	28.3	45.4
Asian American	15.4	25.2	16.2	10.6	25.8	44.1
Native American						
Other	13.3	23.1	12.0	10.1	26.2	40.0

Note: Directly standardized to age 65+ population, U.S. all racial/ethnic groups in 2000, for ages 65–69, 70–74, 75–79, 80–84, 85–89, 90+.

Table 3.5 presents rates for U.S.-born and foreign-born Mexican men compared to non-Hispanic white and all Hispanic men. Mexicans are the only group of older Hispanics for which differences by nativity status are warranted because for most other groups the older population is primarily foreign-born, while for Puerto Ricans, the population is virtually all native-born. We also include U.S.-born all other Hispanic men in this table because our analysis of regional distribution

TABLE 3.4. Age-standardized census disability rates (percent) for U.S. born and foreign-born females 65 and over by race/ethnicity (United States Census, 2000)

Race/ethnicity	Sensory	Physical	Mental	Self-care	Go outside	Any disability
U.S.-born						
Non-Hispanic white	13.3	31.3	12.0	12.6	23.3	42.4
Hispanic	17.4	37.9	18.1	17.2	30.9	53.3
Black	15.2	43.2	19.8	20.5	34.3	57.8
Asian American	12.1	25.9	13.4	12.3	23.9	39.4
Native American						
Other	16.7	36.1	15.8	16.8	25.7	47.1
Foreign-born						
Non-Hispanic white	11.8	27.4	13.2	12.2	25.1	41.5
Hispanic	15.9	28.1	15.8	11.6	27.5	47.7
Black	14.3	33.5	17.5	15.6	30.3	50.3
Asian American	15.2	31.4	20.1	13.8	30.8	48.1
Native American						
Other	16.2	30.7	16.0	14.4	28.2	46.2

Note: Directly standardized to age 65+ population, U.S. all racial/ethnic groups in 2000, for ages 65–69, 70–74, 75–79, 80–84, 85–89, 90+.

TABLE 3.5. Age-standardized census disability rates (percent) for U.S. born and foreign-born males 65 and over by type of Hispanic origin (United States Census, 2000)

Race/ethnicity	Sensory	Physical	Mental	Self-care	Go outside	Any disability
U.S.-born						
Non-Hispanic white	17.4	28.1	11.3	9.5	18.4	42.6
All Hispanic	21.4	33.0	17.2	13.9	26.7	51.6
Mexican	21.8	32.5	17.1	13.6	26.2	51.4
All other Hispanic	21.7	34.9	18.0	14.2	28.3	53.1
Foreign-born						
Non-Hispanic white	13.1	22.6	12.1	9.4	21.2	39.4
All Hispanic	15.9	28.1	15.8	11.6	27.5	47.7
Mexican	18.0	30.1	16.4	12.1	27.4	49.4

Notes: Hispanic origin persons are recoded to a specific group based on ancestry or country of birth if they reported general Hispanic. Directly standardized to age 65+ population, U.S., all racial/ethnic groups in 2000, for ages 65–69, 70–74, 75–79, 80–84, 85–89, 90+.

suggests that this group is primarily composed of older Mexican American-origin men born and living in a Southwestern state who declared a general Hispanic identity without reporting a Mexican ancestry.

The data confirm a general pattern of immigrant advantage, though the differences are substantively small for older Mexican men. Non-Hispanic white men report lower overall disability rates than any other group, as well as the lowest rates of sensory, physical, and self-care disability. Immigrant Hispanics and Mexicans report slightly lower rates of all disabilities except going outside the home disabilities than respective U.S.-born comparison groups. U.S.-born all other Hispanic men have a similar disability profile to U.S.-born Mexican men, with slightly higher rates for each type of disability.

Table 3.6 presents rates for U.S.-born and foreign-born Hispanic women by type of Hispanic origin. Again, foreign-born non-Hispanic white women report

TABLE 3.6. Age-standardized census disability rates (percent) for U.S.-born and foreign-born females 65 and over by type of Hispanic origin (United States Census, 2000)

Race/ethnicity	Sensory	Physical	Mental	Self-care	Go outside	Any disability
U.S.-born						
Non-Hispanic white	13.3	31.3	12.0	12.6	23.3	42.4
All Hispanic	17.4	37.9	18.1	17.2	30.9	53.3
Mexican	17.8	38.1	18.5	17.1	31.1	53.4
All other Hispanic	16.8	38.2	18.0	16.9	30.8	53.7
Foreign-born						
Non-Hispanic white	11.8	27.4	13.2	12.2	25.1	41.5
All Hispanic	15.9	36.0	20.2	15.9	33.0	53.7
Mexican	18.0	37.4	20.1	16.8	33.4	54.4

Notes: Hispanic origin persons are recorded based on ancestry or country of birth, if they report general Hispanic. Directly standardized to age 65+ population, U.S., all racial/ethnic groups in 2000, for ages 65–69, 70–74, 75–79, 80–84, 85–89, 90+.

TABLE 3.7. Odds ratio and 95% confidence intervals from adjusted* logistic regression models for disability by type, immigrant vs. U.S.-born Mexican men and women

Type of disability	Immigrant older Mexican men vs. U.S.-born	Immigrant older Mexican women vs. U.S.-born
Sensory	0.71 (0.66–0.77)	0.92 (0.86–1.00)
Physical	0.79 (0.74–0.84)	0.85 (0.80–0.90)
Mental	0.82 (0.75–0.89)	0.99 (0.92–1.06)
Self-care	0.80 (0.72–0.88)	0.90 (0.83–0.96)
Go outside the home	0.98 (0.92–1.06)	1.02 (0.97–1.09)
Any disability	0.73 (0.67–0.79)	0.84 (0.78–0.90)

*Adjusted for age and education

the lowest overall disability rate, which, much like was the case with men, is driven by low sensory and physical disability. Foreign-born Mexican origin women report slightly more overall disability than U.S.-born Mexican women (54.4 vs. 53.4%). As was the case for men, disability rates are extremely similar for U.S.-born all other Hispanic and U.S.-born Mexican women.

To further explore the contrast in disability rates for immigrant and U.S.-born Mexican American men and women, we estimated logistic regression models for the odds of reporting each type of disability, with adjustment for level of education as well as age. With this adjustment, immigrant men are consistently advantaged for each type of disability except for go-outside-the-home disability (Table 3.7). The immigrant advantage is modest, varying between an odds ratio of 0.71 for sensory disability and 0.82 for mental disability. For women, statistically significant immigrant advantages were found for sensory, physical, self-care, and any disability. In each case, the immigrant advantages are smaller for women than for men, and are substantively small. Gender by immigrant status interactions, were statistically significant ($p < 0.05$) for each type of disability except for go-outside-the-home disability.

3.4. Discussion

The Census disability rates for the year 2000 are generally consistent with what has been found in the literature in smaller national as well as regional studies. Older Hispanics as a group are more disabled than older non-Hispanic whites and slightly less disabled than older Blacks. Older Native Americans report the highest disability rates while Asian Americans are somewhat more disabled than older non-Hispanic whites.

The relatively high disability rates for Asian Americans are a bit of a surprise given previous literature (Hayward and Heron, 1999; Hummer et al., 2004). However, this group's higher—or absence of lower—disability rates may reflect the changing composition of older Asian Americans. As our results by nativity show, immigrant older Asian Americans report more disability that U.S.-born Asian Americans. This is opposite to the pattern found for other groups. However, approximately half of older U.S.-born Asian Americans are of Japanese ancestry, while approximately half of foreign-born older Asian Americans are Filipino, Vietnamese, or other less advantaged origins. Thus, because all subgroups are combined into a single pan-ethnic category, comparisons of foreign-born to U.S.-born older Asian Americans treat members of groups with considerably different composition and experiences as similar.

As expected, among the Hispanic populations older Puerto Rican men and women report the highest disability rates. Mexican, Dominican and other origin men and women report the next highest rates. Wide gender differences are observed in the rates for the largely immigrant populations: older Cuban, Central American, and South American women report disability rates that are considerably higher than rates for non-Hispanic white women. On the other hand, Cuban, Central American, and South American men report disability rates that are comparable to the rates of non-Hispanic white men and considerably lower than the rates for other Hispanic origin men. These gender differences in such overwhelming immigrant populations suggest that women are less selected on health status through immigration than are men.

Is immigration selection also suggested in other groups? Among Mexican origin men and women, disability rates of foreign-born and U.S.-born are not too dissimilar among either men or women. With adjustment for education, a modest advantage of lower rates of most types of disabilities appears for men and, to a far lesser degree, for women. The absence of substantively important differences in disability rates by immigrant status among older Mexican Americans is consistent with what has been found with data from the Hispanic EPESE and other studies (Markides et al., 1997). While there may be nativity differences in mortality among older Mexican Americans, few such differences are observed in disability rates. It should be noted that disability rates for foreign-born black men and women are considerably lower than rates for U.S.-born black men and women. Foreign-born non-Hispanic white men report lower disability rates than U.S.-born white men. The same is not observed among white women, again suggesting the operation of a gender-based migration selection.

In sum, despite data limitations, the Census disability rates corroborate expectations from previous literature on older Hispanics and older Puerto Ricans and Mexican Americans. They also suggest the existence of possible gender-based migration selection favoring men over women in the overwhelmingly immigrant Cuban, Central American, South American, and Dominican groups of older people. The greater overall disability of older Mexican Americans continues to be inconsistent with their supposed lower mortality and suggests that more research is needed in both areas to reconcile this paradox within a paradox.

APPENDIX TABLE. Unweighted frequency counts for race and Hispanic origin subjects 65 years old or older in the Census 2000 five percent public-use micro data file, by gender and immigrant status

	Men			Women		
	U.S.-born	Immigrant	Total	U.S.-born	Immigrant	Total
Hispanic	15,816	20,571	36,387	20,742	29,328	50,070
Mexican	11,216	8,530	19,746	14,291	10,731	25,022
Puerto Rican*	555	3,131	3,686	685	4,640	5,325
Cuban	221	4,756	4,977	326	6,543	6,869
Central American	62	1,188	1,250	89	2,564	2,653
South American	69	1,848	1,917	114	3,022	3,136
Dominican	45	707	752	48	1,268	1,316
Spaniard	212	295	507	266	404	670
Other	3,436	116	3,552	4,923	156	5,079
Non-Hispanic						
White	620,357	31,669	652,026	864,085	50,624	914,709
Black	50,672	2,709	53,381	83,383	4,304	87,687
Asian	4,919	12,771	17,690	5,980	17,858	23,838
AIAN	5,550	85	5,635	7,363	109	7,472
Other	489	360	849	636	491	1,127

*For Puerto Rican, "immigrant" refers to persons born in Puerto Rico or abroad.

References

Crimmins, E. (in press). Compression of morbidity. In K.S. Markides (Ed.), *Encyclopedia of health and aging.* Thousand Oaks, CA: Sage Publications.

Finch, B.K., Hummer, R.A., Reindl, M., and Vega, W.A. (2002). Validity of self-rated health among Latino(a)s. *American Journal of Epidemiology, 155,* 755–759.

Franzini, L, Ribble, J.C., and Keddie, A.M. (2001). Understanding the Hispanic paradox. *Ethnicity and Disease, 11,* 496–518.

Haan, M.N., Mungas, D.M., and González, H.M. (2003). Prevalence of dementia in older Latinos: The influence of Type 2 diabetes mellitus, stroke and genetic factors. *Journal of the American Geriatrics Society, 51,* 160–77.

Hummer, R.A., Benjamins, M.R., and Rogers, R.G. (2004). Racial and ethnic disparities in health and mortality among the U.S. elderly population. In R.A. Bulatao and N.B. Anderson (Eds.), *Understanding racial and ethnic differences in health in late life: A research agenda* (pp. 53–94). Washington, DC: National Academy Press.

John, R. Native Americans. In K.S. Markides (Ed.), *Encyclopedia of health and aging* (pp. xx–yy). Thousand Oaks, CA: Sage Publications.

Manton, K.G., and Wu, S. (2001). Changes in the prevalence of chronic disability in the United States black and non-black population above the age 65 from 1982 to 1999. *Proceedings of the National Academy of Sciences, 98,* 6354–6359.

Markides, K.S., and Coreil, J. (1986). The health of southwestern Hispanics: An epidemiologic paradox. *Public Health Reports, 101,* 253–265.

Markides, K.S., and Eschbach, K. (2005). Aging, migration, and mortality: Current status of research on the Hispanic paradox. *Journal of Gerontology: Social Sciences, 60B,* 68–75.

Markides, K.S., Rudkin, L., Angel, R.J., and Espino, D.V. (1997). Health status of Hispanic elderly in the United States. In L.G. Martin and B. Soldo (Eds.), *Racial and ethnic differences in the health of older Americans* (pp. 285–300). Washington, DC: National Academy Press.

Markides, K.S., and Wallace, S.P. (in press). *Minority elders in the United States: Implications for public policy.*

National Center for Health Statistics, *National Vital Statistics Reports*, Vol. 51, No. 5, March 14, 2003

Palloni, A., and Arias, E. (2004). Paradox lost: Explaining the Hispanic adult mortality advantage. *Demography, 41*, 385–415. National Center for Health Statistics Washington, DC.

Palloni, A., and Morenoff, J. (2001). Interpreting the paradoxical in the "Hispanic Paradox." Demographic and epidemiological approaches. In Weinstein, A. Hermalin, and M. Soto (Eds.), *Population health and aging* (pp. 140–174). New York: New York Academy of Sciences.

Rivera-Batiz, F.L., and Santiago, C.H.A. (1996). *Island paradox. Puerto Rico in the 1990s.* New York: Russell Sage Foundation.

Rudkin, L., Markides, K.S., and Espino, D.V. (1997). Functional disability in older Mexican Americans. *Topics in Geriatric Rehabilitation, 12*, 38–46.

Satariano, W.A. (2006). *Epidemiology of aging.* Boston: Jones and Bartlett.

Sorlie P.D., Rogot E., and Johnson N. Validity of demographic characteristics on the death certificate. *Epidemiology 1992, 3*, 181–184.

Stern, S.M. (2004). Counting people with disabilities: How survey methodology estimates influence estimates in Census 2000 and the Census 2000 supplementary survey. U.S. Bureau of the Census.

Tucker, K.L., Falcon, L.M., Bianchi, L.A., Cacho, E., and Bermudez, O.I. (2000). Self-reported prevalence and health correlates of functional limitation among Massachusetts's elderly Puerto Ricans, Dominicans, and a non-Hispanic neighborhood comparison group. *Journal of Gerontology: Medical Sciences, 55A*, M90–M97.

Wu, J.H., Haan, M.N., and Liang, J. (2003). Diabetes as a predictor of change in functional status among older Mexican Americans: A population-based cohort study. *Diabetes Care, 26*, 314–319.

4
Disability and Active Life Expectancy of Older U.S.-and Foreign-Born Mexican Americans

Karl Eschbach, Soham Al-Snih, Kyriakos S. Markides, and James S. Goodwin

4.1. Introduction

It has long been recognized that nativity status may have a strong relationship to health (Marmot, Adelstein, and Bulusu, 1984; Kasl and Berkman, 1983). Evidence from United States data sources shows that immigrants have generally lower age-specific mortality than members of the same ethnic group born in the United States (Singh and Miller, 2004; Singh and Siahpush, 2001, 2002). This relationship may come about in part because international migration is selected on good health, and return migration on poor health (Palloni and Morenoff, 2004). There is also some evidence that in the U.S. and other settings immigrant/native differentials in health arise because of differences in socio-cultural characteristics and health-related behaviors (Marmot and Syme, 1976; Antecol and Bedard, 2006).

It is particularly important to understand the relationship between immigrant status and health in the case of older Mexican Americans. The 2000 census counted 800,000 Mexican Americans living in the U.S. who were at least 65 years of age. Of these, just less than half had been immigrants to the U.S. at some time in their lives. Of the remainder, nearly 90 percent were born in one of the five Southwestern states that were the historic centers of Mexican American settlement in the U.S.: Texas, California, Arizona, New Mexico, and Colorado.

Immigrant/native differentials in mortality that favor the immigrants have been observed for older Mexican Americans as for other populations (Hummer et al., 2000; Palloni and Arias, 2004; Elo, Turra, and Kestenbaum, 2004). The low age-specific rates of mortality for Mexican Americans compared to non-Hispanic whites—and especially for immigrant Mexican Americans—have sometimes been described as paradoxical because of the high level of socioeconomic risk and reduced access to health care in this population (Markides and Coreil, 1986; Franzini, Ribble, and Keddie, 2001).

While low mortality and extended longevity are obviously important markers of the health and well-being of a population, they do not present a complete portrait. A pressing question that arises in conjunction with the increasing longevity of the modern era concerns the quality of extended life. Are the later years in life plagued by disease and functional dependence, or are they relatively disease free and active?

Viewed from this perspective the lower mortality rates of older U.S.-born Mexican Americans compared to non-Hispanic whites and older Mexican immigrants compared to U.S.-born Hispanics are not complemented by a markedly lower burden of chronic illness and disability. Prevalence data from cross-sectional surveys show that older Hispanics are more likely than non-Hispanic whites to report fair or poor compared to excellent or good self-rated health (Hummer, Benjamins, and Rogers 2004). They report higher rates of diabetes, obesity, and cognitive impairment, and show no advantages in hypertension rates (Markides et al., 1997; Black et al., 1999; Wu, Haan, and Liang, 2003). Data from both the National Health Interview Survey (Hummer, Benjamins, and Rogers, 2004) and the Census show a higher prevalence of limitations in activities of daily living (Markides et al., 2005).

A more mixed picture emerges with respect to comparison of immigrant morbidity and disability profiles with those of natives. In a systematic review of prevalence data for older immigrants and natives in the United States, setting aside ethnicity, Jasso and co-authors report consistently poorer self-rated health for immigrants compared to U.S.-born persons. Lower chronic disease morbidity rates for immigrants narrow or disappear at older ages (Jasso et al., 2004). Census 2000 disability data presented elsewhere in this volume show little difference in the age-standardized disability rates for older U.S. and foreign-born Mexican Americans (Markides et al., 2005).

A related question to that concerning differentials in disability concerns the relationship between age at migration, time in the U.S., acculturation and morbidity, and disability among immigrants. A growing body of recent evidence shows that time in the U.S. among immigrant cohorts is associated with declines in health, health-related behaviors, and functional independence (Antecol and Bedard, 2006; Jasso et al., 2004). This pattern is consistent with a model of declining health as a function of cultural assimilation. However, this explanation of the association cannot be directly inferred because apparent trajectories with increasing settlement durations in cross-sectional data may reflect the effects of selective emigration, and because of the confounding of time in the U.S. with age and time period of migration (Angel and Angel, 1992).

In the current study we draw on data from 7 years of follow-up of the Hispanic Established Population for Epidemiological Studies (HEPESE) cohort to investigate the incidence of disability in the follow-up period, as well as the implications of incident morbidity and mortality for remaining active life expectancy for older Mexican Americans. The HEPESE is a representative population-based cohort of older Mexican Americans living in five Southwestern states in the United States. The HEPESE has been used extensively to study the prevalence and incidence correlates of disability among older Mexican Americans (see, for example Rudkin, Espino, and Markides, 1997; Peek et al., 2004; Peek, Patel, and Ottenbacher 2005; Patel et al., 2006). One of the early findings from the baseline Hispanic EPESE study was that despite the lower mortality of Mexican Americans at older ages, older Mexicans reported a higher rather than lower prevalence of physical disability than older non-Hispanic whites in the baseline survey, and that the prevalence of disability for older Mexican Americans was similar for immigrants and the U.S.-born (Markides et al., 1997). In this paper we extend this work by examining the relationship between nativity status, acculturation, disability, and active life expectancy across a 7-year follow-up period.

4.2. Materials and Methods

4.2.1. Data

The initial wave of the HEPESE was conducted during 1993/1994. There were three subsequent follow-up interviews, in 1995/1996, 1998/1999, and 2000/2001. A fifth interview is currently being fielded, but data from this wave are not included in this study. The median interval between the first and the fourth follow-up interview was 85 months, with 84% of fourth follow-up interviews occurring between 81 and 88 months after the initial interview. The HEPESE baseline cohort of 3,050 older (age 65 or older) Mexican Americans was selected from the five Southwestern states of Texas, California, Arizona, Colorado, and New Mexico using a multistage area probability cluster sample of counties, census tracts, census blocks, and households. The sampling procedure yields a sample that is generalizable to the approximately 500,000 older Mexican Americans living in the Southwest in the mid 1990s. The response rate at baseline (1993/1994) was 83%.

4.2.2. Measures

4.2.2.1. Activity Limitation

Disability was assessed with an Activities of Daily Living (ADL) scale that includes five items from the Katz ADL index (1963) and two items developed by Branch et al. (1983). The Katz ADL index is the most common measure for assessing functioning and has been widely utilized in representative community-dwelling populations. ADLs included walking across a small room, bathing, grooming, dressing, eating, transferring from a bed to a chair, and using the toilet. Respondents were asked to indicate if they could perform the activity without help, with help, or if they were unable to do it. Disability was dichotomized as no help needed versus needing help with or unable to perform one or more of the seven ADLs.

Comparisons of disability rates by immigrant status and ethnicity may be influenced by varying socio-cultural perceptions of disability status. Finch et al. (2002) showed that self-ratings of health are differentially associated with mortality by immigrant status and acculturation among Hispanics. Similar there may be a subjective component to self-reports of physical disability, so that a given level of physical functioning may lead to discrepant rates of reported disability by ethnicity and immigrant status.

4.2.2.2. Chronic Condition

Chronic conditions are self-reported by the respondent at baseline. An additive index was constructed with respect to report of a doctor's diagnosis of cancer, heart attack, diabetes, stroke, or hip fracture.

4.2.2.3. Performance-Oriented Mobility Assessment

The POMA is a direct assessment of a person's physical abilities in the domains of gait, balance, strength, upper arm function, and hand function. Direct timed measurement of these activities gives an objective assessment of a person's abilities and has been found to be a useful predictor of subsequent disability among persons who were not disabled at the time of the test (Guralnik et al., 1995; Ostir et al., 1998).

4.2.2.4. Demographic Characteristics

Age, gender, immigrant status (birth in U.S. or Mexico), education, and language acculturation measures are collected by self-report. Education is reported with respect to years of schooling and entered into models in ordinal form. Language acculturation is measured using a four-value scale constructed from self-report of English-language competence with respect to understanding, speaking, and reading English. *Time from first to last interview,* measured in months, is included in equations with disability at 7-year follow-up dependent.

4.2.2.5. Vital Status

Vital status at fourth follow-up was ascertained by two sources: proxy-report at time of a follow-up interview and a computer-assisted match to the National Death Index (NDI) through 2001. Where date of death disagreed between the NDI and proxy report, the NDI date was accepted as more precise.

4.2.3. Analysis

We compare the distribution of incident disability, mortality, and loss-to-follow-up for immigrant and U.S.-born members of the Hispanic EPESE cohort. We then compare the baseline characteristics of immigrants and natives with respect to social (language acculturation, education) and health (chronic conditions and mobility assessment) risks of disablement and mortality. We then present risk ratios from multinomial logistic regression models for disability and mortality compared to reporting active (non-disabled) status at 7-year follow-up.

We also present estimates of remaining life expectancy and percent active of remaining life expectancy at 5-year age intervals for immigrant and native men and women. These estimates were calculated using the Interpolation of Markov Chains (IMACH) method and software.(Lièvre and Brouard, 2003). This methodology uses data from cross-longitudinal surveys that track disability and vital status over time. Age-sex specific transition probabilities are calculated between each of two

TABLE 4.1. Vital and disability status for the Hispanic EPESE cohort.

At fourth follow-up	Immigrants at baseline		U.S.-born at baseline	
	Not disabled	Disabled	Not disabled	Disabled
Among those enrolled or dead	1,143	198 (15%)	1,477	223 (13%)
% dead	32	66	30	66
% disabled	15	25	14	23
% not disabled	52	9	56	12
% lost to follow-up	17	6	10	10

disability states (non-disabled/disabled, i.e., ADL limitations $> = 1$), and death. The transition probabilities are used to calculate period life table estimates of remaining life expectancy and remaining life expectancy free of disability.

4.2.4. Results

At baseline the prevalence of ADL limitations in the Hispanic EPESE cohort was slightly higher (15%) among persons born in Mexico compared to persons born in the United States (13%) see Table 4.1.). It should be noted that this slight advantage for the U.S.-born members of the cohort reverses with controls for age and gender composition, so that there is a slight immigrant advantage (O.R. = 1.13). In neither the adjusted nor the unadjusted model is the immigrant vs. native difference in baseline disability prevalence statistically significant.

Table 4.1 also shows the distribution of baseline cases, stratified by baseline disability status, to vital and disability status at fourth follow-up, approximately 84 months after baseline. There is little evidence of difference in the trajectory of either disability or mortality comparing the disabled to the active at baseline. Among both the disabled and active at baseline, a slightly higher percentage of persons born in the U.S. was not disabled 7 years later (12 vs. 9% for the baseline disabled; 56 vs. 52% for the baseline active). Again, differences in the distribution to vital or disability status at follow-up are not statistically significant. Non-disabled immigrants were somewhat more likely to be lost to follow-up at 7-year follow-up among those who were not disabled at baseline.

Table 4.2 reports compositional differences in the non-disabled members of the baseline cohort between the Mexico-born and the U.S.-born. The U.S.-born are 1.5 years younger on average. The gender composition of the two groups was identical. As expected, there are sharp differences between immigrants and U.S. natives in educational attainment and in the prevalence of English language proficiency. Even for persons born in the United States, the prevalence of persons with very low levels of schooling and monolingual Spanish-speakers is relatively high.

Table 4.2 also shows the prevalence at baseline of two important risks for subsequent disability: self-reported chronic conditions and performance on a performance-oriented mobility assessment. There is a small difference in the prevalence of at least one self-reported chronic condition favoring the foreign-born

TABLE 4.2. Characteristics of immigrants and natives in the Hispanic EPESE cohort, no ADL limitation at baseline.

		Immigrants	U.S. Born	Sig.
Mean age		72.9	71.4	0.000
	% female	57	57	0.992
Years of schooling				
	% no schooling	19	13	
	% 1–5 years	52	33	
	% 6–11 years	24	39	
	%12+ years	6	15	0.000
English ability				
	% well or very well	18	56	0.000
Chronic conditions				
	% 0	70	64	
	% 1	25	29	
	% 2+	5	8	0.029
Performance-oriented mobility assessment (POMA) (category recode)				
	% can't	5	5	
	% low	10	10	
	% middle	43	46	
	% high	42	39	0.665

(30% for immigrants vs. 36% for the U.S.-born), but not for performance on the POMA.

Table 4.3 reports age-sex adjusted risk ratios for disability and vital status at 7-year follow-up for persons who were not disabled at baseline with respect to immigrant status, schooling, English language ability, chronic conditions at baseline, and baseline POMA score. Age, chronic conditions, and poorer POMA performance all strongly predict both higher levels of disability and mortality at 7-year follow-up, as expected. Female gender predicts lower mortality and higher incident disability, again as expected.

The relationship between immigrant status and both disability and death at 7-year follow-up, if any, is weak and not statistically significant. The relative risk

TABLE 4.3. Risk ratios for ADL limitation/vital status at 7-year follow-up with adjustment for age, sex, and months of follow-up, for non-disabled persons at baseline.

	Disabled	Deceased
Age	1.15 (1.12, 1.19)	1.12 (1.10, 1.14)
Female	1.39 (1.03, 1.88)	0.65 (0.54, 0.78)
POMA category	0.40 (0.32, 0.51)	0.59 (0.50, 0.70)
Chronic condition	2.21 (1.72, 2.83)	1.96 (1.66, 2.32)
Born in U.S.	0.86 (0.62, 1.19)	1.02 (0.85, 1.24)
Years of schooling	0.96 (0.93, 0.99)	0.98 (0.96, 1.01)
Uses English well/very well	0.70 (0.53, 0.93)	0.96 (0.78, 1.19)

TABLE 4.4. Risk ratios for ADL limitation/vital status at 7-year follow-up for non-disabled persons at baseline, Hispanic EPESE cohort.

	Disabled	Deceased	Disabled	Deceased
Born in U.S.	0.85 (0.58,1.24)	0.99 (0.78,1.24)	0.91 (0.68,1.21)	0.98 (0.80,1.22)
Uses English well/				
very well	0.73 (0.53,1.00)	0.94 (0.72,1.22)		
Years of schooling			0.97 (0.94,1.00)	0.98 (0.96,1.01)

Adjusted for age, sex, chronic conditions, performance-oriented mobility assessments and (disability model only) interval in months between first and last interview.

of a U.S.-born person compared to an immigrant reporting a time 4 disability is 0.86, but the 95% confidence band includes 1. The model estimate of relative risk that a U.S. native compared to an immigrant will be deceased during the follow-up period is essentially equal to 1. By contrast, both English language fluency and additional years of schooling are associated with reduced levels of activity impairment at 7-year follow-up.

Considered in a multivariate context these relationships are essentially unchanged, as reported in Table 4.4. Both schooling and English language competence remain associated with lower rates of disablement, even with adjustment for baseline POMA score and reported chronic conditions. Control for schooling and English competence has no discernable impact on the comparison of persons born in Mexico and those born in the United States.

Given the similar pattern of incident disability during the follow-up period already presented, it is not surprising that both life expectancy and active life expectancy are extremely similar for both immigrants and natives in the Hispanic EPESE cohort (Table 4.5). Remaining life expectancy at age 65 is calculated to be 16.4 years for immigrant men and 15.3 years for U.S.-born men in the cohort. For women, life expectancy at age 65 is 19 years for immigrants and 18.2 years for the U.S.-born. At each subsequent five-year age interval, calculated life expectancy is slightly higher for immigrants compared to the U.S.-born. At each age, U.S.-born subjects have a slightly higher expectation for the proportion of life to be spent

TABLE 4.5. Remaining total life expectancy in years and percent active by gender and immigration place of birth.

	Men				Women			
	Born in Mexico		Born in U.S.		Born in Mexico		Born in U.S.	
At age	Total	% Active	Total	% Active	Total	% Active	Total	% Active
65	16.4	83	15.3	85	19.0	75	18.2	77
70	13.4	79	12.5	82	15.7	70	15.0	72
75	10.9	74	10.0	77	13.0	64	12.3	66
80	8.8	69	8.1	71	10.7	57	10.1	59
85	7.2	62	6.5	65	9.0	49	8.4	52
90	5.9	54	5.3	57	7.6	41	7.1	44

without an ADL-limitation. However, none of these differences with respect to either life expectancy or active life expectancy is statistically significant.

4.2.5. Discussion

The longitudinal findings from the H-EPESE cohort corroborate the cross-sectional findings from the baseline cohort (Markides, 1997) and from other cross-sectional sources (such as Census) that show comparable rates of disability between immigrant and U.S.-born Mexican Americans (Markides et al., 2005). There does not appear to be an immigrant advantage in lower rates of either prevalence or incidence of disability in the Hispanic EPESE cohort.

Both language acculturation and additional years of schooling are associated with lower rates of disability at 7-year follow-up. These findings appear in certain respect to cross-cut expectations that increasing assimilation and acculturation will be associated with health declines (Antecol and Bedard, 2006). However, we find that these relationships neither create nor mask differences in incident disability by nativity status. That is, while immigrants in the HEPESE sample report both lower education and less language acculturation compared to their U.S.-born counterparts, statistical control for these characteristics does not reveal significantly lower rates of disability for the foreign-born. Acculturation effects may also be confounded by offsetting socioeconomic risks that may have different implications for morbidity, disablement, and mortality processes (Angel and Angel, 2006; Crimmins, Hayward, and Seeman, 2004).

Rates of activity limitations among older adults that are reported from the Hispanic EPESE and from other sources are higher than those reported for the non-Hispanic white population (Markides et al., 1997; Markides et al., 2006; Crimmins, Hayward, and Seeman, 2004). This finding stands in contrast to the increasingly well-documented finding of low relative age-specific mortality rates for middle age and older foreign-born Mexican Americans (Elo et al., 2004; Singh and Siahpush, 2001, 2002; Hummer et al., 2000). This contrast between lower mortality and higher incident and prevalent rates of disability and physiological risk markers for older Mexican American immigrants increases the sense that the mortality findings present a paradox, as researchers have not yet clearly identified either physiological or socio-cultural mechanisms that explain the lower mortality rates.

References

Angel, J.L., and Angel, R.J. (1992). Age at migration, social connections, and well-being among elderly Hispanics. *Journal of Aging and Health, 4,* 480–499.

Angel, J.L., and Angel, R.J. (2006). Minority group status and healthful aging: Social structure still matters. *American Journal of Public Health, 96,* 1152–1159

Antecol, H., Bedard, K. (2006). Unhealthy assimilation: Why do immigrants converge to American health status levels? *Demography, 43,* 337–360.

Black, S.A., Espino, D.V., Mahurin, R., Lichtenstein, M.J., Hazuda, H., Fabrizio, D., et al. (1999). The influence of non-cognitive factors on the Mini-Mental State Examination in older Mexican-Americans: Findings from the Hispanic EPESE. *Journal of Clinical Epidemiology, 52,* 1095–1102.

Branch, L., Katz, S., Kniepmann, K., and Papidero, J. (1984). A prospective study of health status among community elders. *American Journal of Public Health, 74,* 266–268.

Crimmins, E.M., Hayward, M.D., and Seeman, T.E. (2004). Race/ethnicity, socioeconomic status, and health. In N.B. Anderson, R.A. Bulatao, and B. Cohen (Eds.), *Critical perspectives on racial and ethnic differences in health in late life* (pp. 310–352). Washington, DC: National Academy Press.

Elo, I.T., Turra C.M., Kestenbaum, B., and Ferguson, R.F. (2004). Mortality among elderly Hispanics in the United States: Past evidence and new results. *Demography 2004, 41,* 109–128.

Finch, B.K., Hummer, R.A., Reindl, M., and Vega, W.A. (2002). Validity of self-rated health among Latino(a)s. *American Journal of Epidemiology, 155,* 755–759.

Franzini, L., Ribble, J.C., Keddie, A.M. (2001). Understanding the Hispanic paradox. *Ethnicity and Disease, 11,* 496–518.

Guralnik, J.M., Ferrucci, L., Simonsick, E.M., Salive, M.E., and Wallace, R.B. (1995). Lower-extremity function in persons over the age of 70 years as a predictor of subsequent disability. *New England Journal of Medicine, 332,* 556–561.

Hummer, R.A., Rogers, R.G., Amir, S.H., Forbes, D., and Frisbie, W.P. (2000). Adult mortality differentials among Hispanic subgroups and non-Hispanic whites. *Social Science Quarterly, 81,* 459–476.

Kasl, S.V., and Berkman, L. (1983). Health consequences of the experience of migration. *Annual Review of Public Health, 4,* 69–90.

Katz, S., Ford, A.B., Moskowitz, R.W., Jackson, B.A., and Jaffe, M.W. (1963). Studies of illness in the aged. The index of ADL: A standardized measure of biological and psychosocial function. *JAMA, 185,* 914–919.

Lièvre, A., Brouard, N., and Heathcote, C.R. (2003). Estimating health expectancies from cross-longitudinal surveys. *Mathematical population studies.* 10(4), pp. 211–248.

Markides, K.S., and Coreil, J. (1986). The health of Hispanics in the southwestern United States: An epidemiologic paradox. *Public Health Reports, 101,* 253–265.

Markides, K.S., Eschbach, K., Ray, L., and Peek, M.K. (2005). Old age disability rates by race/ethnicity using the 2000 United States Census. Presented at the 2nd Conference on Aging in the Americas: Key Issues in Hispanic Health and Health Care Policy Research, Austin, TX 2005.

Markides, K.S., Rudkin, L., Angel, R.J., and Espino, D.V. (1997). Health status of Hispanic elderly in the United States. In L.G. Martin and B. Soldo (Eds.), *Racial and ethic differences in the health of older Americans* (pp. 285–300). Washington, D.C.: National Academy Press.

Marmot, M.G., Adelstein, A.M., and Bulusu, L. (1984). Lessons from the study of immigrant mortlity. *Lancet, 30,* 1455–1457.

Marmot, M.G., Syme, S.L. (1976). Acculturation and coronary heart disease in Japanese-Americans. *American Journal of Epidemiology, 104,* 225–247.

Ostir, G.V., Markides, K.S., Black, S.A., and Goodwin, J.S. (1998). Lower body functioning as a predictor of subsequent disability among older Mexican Americans. *Journals of Gerontology, Medical Sciences, 53,* M491–M495.

Palloni, A., and Arias, E. (2004). Paradox lost: Explaining the Hispanic and adult mortality advantage. *Demography, 41,* 385–415.

Palloni, A., and Morenoff, J.D. (2001). Interpreting the paradoxical in the Hispanic paradox: Demographic and epidemiologic approaches. *Annals of the New York Academy of Sciences, 954,* 140–174.

Patel, K.V., Peek, M.K., Wong, R., and Markides, K.S. (2006). Comorbidity and disability in elderly Mexican and Mexican American adults findings from Mexico and the southwestern United States. *Journal of Aging and Health, 18,* 315–329.

Peek, M.K., Ottenbacher, K.J., Markides, K.S., and Ostir, G.V. (2003). Examining the disablement process among older Mexican American adults. *Social Science Medicine, 57,* 413–425.

Peek, M.K., Patel, K.V., and Ottenbacher, K.J. (2005). Expanding the Disablement Process Model Among Older Mexican Americans *Journals of Gerontology Series A: Biological Sciences and Medical Sciences 60:*334–339.

Rudkin, L., Markides, K.S., and Espino, D.V. (1997). Functional disability in older Mexican Americans. *Topics in Geriatric Rehabilitation, 12,* 38–46.

Singh, G.K., Miller, B.A. (2004). Health, life expectancy, and mortality patterns among immigrant populations in the United States. *Canadian Journal of Public Health, 95,* I14–21.

Singh, G.K., Siahpush, M. (2001). All-cause and cause-specific mortality of immigrants and native born in the United States. *American Journal of Public Health, 91,* 392–399.

Singh, G.K., Siahpush, M. (2002). Ethnic-immigrant differentials in health behaviors, morbidity, and cause-specific mortality in the United States: An analysis of two national data bases. *Human Biology, 74,* 83–109.

U.S. Bureau of the Census. (2003). 2000 Census of Population and Housing. Five percent public use micro-data sample. Machine readable file. Produced and distributed by the U.S. Bureau of the Census, Washington, DC, 2003.

Wu, J.H., Haan, M.N., and Liang, J. (2003). Diabetes as a predictor of change in functional status among older Mexican Americans: A population-based cohort study. *Diabetes Care, 26,* 314–319.

5
Predictors of Decline in Cognitive Status, Incidence of Dementia/CIND and All-Cause Mortality in Older Latinos: The Role of Nativity and Cultural Orientation in the Sacramento Area Latino Study on Aging

Mary N. HaanVivian Colon Lopez, Kari M. Moore, Hector M Gonzalez
Kala Mehta, and Ladson Hinton

5.1. Background

Elderly Mexican Americans are one of the fastest growing groups of elderly in the United States. It is estimated that by 2050 older people of Mexican origin will constitute more than one million persons (Day, 1996). The rapid growth of the Mexican American population in the U.S. is due in part to immigration (Stephen, Foote, Hendershot, and Schoenborn, 1994; Larsen, 2004). Many of the elderly in this ethnic group migrated from Mexico in their youth, yet few studies have evaluated the long-term effects of migration on health status in old age. Some research (Angel, Buckley, and Sakamoto, 2001; Buckley, Angel, and Donahue, 2000) has suggested that duration of residence in the U.S. may be associated with changes in health status and that there may be gender differences in the effects of migration. However, migration does not inevitably lead to adoption of new cultural behaviors, and maintenance of cultural forms in the new context is not only possible but common (Benjamin, 1997; Graves et al., 1999). Further, among those born in the U.S., cultural orientation may still be influenced by the cultural heterogeneity of the context in which the individual resides. Continuing connections to the ancestral country in the form of relatives, visits, property ownership, and maintenance of language can also influence cultural change and mediate the effects of nativity on health. The potential for improvement in economic status is a major motivation for many to immigrate. Success in improving economic status is likely to have an important impact on health status that could alter whatever effects nativity and cultural change may have.

This paper focuses on the association between nativity and cultural orientation to three important health outcomes: mortality, cognitive decline, and incidence of dementia and cognitive impairment not meeting dementia criteria (CIND) observed over a four-year time period in a cohort of Mexican Americans aged 60–100 at baseline. The role of socioeconomic status (SES) in mediating the effects of nativity or cultural orientation on these outcomes is examined.

5.2. Methods

The Sacramento Area Latino Study on Aging (SALSA) is an ongoing cohort study, including, at baseline in 1998–99, 1,789 people aged 60–100. Eligibility criteria included age, residency in a selected Census tract, non-institutional living circumstances, and Latino ethnicity. The methods for this study have been extensively published elsewhere (Haan et al., 2003). Sacramento County, California, was the primary recruitment site; an additional five adjacent counties were included. The participants were recruited using a multi-stage sample selection procedure that identified Census tracts from the 1990 and 1998 census that were at least 5% Hispanic. All such tracts were included in a sampling frame, a subset was randomly selected and every household in every tract was enumerated for eligible participants.

All participants have been followed up annually at their homes and at six-month intervals by a short telephone interview. When participants are unable to respond a proxy interview is given. The contents of the interview and clinical assessments have varied over time but have always included a survey of medical history, medications use, functional and cognitive status, socioeconomic factors, and lifestyle factors. The clinical assessment has routinely included blood pressure, anthropometry, physical performance and a blood draw. The general goal of the study is to evaluate cognitive decline and dementia. A second evaluation phase for those at risk for dementia includes intensive clinical assessment for dementia. A vital status follow-up of the SALSA cohort is routinely done to determine all-cause and cause-specific mortality. Mortality ascertainment is ongoing and uses multiple sources for identifying deaths. These include reports of spouses and family members, death certificates, obituaries, state vital statistics, and the National Death Index. Approximately 84.5% (272 of 320) of known deaths have been verified with a death certificate to date. Among those for which it was not possible to obtain the death certificates ($n = 48$), 50.0% died in Mexico or other foreign country and 50% died in the United States. Among the 868 participants born in Mexico, at the fourth year of follow-up 516 remained active in the study, 157 were deceased, 120 had refused, and 108 were lost to follow-up. Among the 847 participants born in the U.S., 524 remained active in the study, 155 were deceased, 120 had refused, and 70 were lost to follow-up. There were no significant differences in attrition (not due to death) by nativity [OR = 0.93 (0.73–1.19)], – after adjustment for age, gender, and education.

5.3. Measurements

Cognitive status for this analysis is always measured using two tests. The first is the modified Minimental State Exam (3MSE), which is a measure of global cognitive function widely used in large-scale epidemiologic studies (Teng and Chui, 1987). The second is the Spanish English Verbal Episodic Memory Test (SEVLT), a test of episodic memory (Mungas, Reed, Haan, and Gonzalez, 2005). These tests are repeated at every follow-up in the study since baseline, when they were field-tested and validated in both English and Spanish. Incidence of dementia and cognitive impairment without dementia (CIND) are combined in these analyses to permit adequate power. Methods for evaluating these clinical outcomes have been published else where (Haan et al., 2003).

At baseline, time elapsed since immigration was ascertained by the question, "How many years ago did you come to the U.S.?" Country of birth was ascertained by the question, "Where were you born?" This was coded as U.S. ($n = 872$), Mexico ($n = 807$) or other Latin American country ($n = 100$). For purposes of analysis this variable was recoded into U.S.-born or birth in Mexico or other Latin American country.

More detailed analysis suggested that most immigrants not born in Mexico were born in Central America. Cultural orientation was measured using an adaptation of the Acculturation Rating Scale for Mexican Americans II, developed by Cuellar (Cuellar, Arnold, and Maldonado, 1995). This is an additive scale with strong metric properties (Cronbach's alpha $= 0.88$) in which a low score indicates Mexican orientation and a high score indicates Anglo orientation. In this analysis, the scale is used both continuously and in categories based on tertiles. Education was measured in years of education. Country of birth was measured by asking the participant where he or she was born. Time since migration among those who were born in Mexico or other Latin American country was measured by comparing the age at which the participant first migrated to the U.S. to their current age. Income was measured in five categories ($<\$1,000$ per month, $\$1,000-\$1500$, $\$1500-\$2000$, $\$2000-\$2500$, $\$2500+$). A measure of socioeconomic status was constructed that combined income and education as: below the median on both, above the median on both, below the median on the first, above the median on the second. Education above 8 years was above the median. A monthly income of $\$1000$ or more was above the median. This was intended to evaluate economic success relative to education and to allow assessment of the effects of education separately from income. It was used as a categorical variable in linear models and as an indicator variable in proportional hazards models where the reference category was the group with high education and high income.

Participants were asked to rate their health as "excellent," "good," "fair," or "poor." Access to medical care was assessed by asking participants about medical insurance and about the source of care (private M.D., public clinic, E.D.). Specific health problems were assessed by a detailed medical history of over 35 conditions and diagnosis. Type-2 diabetes was further ascertained by the use of diabetic medications and a fasting glucose $=>. 126$ mg/dl.

5.4. Statistical Methods

The intent of this analysis is to separately evaluate the influence of nativity and cultural orientation on all-cause mortality, cognitive decline, and combined incidence of dementia and CIND. Annual attrition from the cohort has averaged about 8% per year, split equally between attrition due to death and due to refusals. Both dementia/CIND and mortality rates were calculated using a cumulative incidence density approach with the variable person years as the denominator. Those not known to be dead were categorized as "alive." Those not known to be demented or CIND were categorized as "normal." The association between covariates and dichotomous outcomes was assessed using proportional hazards models with a series of adjustments for potential confounders or variables on the pathway.

Change over time in the 3MSE or the SEVLT was assessed using mixed linear models with a repeated measures approach, random intercept, and an autoregressive option. For each outcome we tested for change over time by including a time*variable term. If the term was not significant, it was dropped from subsequent models. Because time since migration did not demonstrate any association with any outcomes, the results from that analysis are not shown here. The 3MSE test was reverse-coded, so that the score represented the number of errors on the test, and then log-transformed to meet the assumptions of the linear mixed model. The SEVLT was transformed by square root to improve model fit in linear mixed models.

5.5. Results

Table 5.1 displays descriptive statistics by country of birth for demographic, economic, cultural, and health care access factors. Participants born in Mexico or other Latin American countries were significantly older than those born in the U.S. They were less likely to have attended any school and had significantly fewer years of education than their U.S.-born counterparts. Just 6% of Mexican/Other Latin American-born, compared to nearly 22% of U.S.-born participants had more than a high school education. Twenty one percent of Mexican/Other Latin American-born compared to 5% of U.S.-born had no formal education. Although there were no differences by nativity in those currently working, those born in Mexico/Latin America were more likely to have an income less than $1,000 per month, and slightly more likely to have wage or salary income. More than half of the Mexican/Latin-born participants had incomes and education below the median for the entire sample compared to almost half of the U.S.-born participants. Slightly more Mexican/Other Latin American-born than U.S.-born participants had high income and low education and the percent with low income and high education was approximately equal by nativity. Figure 5.1 shows the mean years of education by income level by country of birth. The between-country discrepancy

TABLE 5.1. Descriptive socio-demographic and health status at baseline by country of birth in study participants.

| Covariate | Country of birth | | Total (1789) | p values |
	U.S. 49.4% (847)	Mexico/other LatinAmerican Born 50.6% (868)		
	Demographic and socioeconomic factors			
Age (mean/SE)	70.0 (.24)	71.3 (.24)	70.65	0.0004
Gender (% female)	46.5 (493)	52.5 (545)	58.3 (1038)	0.13
Attended school (%yes)	94.5 (826)	79.7 (722)	87.0 (1548)	<0.0001
Years of education (mean/SE)	9.58 (0.16)	4.99 (0.16)	7.24 (0.16)	<0.0001
Currently working (% Yes)	17.4 (147)	16.7 (145)	16.4 (292)	0.72
Monthly household income <$1K (%)	29.75 (257)	59.5 (527)	44.8 (784)	0.002
Wage or salary income (self or spouse)	18.8 (163)	19.8 (178)	19.3 (341)	0.035
Monthly income <$1,000 and education <= 8years	20.1 (175)	52.2 (473)	36.42 (648)	<0.0001
Monthly income <$1,000 and education > 8 years	10.3 (90)	8.3 (75)	9.3 (165)	
Monthly income => $1,000 and education <= 8 years	21.3 (186)	27.5 (249)	24.45 (435)	
Monthly income => $1,000 and education => 8 years	48.3 (421)	12.1 (110)	29.9 (531)	
Own your house (% yes)	77.7 (676)	56.12 (509)	66.7 (1185)	0.04
	Cultural factors			
Cultural scale (mean/SE)	30.86 (0.34)	13.6 (0.33)	22.11	<0.0001
Spanish speaking (%)	31 (225)	88.6 (804)	51% (907)	<0.0001
Live in a neighborhood that is 50%+ Latino	28.4 (184)	42.2 (239)	34.8 (423)	<0.0001
Years lived in neighborhood	25.13 (0.56)	18.96 (.59)	22.2 (57)	<0.0001
	Health status, insurance, and access to care			
Self-rated health (fair or poor) (%/N)	41.2 (349)	56.7 (866)	49.0 (840)	<0.001
Prevalence of type-2 diabetes (%/N)	32.8 (286)	25.3 (229)	28.9 (515)	0.0005
Prevalence of stroke (%/N)	11.9 (104)	7.1 (64)	9.4 (168)	0.0005
Baseline Modified Mini-Mental State Exam* (Mean/SE)	84.7 (0.43)	83.7 (0.41)	84.0 (0.42)	0.11
Baseline word list* (Mean/SE)	8.4 (0.09)	8.3 (0.09)	8.33 (0.09)	0.45
No medical insurance (%/N)	2.53 (22)	15.8 (143)	9.3 (165)	<0.001
Source of medical care not private clinic or doctor (%/N)	38.0 (336)	44.0 (396)	41.4(732)	0.45

* Adjusted for gender, age and education

in education diminishes with increasing income ($p > 0.0001$). Figure 5.2 shows the joint distribution of income and education by nativity for the study. A much greater percentage of Mexican/Other LatinAmerican-born falls into the lowest SES category compared to U.S.-born.

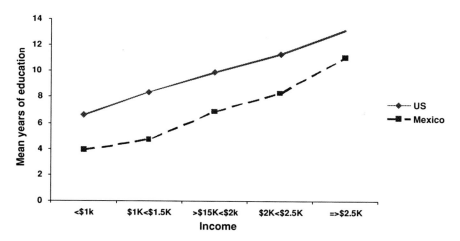

FIGURE 5.1. Mean years of education by total household income by country of birth.

More of the U.S.-born participants owned a house compared of the Mexican/Latin-born participants. Mexican/Other Latin American-born participants had lived an average of 37 (SE: 19.6) years in the U.S. Their mean cultural orientation score was lower than U.S.-born and they were significantly more likely to be monolingual in Spanish. Mexican/Latin-born were more likely than U.S.-born participants to live in a neighborhood that was at least 50% Latino and they had lived less time in that neighborhood.

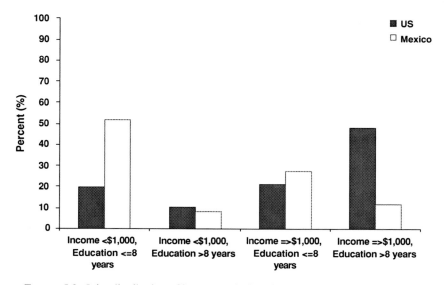

FIGURE 5.2. Joint distribution of income and education (<=> median) by nativity.

Mexican/Latin-born participants were more likely to report their health status as only "fair" or "poor" than U.S.-born. Among those born in Mexico/Latin America, fewer had type-2 diabetes at baseline compared to those born in the U.S. Stroke at baseline was much lower among Mexican/Other Latin American-born participants compared to those born in the U.S. At baseline, Mexican/Other Latin American-born did not differ from U.S.-born on the 3MSE or the verbal recall test (SEVLT). They were less likely to have any medical insurance and more of them reported not using a private clinic or doctor for medical care.

5.6. All-Cause Mortality

Among those born in Mexico or other Latin American countries, the crude mortality rate was 38.23 per 1,000 person-years compared to U.S.-born participants, for whom the mortality rate was 36.61. There were 157 deaths among Mexican/Latin-born and 160 among U.S.-born participants. Examination of the influence of nativity on mortality by gender revealed that crude rates among women were lower in Mexican/Latin-born women than in U.S.-born women. A series of proportional hazards models were used to evaluate the potential interaction between age and nativity by gender (Table 5.2). A significant interaction term ($p = 0.009$) was found between nativity and gender so the subsequent models were stratified on gender. The initial models included nativity, age, and an interaction term between nativity and age. For men, this interaction term was not significant and was dropped from the model. Among women, there was a significant interaction between age and nativity so this term was retained in all subsequent models. Mexican/Other Latin American-born women had a lower risk of dying than U.S.-born women, which was modified by age such that the advantage among Mexican/Latin-born women disappeared with increasing age. Adjustment for SES, stroke, and diabetes at baseline did not affect this finding. Among men, adjustment for SES, type-2 diabetes, and stroke slightly increased the magnitude of the association between the cultural scale and both nativity and mortality, but it was still not significant. For both men and women, lowest SES was associated with greater risk of death compared to highest SES. Those with higher income and lower education experienced a higher risk of death although among men this was not significant. Those with lower income and higher education did not differ from those higher on both. For each gender, type-2 diabetes at baseline was a significant predictor of mortality but stroke was not.

5.7. Cognitive Decline: Modified Mini-Mental State Exam

5.7.1. Nativity

Change over time in the 3MSE was assessed in relation to nativity and cultural orientation (Table 5.3). Overall, nativity in Mexico was associated with more errors

TABLE 5.2. Association between nativity or cultural orientation and all-causes mortality by gender from a series of proportional hazards models with adjustments for demographic and health covariates.

	Men		Women	
	Deaths by gender and nativity			
U.S.	76/379		89/493	
Mexico or other Latin American country	71/362		81/545	
	Model 1			
−2 LL (p-value)	5418		2293.73 (0.45)	
	Hazard Ratio	95% CL	Hazard Ratio	95% CL
Nativity (U.S. (0) vs. Mexico (1))	1.21	0.88–1.68	0.89	0.66–1.21
	Model 2			
−2 LL	1814.17		2126.6	
Nativity	3.27	0.16–80.19	0.03*	0.001–0.58
Age	1.08**	1.05–1.12	1.09**	1.05–1.12
Nativity*age			0.025[1]	0.001–0.61
	Model 3			
−2 LL	3329.82		3327.17	
Nativity	0.97	0.70–1.34	0.11	0.01–1.55
Age	1.08	1.06–1.10	1.06	1.03–1.09
Nativity*age			0.02[1]	0.001–0.68
Stroke	1.41	0.97–2.05	1.40	0.96–2.04
Diabetes	2.89	2.23–3.72	2.82	2.18–3.66
Monthly income[1] < $1,000 and education > 8 years	0.89	0.41–1.90	0.63	0.38–1.06
Monthly income =>$1,000 and education <= 8 years	0.85	0.56–1.29	0.77	0.53–1.18
Monthly income =>$1,000 and education > 8 years	0.72	0.46–1.14	0.43	0.26–0.71

* $p < 0.05$, ** $p < 0.00001$

[1] Calculated as the effect of 5-year difference in age on the association between nativity and mortality. $p = 0.003$. Nativity*age was not significant in model for men.

[2] Reference is monthly income <$1,000 and education < = 8 years.

on the 3MSE in models 1 and 2. The time*nativity interaction term was significant in the unadjusted and adjusted models ($p = 0.03$ and 0.005), suggesting that being born in Mexico was associated with more rapid decline in cognitive function over the four years of follow-up. Adjustment for covariates shown in the table reduced the overall association substantially (about 67%) but did not affect significance. The mean errors on the 3MSE among Mexican/Latin-born compared to U.S.-born participants were 1.12, 1.2, 1.21, and 1.35 points at the first, second, third, and fourth follow-up, respectively, compared to the baseline difference.

5.7.2. Cultural Orientation

Anglo orientation on the cultural scale (a higher number) was significantly associated overall with fewer errors on the 3MSE in both models (Table 5.3). The

TABLE 5.3. Nativity and cultural orientation as predictors of change in cognitive function from a series of mixed linear models.

	Modified Mini-Mental State Exam (log of errors)[1]		Delayed Word List Recall (square root)	
	Model 1	Model 2	Model 1	Model 2
Factor	Nativity		Nativity	
AIC	14036.8	12711.0	7141.1	6536.5
	Coefficients (p value)		Coefficients (p value)	
Intercept	2.25*	0.43*	2.91*	4.74*
Nativity (US (0) vs. Mexico (1))	0.49*	0.16*	–0.18*	–0.05
Time 4	–0.31*	0.24*	–0.05 ($p = 0.002$)	–0.06*
3	–0.33*	–0.29*	0.02	0.014
2	–0.36*	–0.35*	0.01	0.02
1	0.02 (ns)	–0.03 ns	–0.086*	–0.08*
Time*nativity (p-value for fixed effects)	$p = 0.03$	$p = 0.005$	$p = 0.73$	–
Age		0.033*		–0.03*
Gender		–0.009		0.25*
Type-2 diabetes		0.19*		–0.12*
Stroke		0.27*		–0.18*
Monthly income <$1,000 and education <= 8 years[2]		–0.91*		0.35*
Monthly income <$1,000 and education > 8 years		–0.32*		0.14*
Monthly income => $1,000 and education <= 8 years		–0.67*		0.23*
Factor	Cultural orientation		Cultural orientation	
AIC	12,902.6	11,945.4	6348.2	5,898.9
Intercept	3.01	0.76*	2.66*	4.33
Culture	–0.03*	–0.01*	0.009 *	0.004 (0.004)
Time 4	–0.04	0.003	–0.04	–0.06*
3	–0.24*	–0.24*	0.02	0.013
2	–0.27*	–0.27*	0.02	0.011
1	–0.03	–0.03	–0.11*	–0.084
Culture*time (p-value)	0.0006	0.0004	$p = 0.27$	–
Age		0.03*		–0.03*
Gender		–0.006		0.25
Type-2 diabetes		0.17*		–0.09 *
Stroke		0.19*		–0.12*
Monthly income <$1,000 and education <= 8 years[2]		–0.75*		0.28*
Monthly income <$1,000 and education > 8 years		–0.26*		0.11*
Monthly income =>$1,000 and education <= 8 years		–0.54*		0.18*

[1] Formula: log (101 – 3MSE).
[2] Reference is participants with high income and high education
* $p < 0.01$

time*cultural scale term was significant in both models. Adjustment for covariates in model 2 substantially affected the magnitude of the time*cultural scale coefficient, but it remained significant. A 1 SD (12.97) difference in the cultural scale resulted in 1.2 fewer errors at each follow-up in model 3. Type-2 diabetes, stroke, gender, age, and socioeconomic status were all significantly associated with the 3MSE in all models.

5.8. Spanish English Verbal Learning Test

Cultural orientation was weakly but significantly associated with the SEVLT overall score. The time*cultural score term was not significant.

5.9. Health Status

Type-2 diabetes and stroke were both associated with more errors on the 3MSE and a lower score on the SEVLT. These adjustments did not influence the associations between nativity or cultural orientation and the cognitive outcomes.

5.10. Cognitive Decline and Socioeconomic Status

Overall, SES was associated with significantly more errors on the 3MSE: compared to those with both income and education below the median, those who were high on both measures had the least number of errors, followed by those with low income and high education, then by those with high income and low education. The time* SES interaction term was significant ($p < 0.0001$), indicating that errors on the 3MSE increased more rapidly among those in the lowest SES group compared to the other groups (data not shown). Those with high income and low education were the least different for a change in errors on the 3MSE from the lowest SES group, followed by those with low income and high education.

For the SEVLT, compared to those in the highest SES group, those in the lowest group also scored lowest on the test, followed by those in the high income, lower education group, then followed by those in the lower income, higher education group. For the SEVLT the time*SES interaction term was not significant, indicating that SES was not associated with decline in the SEVLT scores.

5.11. Dementia/CIND Incidence

The incidence rates were calculated for combined dementia and CIND through the fourth follow-up year by nativity (Table 5.4). The cultural orientation score was categorized into three groups (terciles), and incidence rates were also computed. Table 5.4 shows the dementia/CIND rates by Nativity. Rates were higher

TABLE 5.4. Incidence of dementia/CIND by nativity and cultural orientation from a series of proportional hazards models.

	Nativity		
	Mexico		United States
	Crude incidence rate of dementia/CIND		
Rates in person years			
Crude rate (N)	15.43 (60)		12.35 (51)
Total person yrs (py)	3887.77		4128.60
Total N	846		828
	Proportional hazards models[1] cultural orientation score		
Hazard ratio (95% CL)	Mexico (1)		0.64 (0.52–0.78)[2]
	USA (0)		0.97 (0.94–1.01)[3]
	Cultural orientation score (tertiles)		
	Mexican	Middle	Anglo
Rate per 1,000 py	17.88	16.58	8.29
Number of cases	40	43	24
Total number	510	542	557
Total person years	2237.62	2593.19	2893.9
	Cultural orientation score: Proportional hazards models		
	Unadjusted	Adjusted[4]	Full Model[5]
Hazard ratio (95% CL)	0.98 (0.96–0.99)	0.98 (0.96–0.99)	0.99 (0.97–1.01)

[1] Adjusted for age, type-2 diabetes, stroke, gender, combined education and income. Cultural score continuous in model
[2] Calculated as the effect of one-year increase in age on the association between a one-unit difference in cultural score and dementia/CIND. As age increases, the protective effect of a higher cultural score decreases.
[3] Culture*age interaction term not significant and was dropped from the model.
[4] Adjusted for age and gender, baseline diabetes and stroke.
[5] Adjusted for age, gender, baseline diabetes and stroke, SES.

among those born in Mexico or another latin American country than among those born in the US. The association between cultural orientation score and rate of dementia/CIND was examined in models stratified by nativity. Among those born in Mexico, a higher cultural orientation score was associated with a lower rate of dementia/CIND. Among those born in the US, cultural orientation score had no association with dementia/CIND.

Cultural orientation score was divided into tertiles. Those in the lowest and middle tertiles (most Mexican oriented) had higher rates of dementia/CIND than those in the most Anglo oriented group. Cultural orientation score as a continuous variable was examined in relation to dementia/CIND. Adjustment for age, gender, diabetes, stroke and education was included in the full model. Cultural orientation in this model had only a slight effect on the outcome and was attenuated by adjustment for covariates. Anglo orientation was associated with a significantly lower risk of dementia/CIND compared to Mexican orientation ($p = 0.005$). Two proportional hazards models were fit-stratified on nativity. Effect modification of the association between cultural orientation and age was tested in each model. The term was not significant in the U.S.-born model and was dropped from analyses. The term was significant in the Mexican/Latin-born model ($p = 0.003$) and was

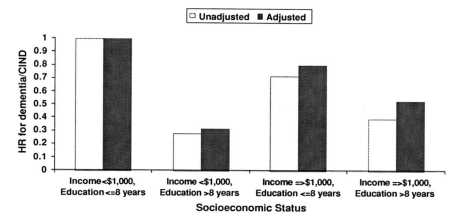

FIGURE 5.3. Risk of dementia/CIND associated with combined income and education in study participants from proportional hazards models.

retained. For a 5-year difference in age the protective effect of cultural orientation for dementia/CIND was diminished by about 6%. Adjustment for age, education, type-2 diabetes, and stroke and waist circumference did not substantially influence this association. Among the U.S.-born, cultural orientation was not associated with dementia/CIND incidence even after adjustment for covariates.

Proportional hazards models were used to test the association between the continuous forms of the cultural orientation score. A one standard deviation (SD = 1) difference in cultural orientation was associated with a 6% lower risk of dementia/CIND. Adjustment for age and gender, diabetes, or stroke did not influence this association. Inclusion of the SES variable reduced the risk difference to 1% per unit of the score and the hazard ratio became non-significant.

5.11.1. Socioeconomic Status

Figure 5.3 shows the results of a proportional hazards model, including the SES measure as indicator variables with the lowest SES as a reference. In both the unadjusted and adjusted models, those with lower income and higher education were at significantly lower risk for dementia/CIND compared to those in the lowest SES group, as were those with both higher income and higher education. Those with higher income but lower education were not significantly different in the unadjusted model from those in the lowest SES group. In the adjusted model, age, baseline diabetes, and stroke were significant predictors of dementia/CIND incidence. There was no gender difference.

5.12. Discussion

We found substantial differences in social, economic, cultural, and health factors by nativity in this study. Most of these would increase the risk of mortality, cognitive decline, and dementia/CIND for those born in Mexico or Latin America compared to those born in the United States. However, our results suggest that there are no overall mortality differences by nativity or by cultural orientation. Our findings for the effects of nativity on mortality among women suggest that Mexican/Latin-born women have an advantage compared to U.S.-born women; but this advantage weakens with increasing age. Mexican/Latin nativity was associated with worse cognitive scores that were largely explained by SES and health status at baseline. Cultural orientation is also strongly correlated with socioeconomic factors, but weakly predicts cognitive outcomes. Cultural orientation is associated with dementia/CIND, but this association is explained by adjustment for socioeconomic status. Those with lower income and education were at highest risk for dementia/CIND.

Our assumption that those lost to follow-up are alive and/or not demented represents a conservative bias in our results. However, we have not found a difference in loss to follow-up or in refusals to continue in the study by nativity. Attrition from the cohort due to causes other than death accounts for about 24% of the original cohort, or 6% per year. We have not found evidence that remigration to Mexico has created an underestimate of mortality rates or of dementia rates. The use of complex longitudinal models helps us include all information for follow-up years that is available to us. Since participants who were immigrants arrived in the U.S. on average 37 years before the study (Range 1–85), selective effects due to the "healthy" migrant' effect (Abraído-Lanza, Dohrenwend, Ng-Mak, and Turner, 1999; Franzini, Ribble, and Keddie, 2001) probably have little impact on the participants' health status today.

Other studies have reported similar findings regarding predictors of decline in cognitive outcomes (Raji, Al Snih, Ray, Patel, and Markides, 2004; Raji, Ostir, Markides, and Goodwin, 2002). There are few other population-based studies that have published reports to evaluate the influence of immigration status or cultural orientation on dementia incidence or cognitive decline in an elderly population. A notable exception to this is the work by Angel (Angel et al., 2001) and the work by Black (Black and Rush, 2002) and by Nguyen et al. (Nguyen, Black, Ray, Espino, and Markides, 2003); both reports had similar findings for cognitive outcomes.

Our work strongly supports the notion that immigrants continue to experience worse health status than non-immigrants. To some extent this may be mitigated by the degree to which new cultural forms are adopted, but education and current socioeconomic factors appear to be more important. As in other studies, stroke and type-2 diabetes are strong predictors of decline on these measures and serve as pathways by which social disadvantage attached to migration has its impact on health of older Mexican Americans. Major chronic conditions in this group are worse among those at lower SES but also predict mortality, cognitive decline, and dementia/CIND independently of SES.

Socioeconomic status has a strong influence on the health of older Mexican Americans. Those with higher education generally had better health outcomes regardless of income status than those with lower education. However, the combination of high education and income provided substantially more protection for health than high education alone. Education is usually attained in early life and reflects childhood socioeconomic circumstances. These early exposures influence disease risk via biological pathways and via later opportunities for gaining material wealth. The category "Years of education" predicts economic success in immigrants. Among those who are discrepant on income and education the group with low income and higher education is at lower risk for cognitive impairment, dementia/CIND, and mortality compared to the lowest group. Those who fall into the group with higher income and lower education have worse health outcomes in our analyses than those in the other discrepant group. In summary, the effects of cultural orientation and nativity on health are likely to be mediated or explained by socioeconomic status.

References

Abraído-Lanza, A.F., Dohrenwend, B.P., Ng-Mak, D.S., and Turner, J.B. (1999). The Latino mortality paradox: A test of the "salmon bias" and healthy migrant hypotheses. *American Journal of Public Health, 89*(10), 1543–1548.

Angel, J.L., Buckley, C.J., and Sakamoto, A. (2001). Duration or disadvantage? Exploring nativity, ethnicity, and health in midlife. *Journals of Gerontology Series B—Psychological and Social Sciences, 56*(5), S275–284.

Benjamin, B.J. (1997). Normal aging and the multicultural population. *Seminars in Speech and Language, 18*(2), 127–134.

Black, S.A., and Rush, R.D. (2002). Cognitive and functional decline in adults aged 75 and older. *Journal of American Geriatrics Society, 50*(12), 1978–1986.

Buckley, C., Angel, J.L., and Donahue, D. (2000). Nativity and older women's health: Constructed reliance in the health and retirement study. *Journal of Women and Aging, 12*(3–4), 21–37.

Cuellar, I., Arnold, B., and Maldonado, R. (1995). Acculturation rating scale for Mexican Americans-II: A revision of the original ARMSA scale. *Hispanic Journal of Behavioral Sciences, 17*, 275–304.

Day, J. (1996) Population projections of the United States by age, sex, race, and Hispanic origin: 1995 to 2050. *Current Population Reports, Series P-25-1130*, U.S. Census Bureau. Washington, DC.

Franzini, L., Ribble, J.C., and Keddie, A.M. (2001). Understanding the Hispanic paradox. *Ethnicity and Disease, 11*(3), 496–518.

Graves, A.B., Rajaram, L., Bowen, J.D., McCormick, W.C., McCurry, S.M., and Larson, E.B. (1999). Cognitive decline and Japanese culture in a cohort of older Japanese Americans in King County, WA: The Kame Project. *Journals of Gerontology Series B—Psychological and Social Sciences, 54*(3), S154–61.

Haan, M.N., Mungas, D.M., Gonzalez, H.M., Ortiz, T.A., Acharya, A., and Jagust, W.J. (2003). Prevalence of dementia in older Latinos: The influence of type 2 diabetes mellitus, stroke and genetic factors. *Journal of American Geriatrics Society, 51*(2), 169–77.

Larsen, L.J. (2004) *Foreign-born population in the United States.* 1–9. *Current Population Reports*, Series P-20-551, U.S. census Bureau: Washington, DC.

Mungas, D., Reed, B.R., Haan, M.N., and Gonzalez, H. (2005). Spanish and English neuropsychological assessment scales: Relationship to demographics, language, cognition, and independent function. *Neuropsychology, 19*(4), 466–75.

Nguyen, H.T., Black, S.A., Ray, L.A., Espino, D.V., and Markides, K.S. (2003). Cognitive impairment and mortality in older Mexican Americans. *Journal of American Geriatrics Society, 51*(2), 178–83.

Raji, M.A., Al Snih, S., Ray, L.A., Patel, K.V., and Markides, K.S. (2004). Cognitive status and incident disability in older Mexican Americans: Findings from the Hispanic established population for the epidemiological study of the elderly. *Ethnicity and Disease, 14*(1), 26–31.

Raji, M.A., Ostir, G.V., Markides, K.S., and Goodwin, J.S. (2002). The interaction of cognitive and emotional status on subsequent physical functioning in older Mexican Americans: Findings from the Hispanic established population for the epidemiologic study of the elderly. *Journals of Gerontology Series A—Biological and Medical Sciences, 57*(10), M678–82.

Stephen, E.H., Foote, K., Hendershot, G.E., and Schoenborn, C.A. (1994). Health of the foreign-born population: United States, 1989–90. *Advance Data,* (241), 1–12.

Teng, E.L., and Chui, H.C. (1987). The Modified Mini-Mental State (3MS) examination. *Journal of Clinical Psychiatry, 48*(8), 314–318.

6
Education and Mortality Risk Among Hispanic Adults in the United States*

Sarah A. McKinnon and Robert A. Hummer

6.1. Introduction

In the United States the impact of socioeconomic status on racial and ethnic mortality differentials is undeniable. Researchers have consistently found that groups with higher levels of socioeconomic status, namely whites and Asian Americans, have better mortality outcomes than those with lower overall socioeconomic status, such as blacks and American Indians (e.g., Elo and Preston, 1996; Hayward, Miles, Crimmins, and Yang, 2000; Hummer, Benjamins, and Rogers, 2004; Kitagawa and Hauser, 1973; Molla, Madans, and Wagener, 2004; Pappas, Queen, Hadden, and Fisher, 1993; Rogers, Hummer, and Nam, 2000). Indicators of socioeconomic status including income, education, employment status, occupation, and wealth have been found to be strongly associated with mortality risks among a variety of different populations and across time. Yet, for one racial/ethnic group, Hispanics, the link between socioeconomic status and mortality is not as apparent. For at least two decades now, researchers have noted that, in spite of an overall socioeconomic status that is similar to that of blacks, Hispanics tend to exhibit mortality rates much more like those of whites (Markides and Coreil, 1986). Termed the "epidemiologic paradox," this phenomenon has received a substantial amount of research interest. Initially, most studies demonstrated support for the paradox and concluded that Hispanics appeared to have similar mortality risks when compared to non-Hispanic whites and/or more favorable risks compared to non-Hispanic blacks (Abraído-Lanza, Dohrenwend, Ng-Mak, and Turner, 1999; Elo, Turra, Kestenbaum, and Ferguson, 2004; Hummer, Rogers, Amir, Forbes, and Frisbie, 2000; Rosenwaike, 1987). Subsequent research has been more critical of the observed findings and has offered a variety of other potential explanations (e.g., immigrant selectivity, return migration selectivity, and racial/ethnic misclassification) for the apparent

* We thank Starling Pullum for her computing assistance. This research was supported, in part, by a population center infrastructure grant from the National Institute of Child Health and Human Development (Grant # 5 R24 HD 42849) and by grant # 1 R01 HD 053696 from the National Institute of Child Health and Development. Please contact Sarah McKinnon at sarah@prc.utexas.edu with any questions or comments.

paradoxical situation for Hispanics (Palloni and Arias, 2004; Palloni and Ewbank, 2004; Palloni and Morenoff, 2001).

Although both earlier and later studies have contributed to a greater knowledge of overall Hispanic mortality patterns, analyses of the relationship between socioeconomic status and mortality patterns have been lacking. As the definition of the epidemiologic paradox is based on the unexpectedly low mortality rates of Hispanics given their relatively low socioeconomic status level, thus implying that their socioeconomic status is not impacting their mortality in the expected direction, we believe that this is a critical omission that demands research attention. Therefore, our primary goal in this chapter is to advance our understanding of the relationship between socioeconomic status and mortality for Hispanics as compared to other population groups in the United States. We select education as our measure of socioeconomic status for several important reasons. First, it is fairly constant, changing very infrequently once individuals become adults. Second, it has been strongly linked to adult health and mortality outcomes in the U.S. population and among other racial/ethnic subgroups. And, finally, it is often considered to be the driving socioeconomic variable in differentiating health and mortality patterns across the life course (Crimmins, 2005; Mirowsky and Ross, 2003; Smith, 2005). In this chapter we specifically address three questions: 1) To what extent do differences in education influence mortality differentials between Hispanics and whites and between Hispanics and blacks? 2) Does the education-mortality relationship differ for Hispanic groups in comparison to blacks and whites? 3) How does the education-mortality relationship among Hispanics differ when separate age and nativity groups of adults are considered? To best answer these questions we utilize data from multiple years of the National Health Interview Survey (NHIS) linked with the Multiple Cause of Death File (MCD), hereafter referred to as the NHIS-MCD. Specific age subpopulations (25–59 and 60+) and nativity groups (foreign-born and native-born) will be separately examined in part of this work in order to most appropriately address these questions.

6.2. Background

6.2.1. General Studies of Educational Differences in Mortality

The relationship between educational attainment and mortality risk is well established. One of the earliest and most comprehensive studies in the U.S. was conducted by Kitagawa and Hauser (1973), using national and metropolitan (Chicago) mortality data from 1960. The researchers found that higher levels of education were associated with lower overall mortality, infant mortality, and cause-specific mortality risks. They determined that the association was similar for both whites and non-whites; at that time, however, the non-white population of the U.S. was mostly black. In 1993, Pappas et al. replicated Kitagawa and Hauser's study using the National Mortality Followback Survey and the NHIS from 1986. They found that lower education continued to be associated with higher mortality rates for whites and blacks and for both genders. In addition, they concluded that the relationship between education and mortality had actually strengthened since Kitagawa and Hauser's analysis of 1960 data.

Successive research using a variety of data sources has continued to provide support for the strong association between education and adult mortality. Elo and Preston (1996) examined National Longitudinal Mortality Survey data and found sizable education-mortality differentials among the U.S. adult population even after controlling for income, marital status and current place of residence. Rogers, Hummer, and Nam (2000) used NHIS-MCD data and reported that adults with less education had substantially higher mortality risks. Again, this relationship remained consistent even with controls for other socioeconomic factors as well as perceived health status. Likewise, Molla, Madans, and Wagener (2004), analyzing data from the U.S. Census and the National Center for Health Statistics, were able to demonstrate that death rates are lowest and life expectancy is highest among individuals with the greatest educational attainment.

A recent study conducted by Smith (2005) proposes that studies of socioeconomic status and mortality risk should focus on education rather than income. Smith found consistent evidence that poor adult health can and does impact income by reducing one's ability to work. Therefore, those at greater risk of mortality would also be more likely to have a lower income due to these prior health conditions. In contrast, adult levels of education are rarely affected by adult health because, for the most part, individuals complete their education prior to turning 25 years of age. Smith also found education to be a very strong predictor of health change among persons aged 51 and older; that is, persons with lower levels of education were much more likely to experience negative health during the follow-up period. Moreover, sociological work has strongly emphasized the importance of education for the health outcomes of adults across the life course. Mirowsky and Ross (2003) most forcefully argue that higher levels of education impact individuals' sense of control over their lives, which helps to encourage healthy living. Further, higher levels of education can directly lead to better and more stable employment, higher incomes, and an overall higher standard of living, all of which are associated with better health and lower mortality.

6.2.2. The Epidemiologic Paradox

Markides and Coreil (1986) first introduced the concept of an epidemiologic paradox for Hispanic health, basing their ideas upon a review of studies that examined the health and mortality of Hispanics residing in the southwestern region of the United States. In terms of mortality, Markides and Coreil discovered that life expectancy rates among southwestern Hispanics were comparable to those of whites and substantially higher than those of blacks. In addition, they also found that Hispanic mortality rates from cardiovascular disease and some cancers were significantly lower than rates of non-whites and more comparable to those of whites. After a thorough review of the studies, the authors concluded that, indeed, a paradox does seem to exist in which "the health status of Hispanics in the Southwest is much more similar to the health status of other Whites than that of Blacks although socioeconomically, the status of Hispanics is closer to that of Blacks" (p. 253). However, this review did not include any studies that specifically examined the association between socioeconomic status (or, more specifically, education) at the individual level and Hispanic mortality risk. In addition, few of the early studies on Hispanic health and mortality differentiated between persons born in the U.S. and those born outside the U.S. and, with their regional focus on the Southwest, the authors were primarily presenting information on Hispanics of Mexican origin.

Most demographic research work in this area has since focused more on the technical aspects of Hispanic mortality estimates, for example trying to account for selective out-migration, making corrections for questionable data, and dealing with the issue of racial/ethnic reporting (Elo et al., 2004; Hummer et al., 2004; Palloni and Arias, 2004; Turra, Elo, Kestenbaum, and Ferguson, 2004). Although this whole set of studies agrees that officially reported Hispanic mortality rates are too low because of data quality issues, there is not yet consensus with regard to whether or not Hispanic adult mortality rates are actually slightly lower than whites, the same as whites, or slightly higher than whites. The answer to this question probably depends on the Hispanic subgroup in question, whether or not immigrant Hispanics are separated out from native-born Hispanics, the specific causes of death and age groups that are examined, and the ever-present data-quality issues that are particularly important when studying a very mobile population like Hispanics.

At the same time, relatively few studies have directly examined the education-mortality relationship among Hispanics vis-à-vis non-Hispanic blacks and whites. In fact, most studies of Hispanic mortality that include education do so as a control rather than a primary variable of interest. For example, Hummer, Rogers, Nam, and Le Clere (1999) used 1989–1994 NHIS data linked to 1989–1995 National Death Index data to examine racial/ethnic/nativity differentials in adult mortality in the U.S. They found that overall mortality risk was significantly higher among non-Hispanic blacks than among non-Hispanic whites, yet found no significant difference between whites and Mexican Americans or other Hispanics. However, once they controlled for nativity status the Hispanic groups demonstrated significantly higher mortality risks. In addition, the researchers conducted regression analyses using racial/ethnic groups that had been differentiated into foreign- and native-born subgroups. They found that foreign-born individuals tended to have lower mortality risks than native-born individuals within most groups, especially among the older adults. These findings highlight the importance of differentiating both age and nativity status groups in studies of Hispanic mortality. Although this study did not specifically address the relationship between mortality and socioeconomic status, socioeconomic controls (including education) were included in all models. The inclusion of these controls resulted in decreases in mortality risks for all racial/ethnic groups and, in fact, for Mexican Americans and other Hispanics resulted in mortality risks that no longer differed from those of non-Hispanic whites.

In a subsequent analysis that further differentiated among the Hispanic subgroups, Hummer et al. (2000) found that Mexican Americans, Cubans, and other Hispanics had adult mortality risks that did not significantly differ from those of white non-Hispanics. In contrast, Central/South Americans had significantly lower risks, whereas Puerto Ricans had higher risks of mortality. By separate age and sex groups, they found higher mortality risk among Mexican American and Puerto Rican males and females between the ages of 18 and 44 and Cuban and other Hispanic men between the ages of 18 and 44. For all other age and sex groups, mortality risks were equal to or lower for Hispanics than for whites. Once again, education was included as a control variable along with other socioeconomic variables (income and employment). The authors find that once these controls are included in the model, mortality risks decrease for all groups, with the strongest declines for Mexican Americans, Puerto Ricans, and Central/South Americans. However, they did not formally test interaction terms that could have determined whether or not socioeconomic factors differed in their effects on mortality across the racial/ethnic groups.

Lin, Rogot, Johnson, Sorlie, and Arias (2003) published a study that included a more thorough analysis of the education-mortality relationship of Hispanics as compared to non-Hispanic whites and blacks. In this study, they found that Hispanics demonstrated the smallest differences in life expectancy across education groups. For example, life expectancy at age 25 increased by 3 years for both white and black males with 12 years of education as compared to those with less than 12 years of education. However, for Hispanic males, life expectancy only increased by half a year. Unfortunately, due to small sample sizes this study did not include information on life expectancies of Hispanics with 13 or more years of education nor are Hispanics differentiated by immigrant status.

In one of the most recent studies on this topic, Hummer et al. (2004) examined mortality differentials among elderly racial/ethnic subpopulations in the U.S. using NHIS–MCD data from 1989–1997 and, once again, concluded that mortality rates were more favorable for both Mexican-origin and other Hispanic populations when compared to whites and blacks. The authors also determined that nativity status and length of residence in the United States impacted mortality risk, although controlling for nativity did not completely explain the advantageous Hispanic mortality rates. They found that education had only a modest impact on mortality risk among the elderly Hispanic population. Indeed, after adding a control for education into their model, the hazard ratios for Mexican Americans and other Hispanics (compared to whites) only decreased by 0.05 and 0.01, respectively.

Although the findings of the above studies vary to some degree, there is some evidence that socioeconomic status impacts Hispanic adult mortality in the expected direction, but perhaps with weaker effects than for other racial/ethnic subgroups. However, most of the studies merely included education as a control variable and, many times, did not distinguish education from other socioeconomic controls. Also, many of the studies to date have not differentiated foreign-born Hispanics from native-born Hispanics, which may be very critical in examining socioeconomic differences in mortality. Educational levels are much lower among foreign-born Hispanics in comparison to native-born Hispanics, especially for Mexican Americans (Saenz 2004). Further, educational changes in the U.S. have been so profound over the past 40 years that older persons need to be separated from younger persons if at all possible when conducting education–mortality analyses. In order to better understand the degree to which the socioeconomic status of Hispanic groups is related to their mortality risk relative to both whites and blacks, careful comparisons across age and nativity groups must be made. This study attempts to address these issues by providing analyses of educational differences in adult mortality risk for Hispanic subgroups as well as for non-Hispanic blacks and whites. Age subgroups and nativity subgroups are separately examined as an important part of this effort.

6.3. Data, Measurement, and Methods

6.3.1. Data Set

This study uses data from the National Health Interview Survey (NHIS) for the years 1986–1994 linked to the National Death Index (NDI) for the years 1986–1997 (NHIS–MCD) to analyze educational differences in Hispanic adult mortality. The NHIS is a cross-sectional, nationally representative survey of the civilian non-institutionalized population residing in the United States (NCHS, 1997). Although the primary purpose of the NHIS is to provide information about health, mortality information is available from linkages made from identifying information of adults 18 and older to the National Death Index. The linkages are made utilizing the following criteria: social security number, first and last name, middle initial, father's surname, month and year of birth, age, sex, race, state of birth, and state of residence. The criteria are weighted and summed to determine the quality of potential matches. Although it is not possible to link every individual, we use the links recommended by the National Center for Health Statistics that is believed to correctly match 97–98% of individuals (NCHS, 1997). Individuals with poor identifying information, termed "ineligibles," are dropped from the study. The advantages of this data set include the use of self-reported race/ethnicity from the NHIS and a very large sample size as a result of the merging of 9 years of survey data with up to 12 years of mortality follow-up data. Further, there is detailed information on a number of demographic and socioeconomic factors that allow for in-depth adult mortality analyses (i.e., Rogers et al., 2000).

6.3.2. Measurement

Race/ethnicity was constructed using two questions in the NHIS. The first, Hispanic Origin, was used to identify the specific Hispanic ethnic groups: Mexican origin, Puerto Rican, and other Hispanics (Cuban, other Latin American, Spanish, and Hispanic but unsure of origin). For persons who identified themselves as non-Hispanic, the second question, Main Racial Background, was used to identify non-Hispanic blacks and whites. Because the focus of our study was limited to

Hispanics and non-Hispanic whites and blacks, we excluded the following categories: Aleut, Eskimo, or American Indian; Asian/Pacific Islander; other; multiple races; and unknown. We selected "completed years of education" as our education measure. We categorized education into: a) less than 11 years; b) 12 years; and c) 13 or more years. Demographic variables included age, sex, marital status, and nativity. In order to allow for sufficient time for individuals to complete their education, we excluded all individuals below the age of 25 from our analyses. In our analyses we used age as both a continuous control variable as well as a stratifying variable (25 to 59, and 60 and above). Marital status was measured as a dichotomous variable with currently married individuals differentiated from widowed, divorced, separated, and never married individuals. After 1988 NHIS included a question regarding the number of years an individual has resided in the United States. As this question was only asked of individuals who were not born in the U.S., nativity status could be ascertained. Thus, for those surveyed after 1988 we differentiated the foreign-born from the native-born. For those surveyed earlier, we denote their nativity status as missing.

6.3.3. Methodology

Descriptive analyses are provided for each racial/ethnic group in our sample. Mortality risks are analyzed using proportional hazard models (Allison 1984). In our first model we examine mortality risk for Hispanic adults of all ages compared to whites and blacks controlling only for age, sex, and marital status. The next model includes the addition of nativity status and the third model includes both nativity status and education. In order to specify the relationship between education and each racial/ethnic group's mortality risks we add a race/ethnicity-by-education interaction term in our final model. Several of the studies we review above suggest that relationships between education and mortality for the racial/ethnic groups are differentiated by age. Therefore, we specify proportional hazard models for all ages (25 and above) as well as for the age groups of 25 to 59, and 60 and above. Finally, in our last set of models, each racial/ethnic group is considered separately, with subgroups by nativity also separated out, to best determine if the relationship between education and mortality risk differs by race/ethnicity and nativity. Again, we not only estimate hazard models for all ages but also for younger and older adult age groups separately. All regression coefficients are reported as hazard ratios throughout the analyses. We weight the data according to NHIS-provided weights. Although descriptive analyses were conducted using SAS version 9.1, we use Sudaan version 9.0.1 to adjust for the design effects of the NHIS in our regression analyses.

6.4. Results

6.4.1. Descriptive Results

Table 6.1 provides descriptive statistics for the demographic and mortality outcome variables for each racial and ethnic group. Whites and blacks constitute the

TABLE 6.1. Descriptive statistics of mortality risk factors and outcomes by race/ethnicity: U.S., 1986–1994

	Mexican American	Puerto Rican	Other Hispanic	Non-Hispanic black	Non-Hispanic white
Demographic variables					
Age[+]					
Mean in years (sd)	41.3 (13.8)	42.4 (13.7)	44.2 (15.0)	46.8 (15.9)	48.8 (16.6)
Sex					
% male	49.6	43.9	45.4	44.1	47.7
% female	50.4	56.1	54.6	55.9	52.3
Nativity					
% Foreign-born	34.9	45.7	47.4	4.4	3.2
% U.S.-born	37.5	24.6	22.4	64.2	64.1
% Missing	27.6	29.7	30.2	31.5	32.7
Marital status					
% Married	74.4	61.5	68.0	51.0	73.0
% Not married	25.6	38.5	32.0	49.0	27.0
Socioeconomic variables					
Education					
% Less than 12 years	52.7	41.5	31.1	32.4	18.8
% 12 years	27.4	31.7	30.7	37.4	39.0
% 13 or more years	19.9	26.8	38.3	30.2	42.3
Outcome variables					
% died during follow-up	4.6	5.4	5.2	9.3	9.0
% survived during follow-up	95.4	94.6	94.8	90.7	91.0
Group N[+]	21,993	4,788	14,274	77,366	466,100

[+] Not weighted

majority of the sample, followed by Mexican Americans and other Hispanics. The non-Hispanic groups also tend to be older and have a substantially lower percentage of foreign-born persons. Blacks have, by far, the lowest percentage of currently married individuals (51.0%) followed by Puerto Ricans (61.5%). Mexican Americans and Puerto Ricans have the highest percentage of people with less than 12 years of education (52.7% and 41.5%, respectively) and the lowest percentage of people with 13 or more years of education (19.9% and 26.8%, respectively). Consistent with their age compositions, mortality during the follow-up period was highest among non-Hispanic whites (9.0%) and blacks (9.3%) and lowest among Mexican Americans (4.6%).

6.4.2. Hazard Models of Overall Adult Mortality

Table 6.2 presents the mortality risks of each Hispanic subgroup as well as non-Hispanic blacks compared to non-Hispanic whites for ages 25 and above. As seen in the first model controlling for demographic variables, only blacks have a significantly higher adult mortality risk than whites. In contrast, Mexican-Americans and Puerto Ricans do not differ from whites while the mortality of other Hispanics

TABLE 6.2. Overall adult mortality risks for Hispanic subgroups and non-Hispanic blacks relative to non-Hispanic whites: ages 25 and above ($n = 584{,}521$)

| | Hazard ratios | | | |
	Model 1	Model 2	Model 3	Model 4
Ethnicity (non-Hispanic white)				
Mexican American	0.97	1.01	0.91*	1.03
Puerto Rican	1.11	1.21**	1.10	1.05
Other Hispanic	0.83**	0.88**	0.86**	0.99
Non-Hispanic black	1.30**	1.30**	1.21**	1.38**
Demographic variables				
Age (continuous)	1.09**	1.09**	1.08**	1.08**
Sex (female)				
Male	1.87**	1.87**	1.88**	1.88**
Marital status (married)				
Not married	1.31**	1.31**	1.30**	1.30**
Nativity (U.S.-born)				
Immigrant		0.81**	0.80**	0.80**
Missing		0.99	0.98*	0.98*
Socioeconomic variables				
Education (13+ years)				
Less than 12 years			1.45**	1.49**
12 years			1.22**	1.23**
Interaction variables				
Race/ethnicity*education				
Mexican American*less than 12 years				0.90
Mexican American*12 years				0.73+
Puerto Rican*less than 12 years				0.99
Puerto Rican*12 years				1.24
Other Hispanic*less than 12 years				0.80**
Other Hispanic*12 years				0.87
Black*less than 12 years				0.81**
Black*12 years				0.98
–2*log-likelihood	710,370	710,240	709,296	709,237
Degrees of freedom	7	9	11	19

Source: 1986–1994 National Health Interview Survey–Multiple Cause of Death Linked File (NCHS, 1997)**$p < .01$; *$p < 0.05$; +$p < 0.10$

is significantly lower than whites. With the inclusion of nativity status, both blacks and Puerto Ricans now demonstrate significantly higher mortality risks while the risk among other Hispanics remains significantly lower than that of whites. Mexican Americans continue to not demonstrate any statistically significant differences from whites in Model 2, thus, substantiating the assertion of an epidemiologic paradox spelled out by Markides and Coreil (1986).

We include our key measure of education in Model 3. Consistent with previous research (Hummer et al., 2000; Liao, Cooper, Cao, Durazo-Arvizu, Kaufman, Luke, and McGee, 1998), controlling for education leads to significantly lower mortality risk among Mexican Americans vis-à-vis whites. In addition, Puerto Ricans no longer demonstrate a significant difference from whites while the mortality

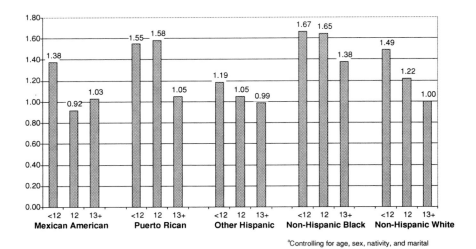

^aControlling for age, sex, nativity, and marital

FIGURE 6.1. Hazard ratios depicting educational differences in adult mortality by race/ethnicity group compared to non-Hispanic whites with 137 years of education.[a]

risk of blacks diminishes somewhat (although remains significant) in comparison to whites. In Model 4, interaction terms for the multiplicative effects of education by race/ethnicity are included, thus allowing us to identify a possible differential impact of education for the specific racial/ethnic groups. For ease in interpreting this model we have graphed the results of the relationship between education and mortality risk for each racial/ethnic group (Figure 6.1). The figure demonstrates that whites with less than 12 years of education experience a 1.49 times greater risk of death than whites with 13 or more years of education. However, Mexican Americans and other Hispanics with less than 12 years of education demonstrate hazard ratios that are modestly lower than the hazard ratios of whites, as they experience 1.38 and 1.19 times greater mortality risk than whites with 13 or more years of education. In contrast, Puerto Ricans and blacks with the lowest educational attainment experience the highest mortality risks (1.55 and 1.67, respectively). Figure 6.1 also shows that whites with 12 years of education experience a risk of death 1.22 times higher than whites with 13 or more years of education. Yet, for both Mexican Americans and other Hispanics, mortality risks for persons with 12 years of education are more similar to those of whites with high education.

As there is sufficient prior evidence that associations between education and mortality vary extensively by age, we next extend our analysis by estimating hazard models of race/ethnicity, education, and mortality for specific age groups. Table 6.3 focuses on adults aged 25 through 59. In the first model, we see that younger adult Mexican Americans, Puerto Ricans, and non-Hispanic blacks all have significantly elevated mortality risks when compared to younger adult non-Hispanic whites. Only one group, other Hispanics, demonstrates a mortality risk that does not differ significantly from that of whites. The inclusion of nativity in Model 2 results in

TABLE 6.3. Overall adult mortality risks for Hispanic subgroups and non-Hispanic blacks relative to non-Hispanic whites: ages 25 to 59 ($n = 428,778$)

	Hazard ratios			
	Model 1	Model 2	Model 3	Model 4
Ethnicity (on-Hispanic white)				
Mexican American	1.26**	1.32**	1.05	1.30+
Puerto Rican	1.51**	1.65**	1.38**	1.09
Other Hispanic	0.93	1.01	0.94	1.19
Non-Hispanic black	1.70**	1.71**	1.50**	1.76**
Demographic variables				
Age (continuous)	1.09**	1.09**	1.08**	1.08**
Sex (Female)				
Male	1.81**	1.81**	1.83**	1.83**
Marital status (married)				
Not married	1.77**	1.77**	1.73**	1.73**
Nativity (U.S.-born)				
Immigrant		0.77**	0.74**	0.74**
Missing		1.02	0.98	0.98
Socioeconomic variables				
Education (13+ years)				
Less than 12 years			2.05**	2.23**
12 years			1.37**	1.38**
Interaction variables				
Race/ethnicity*education				
Mexican American*less than 12 years				0.77+
Mexican American*12 years				0.72+
Puerto Rican*less than 12 years				1.14
Puerto Rican*12 years				1.60
Other Hispanic*less than 12 years				0.60**
Other Hispanic*12 years				0.86
Black*less than 12 years				0.72**
Black*12 years				0.93
−2*log-likelihood	203,245	203,217	202,381	202,332
Degrees of freedom	7	9	11	19

Source: 1986–1994 National Health Interview Survey–Multiple Cause of Death Linked File (NCHS, 1997) ** $p < 0.01$; * $p < 0.05$; + $p <. 010$

even higher relative risks for Mexican Americans and Puerto Ricans compared to whites. As with Table 6.2, once education is included in Model 3, the hazard ratios decrease for every group. This is especially the case among Mexican Americans, who no longer demonstrate significantly higher mortality risks than whites with the inclusion of education in the model.

Once again, we have graphed the results of the final model from Table 6.3, which includes the race/ethnicity by education interaction terms. Figure 6.2 demonstrates that educational differences in younger adult mortality for every race and ethnic group are wider than when we included all adult ages. There are also differences across racial/ethnic groups. Whereas whites at the lowest educational level

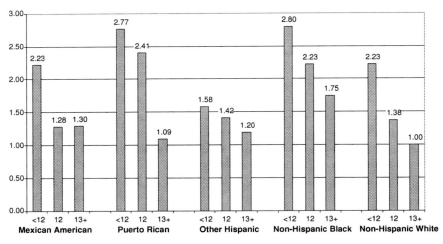

FIGURE 6.2. Hazard ratios depicting educational differences in adult mortality by race/ethnicity group compared to non-Hispanic whites with 13 or more years of education: ages 25 to 59[a]

experience a mortality risk that is 2.23 times greater than whites at the highest educational level, other Hispanics experience only a 1.58 times greater risk compared to whites at the highest educational level. In contrast, the risk of mortality among Mexican Americans with less than 12 years of education appears to be quite similar to the risk to their white counterparts. Consistent with the previous figure for all ages, Puerto Ricans and blacks demonstrate a much higher penalty for low education, with hazard ratios of about 2.8 compared to highly educated whites. Among whites with 12 years of education the mortality risk is 1.38 times greater than the mortality risk of whites with 13 or more years of education. With hazard ratios of 1.28 for Mexican Americans and 1.42 for other Hispanics, the effect of education at this age group and educational level is actually quite similar to what we see with the whites. Finally, at the highest educational level every non-white group experiences elevated mortality risk compared to highly educated whites.

Table 6.4 turns to an identical set of models for adults 60 and older. Here, racial/ethnic and educational differences in mortality are less pronounced in comparison to younger adults. This finding has been documented previously and is believed to be most likely due to mortality selection (Crimmins, 2005; Hummer et al., 2004). Model 1 shows that for individuals aged 60 and above, non-Hispanic blacks are the only group to demonstrate a significantly higher mortality risk than non-Hispanic whites. Moreover, older Mexican American and other Hispanics display a significantly lower mortality risk than older whites at these ages (with Puerto Ricans approaching significance as well). This pattern remains consistent in the next model, even net of the protective effect of foreign-born nativity. Controlling for education in Model 3 results in a modest reduction in the hazard ratios

TABLE 6.4. Overall adult mortality risks for Hispanic subgroups and non-Hispanic blacks relative to non-Hispanic whites: ages 60 and above ($n = 155,743$)

	Hazard ratios			
	Model 1	Model 2	Model 3	Model 4
Ethnicity (on-Hispanic white)				
Mexican American	0.83**	0.86*	0.80**	0.93
Puerto Rican	0.85+	0.93	0.86+	1.31
Other Hispanic	0.79**	0.84**	0.82**	0.93
Non-Hispanic black	1.14**	1.13**	1.07**	1.15*
Demographic variables				
Age (continuous)	1.08**	1.08**	1.08**	1.08**
Sex (female)				
Male	1.85**	1.85**	1.85**	1.85**
Marital status (married)				
Not married	1.22**	1.22**	1.21**	1.21**
Nativity (U.S.-born)				
Immigrant		0.83**	0.82**	0.82**
Missing		0.98	0.97*	0.97*
Socioeconomic variables				
Education (13+ years)				
Less than 12 years			1.31**	1.32**
12 years			1.14**	1.15**
Interaction variables				
Race/ethnicity*education				
Mexican American*less than 12 years				0.88
Mexican American*12 years				0.65*
Puerto Rican*less than 12 years				0.63
Puerto Rican*12 years				0.65
Other Hispanic*less 12 years				0.86
Other Hispanic*12 years				0.86
Black*less than 12 years				0.91
Black*12 years				0.97
–2*log-likelihood	516,138	516,088	515,655	515,641
Degrees of freedom	7	9	11	19

Source: 1986–1994 National Health Interview Survey–Multiple Cause of Death Linked File (NCHS, 1997) ** $p < 0.01$; * $p < 0.05$; + $p < 0.10$

for each of the minority groups in comparison to whites. However, the mortality risk for blacks remains statistically higher, while the risks for Mexican Americans and other Hispanics remain statistically lower than those of non-Hispanic whites.

Once the interaction terms for race/ethnicity by education are included (Figure 6.3), we can see clear evidence that education has a much weaker effect on all racial/ethnic groups aged 60 and above. For older white adults with less than 12 years of education, the risk of death is only 1.32 times greater the risk for older white adults with 13 or more years of education. Among all Hispanic groups the risk is even less with a hazard ratio of approximately 1.1 for each group. This pattern is evident in the next educational level as well. Whereas whites with

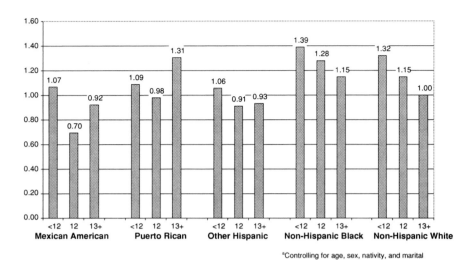

^aControlling for age, sex, nativity, and marital

FIGURE 6.3. Hazard ratios depicting educational differences in adult mortality by race/ethnicity group compared to non-Hispanic whites with 13 or more years of education: ages 60 and above[a]

12 years of education have a hazard ratio of 1.15, the hazard ratios of Mexican Americans, Puerto Ricans, and other Hispanics with 12 years of education are 0.70, 0.98, and 0.91, respectively. The overall pattern in Figure 6.3 suggests that educational differences in mortality risks for older U.S. adults are not consistent across racial/ethnic groups and, in fact, appear weaker among older Hispanics, especially Mexican Americans.

Based upon our analyses in the previous tables and accompanying figures, educational differences in mortality appear to be somewhat weaker for non-whites. In Table 6.5, we further analyze the education-mortality relationship by estimating models separately for each racial/ethnic group. Furthermore, we split Mexican Americans and other Hispanics into foreign- and native-born subgroups because of the high percentage of immigrants within each of those groups. Thus, we present separate proportional hazard models for the following groups: foreign-born Mexican Americans, U.S.-born Mexican Americans, Puerto Ricans, foreign-born other Hispanics, U.S.-born other Hispanics, non-Hispanic blacks, and non-Hispanic whites. We do not differentiate between island- and mainland-born Puerto Ricans because of the small sample size for Puerto Ricans.

Educational differences in mortality displayed in Table 6.5 for all adults aged 25 and over separately by racial/ethnic and nativity group are strong and in the expected direction for non-Hispanic blacks and non-Hispanic whites. That is, both blacks and whites with lower levels of education have higher mortality risks than blacks and whites with higher educational levels, respectively. However, within the Mexican American and other Hispanic groups, educational differences in mortality risk are statistically significantly only among U.S.-born individuals with less than

TABLE 6.5. Overall adult mortality risks by age, sex, nativity, and education for Hispanic ethnicities and non-Hispanic blacks and whites: all ages, both sexes

	Hazard ratios						
	Mexican American foreign-born	Mexican American missing or U.S.-born	All Puerto Ricans	Other Hispanic foreign-born	Other Hispanic missing or U.S.-born	Black non-Hispanic	White non-Hispanic
	($n = 7,885$)	($n = 14,083$)	($n = 4,786$)	($n = 6,912$)	($n = 7,351$)	($n = 77,366$)	($n = 466,100$)
Demographic variables							
Age (continuous)	1.06**	1.07**	1.06**	1.09**	1.08**	1.07**	1.09**
Sex (male)							
Female	1.59**	1.98**	1.68**	1.75*	1.81*	1.87**	1.88**
Marital status (married)							
Not married	1.35+	1.48**	1.26	0.95	1.07	1.28**	1.29**
Socioeconomic variables							
Education (13+ years)							
Less than 12 years	1.29	1.51**	1.99**	1.20	1.28*	1.50**	1.44**
12 years	0.75	0.96	1.61	1.12	1.08	1.23**	1.20**
−2*log-likelihood	4,265	10,859	3,911	3,675	7,131	101,251	578,132
Degrees of freedom	5	5	5	5	5	5	5

Source: 1986–1994 National Health Interview Survey–Multiple Cause of Death Linked File (NCHS, 1997) ** $p < 0.01$; * $p < 0.05$; + $p < 0.10$

12 years of education. For Mexican Americans and other Hispanics with 12 years of education there is no significant association between education and mortality risk. Among all the Hispanic subgroups education appears to have, by far, the greatest impact on Puerto Ricans. Puerto Ricans with less than 12 years of education have twice the mortality risk than Puerto Ricans with 13 or more years of education.

To incorporate differential effects by age, Tables 6.6 and 6.7 replicate the above results for those aged 25–59 and those aged 60 and above, respectively. In Table 6.6 we see that both black and white younger adults with lower educational attainment have much higher mortality than their highly educated counterparts. Once again, education appears to have a weaker effect on mortality for both Mexican Americans and other Hispanics; for both of these groups, regardless of nativity, there are no mortality differences between persons who have 12 years of education in comparison to 13 or more years of education. Puerto Ricans are more similar to whites and blacks, as lower education is substantially and significantly associated with higher mortality risks.

Table 6.7, which includes only individuals aged 60 and above, finds relatively weak relationships between education and mortality risk, particularly among Hispanics. In fact, of all the racial/ethnic groups, only non-Hispanic whites display a consistent graded relationship between educational level and mortality risk. Whereas blacks with less than 12 years of education do have a significantly higher mortality risk than their highly educated counterparts, blacks with 12 years of education do not. Of all Hispanic-education groups, only one (Mexican Americans with 12 years of education) demonstrates a significant association between education and mortality. However, the coefficient actually demonstrates lower mortality risk among individuals with 12 years of education compared to individuals with 13 or more years of education; caution is urged, however, because of very small sample sizes in the Mexican immigrant high education groups at these older ages.

6.5. Conclusion

The purpose of this chapter is to further our understanding of the epidemiologic paradox, namely, that the socioeconomic status of Hispanics does not seem to have the same strong relationship with mortality risk when compared to non-Hispanic blacks and whites. To accomplish this we attempted to quantify the individual-level effect that education has on Hispanic mortality vis-à-vis the effect that it has on non-Hispanic whites and blacks. By using interaction terms and creating separate models for each racial/ethnic group, we are able to demonstrate that, indeed, education does not seem to have the same impact for some of the Hispanic subgroups as it does for non-Hispanic whites and blacks. Our results suggest that educational differences in mortality risk for both Mexican Americans and other Hispanics are somewhat smaller than those observed for non-Hispanic whites, both for overall mortality and for the separate younger and older adult age groups that were specified. Further, the weakest relationships between educational attainment

TABLE 6.6. Overall adult mortality risks by age, sex, nativity, and education for Hispanic ethnicities and non-Hispanic blacks and whites: ages 25 to 59, both sexes.

	Hazard ratios						
	Mexican American foreign-born	Mexican American missing or U.S.-born	All Puerto Ricans	Other Hispanic foreign-born	Other Hispanic missing or U.S.-born	Black non-Hispanic	White non-Hispanic
	(n = 7,130)	(n = 12,077)	(n = 4,142)	(n = 5,779)	(n = 5,965)	(n = 59,006)	(n = 334,636)
Demographic variables							
Age (continuous)	1.04**	1.07**	1.04**	1.07**	1.06**	1.07**	1.09**
Sex (male)							
Female	1.52**	2.36**	1.82**	1.52	2.00**	1.83**	1.81**
Marital status (married)							
Not married	0.87	2.02**	1.50	1.37	1.57**	1.54**	1.82**
Socioeconomic variables							
Education (13+ years)							
Less than 12 years	1.41	1.67**	2.91**	1.58$^+$	1.44	1.76**	2.17**
12 years	1.10	0.98	2.24*	1.40	1.18	1.31**	1.37**
−2*log-likelihood	2,695	5,287	2,315	1,222	3,031	41,645	147,620
Degrees of freedom	5	5	5	5	5	5	5

Source: 1986–1994 National Health Interview Survey–Multiple Cause of Death Linked File (NCHS, 1997) **$p < 0.01$; *$p < 0.05$; $^+p < 0.10$

TABLE 6.7. Overall adult mortality risks by age, sex, nativity, and education for Hispanic ethnicities and non-Hispanic blacks and whites: ages 60 and above, both sexes.

	Hazard ratios						
	Mexican American foreign-born	Mexican American missing or U.S.-born	All Puerto Ricans	Other Hispanic foreign-born	Other Hispanic missing or U.S.-born	Black non-Hispanic	White non-Hispanic
	(n = 727)	(n = 1,782)	(n = 616)	(n = 1,115)	(n = 1,386)	(n = 18,344)	(n = 131,464)
Demographic variables							
Age (continuous)	1.09**	1.07**	1.05**	1.08**	1.10**	1.06**	1.08**
Sex (male)							
Female	1.73*	1.50**	1.51	1.71+	1.58**	1.86**	1.86**
Marital status (married)							
Not married	1.56+	1.04	0.99	0.87	0.74*	1.16**	1.21**
Socioeconomic variables							
Education (13+ years)							
Less than 12 years	0.88	1.22	0.92	1.06	1.24	1.26**	1.31**
12 years	0.10*	0.81	0.81	0.98	1.06	1.09	1.15**
−2*log-likelihood	1,461	4,628	1,376	2,228	4,044	65,513	435,756
Degrees of freedom	5	5	5	5	5	5	5

Source: 1986–1994 National Health Interview Survey–Multiple Cause of Death Linked File (NCHS, 1997) **$p < 0.01$; *$p < 0.05$; +$p < 0.10$

and mortality risk were exhibited among foreign-born Mexican Americans and other Hispanics.

In sum, our research suggests that educational differences in adult mortality are smaller for Mexican Americans and other Hispanics compared to non-Hispanic whites. This is consistent with the idea of the epidemiologic paradox among Hispanics in the United States. Future research could build upon this work by estimating educational differentials in cause-specific mortality among these Hispanic groups and, perhaps most important, by determining why educational differences in mortality seem to be weaker among these Hispanic groups compared to whites. The recently expanded NHIS–MCD data set should allow for more in-depth examinations of this topic. Further, it will be important to continue to be cautious regarding the findings of studies such as ours, given that statistically based linkages between surveys like the NHIS and the NDI may underestimate the mortality of Hispanics compared to whites due either to the more frequent out-migration of Hispanics compared to whites or to lower quality data among Hispanics. Nevertheless, our results add to the growing body of literature on the uniqueness of Hispanic mortality by showing somewhat weaker educational differences in adult mortality when compared to the non-Hispanic white population.

References

Abraído-Lanza, A.F., Dohrenwend, B.P., Ng-Mak, D.S., and Turner, J.B. (1999). The Latino mortality paradox: A test of the "Salmon bias" and healthy migrant hypotheses. *American Journal of Public Health, 89*, 1543–1548.

Allison, P. (1984). *Event history analysis*. Beverly Hills, CA: Sage Publications.

Crimmins, E.M. (2005). Socioeconomic differentials in mortality and health at the older ages. *Genus, LXI*(1), 163–178.

Elo, I.T., Turra, C.M., Kestenbaum, B., and Ferguson, B.R. (2004). Mortality among elderly Hispanics in the United States: Past evidence and new results. *Demography, 41*, 109–128.

Elo, I.T., and Preston, S.H. (1996). Educational differentials in mortality: United States, 1979–85. *Social Science Medicine, 42*, 47–57.

Hayward, M.D., Miles, T.P., Crimmins, E., and Yang, Y. (2000). The significance of socioeconomic status in explaining the racial gap in chronic health conditions. *American Sociological Review, 65*, 910–930.

Hummer, R.A., Benjamins, M.R., and Rogers, R.G. (2004). Racial and ethnic disparities in health and mortality among the U.S. elderly population. In N. Anderson, R. Bulatao, and B. Cohen (Eds.), *Critical perspectives on racial and ethnic differences in health in late life* (Chapter 3). Washington, DC: National Academies Press.

Hummer, R.A., Rogers, R.G., Nam, C.B., and Le Clere F.B. (1999). Race/ethnicity, nativity, and U.S. adult mortality. *Social Science Quarterly, 80*, 136–153.

Hummer, R.A., Rogers, R.G., Amir, S.H. Forbes, W.D., and Frisbie, P. (2000). Adult mortality among Hispanic subgroups and non-Hispanic whites. *Social Science Quarterly, 81*, 459–476.

Kitagawa, E.M. and Hauser, P.M. (1973). *Differential mortality in the United States: A study in socioeconomic epidemiology*. Cambridge, MA: Harvard University Press.

Liao, Y., Cooper, R.S., Cao, G., Durazo-Arvizu, R., Kaufman, J.S., Luke, A., and McGee, D.L. (1998). Mortality patterns among adult Hispanics: Findings from the NHIS, 1986 to 1990. *American Journal of Public Health, 88*, 227–232.

Lin, C.C., Rogot, E., Johnson, N.J., Sorlie, P.D., and Arias, E. (2003). A further study of life expectancy by socioeconomic factors in the National Longitudinal Mortality Study. *Ethnicity and Disease, 13*, 240–247.

Markides, K, and Coreil, J. (1986). The health of Hispanics in the southwestern United States—An epidemiologic paradox. *Public Health Reports, 3*, 253–265.

Mirowsky, J., and Ross, C.E. (2003). *Education, social status, and health.* New York: Aldine De Gruyter.

Molla, M.T., Madans, J.H., and Wagener, D.K. (2004). Differentials in adult mortality and activity limitation by years of education in the United States at the end of the 1990s. *Population and Development Review, 30*, 625–646.

National Center for Health Statistics. (1987).

National Health Interview Survey. (1987). [Computer file]. Washington, DC: U.S. Department of Health and Human Services, National Center for Health Statistics [producer], 1987. Ann Arbor, MI: Inter-University Consortium for Political and Social Research [distributor].

Palloni, A. and Arias, E. (2004). Paradox lost: Explaining the Hispanic adult mortality advantage. *Demography, 41*, 385–415.

Palloni, A., and Ewbank, D.C. (2004). Selection processes in the study of racial and ethnic differentials in adult health and mortality. In *Critical perspectives on racial and ethnic differences in health and late life.* Washington, DC: The National Academies Press.

Palloni, A., and Morenoff, J.D. (2001). Interpreting the paradoxical in the Hispanic paradox: Demographic and epidemiologic approaches. *Population Health and Aging Annals of the New York Academy of Sciences, 954*, 140–174.

Pappas, G., Queen, S., Hadden, W., and Fisher, G. (1993). The increasing disparity in mortality between socioeconomic groups in the United States, 1960 and 1986. *New England Journal of Medicine, 329*, 103–109.

Rogers, R.G., Hummer, R.A., and Nam, C.B. (2000). *Living and dying in the USA: Behavioral, health, and social differentials of adult mortality.* San Diego, CA: Academic Press.

Rosenwaike, I. (1987). Mortality differentials among persons born in Cuba, Mexico, and Puerto Rico residing in the United States. *American Journal of Public Health, 77*, 603–606.

Saenz, R. (2004). *Latinos and the changing face of America.* New York, NY: Russell Sage Foundation.

Singh, G.K., and Siahpush, M. (2002). Ethnic-immigrant differentials in health behaviors, morbidity, and cause-specific mortality in the United States: An analysis of two national data bases. *Human Biology, 74*, 93–109.

Smith, J.P. (2005). Unraveling the SES-health connection. *Population Development and Review, 30*, 108–132.

Turra, C.M., Elo, I.T., Kestenbaum, B., and Ferguson, B.R. (2005). The impact of Salmon bias on the Hispanic mortality advantage: New evidence from social security data. Paper presented at the annual meeting of the Population Association of America, Philadelphia, PA.

7
Does Longer Life Mean Better Health? Not for Native-Born Mexican Americans in the Health and Retirement Survey*

Mark D. Hayward, David F. Warner, and Eileen M. Crimmins

7.1. Introduction

Race/ethnic differences in mortality constitute a fundamental form of inequality in the United States. Black Americans live fewer years than whites, and the disadvantage for blacks in chances for long life is a long-term historical pattern. National mortality estimates for Hispanics show that longevity of individuals in this group is equivalent to and sometimes exceeds those for whites, despite Hispanics' stark socioeconomic disadvantages (National Center for Health Statistics 2001). Trend data for Hispanics spanning several decades are not available.

Prior studies suggest that blacks' truncation of life relative to whites is accompanied by inequality of another sort—an extended period of disability relative to whites (Crimmins, Hayward and Saito, 1996; Crimmins and Saito, 2001; Geronimus et al., 2001; Hayward and Heron, 1999). The pattern is consistent with the basic idea that mortality advantages are brought about through the postponement of disease and disability. How Hispanics fare is less clear. Is their relatively lengthy life expectancy accompanied by a compressed period of disability similar to whites? Or, do Hispanics' socioeconomic disadvantages translate into more years of disabled life? As shown by Hayward and Heron (1999), the association between life expectancy and disabled life expectancy is not straightforward. Longer life means a compacted period of disability for some race/ethnic groups— but not others. Considerable uncertainty about the quality of mortality estimates

*Partial support for this research was provided by a research grant from NIA (1 R55 AG19311) and two population center grants from NICHD (1 R24 HD41025 and 5 R24 HD42849). The paper is a preliminary draft and is not for quotation without permission of the first author. For additional information, please contact Mark Hayward, Population Research Center, 1800 Main, University of Texas at Austin, Austin, TX 78712. Mhayward@prc.utexas.edu

for Hispanics contributes to the difficulty in anticipating how mortality is associated with disability among Hispanics. Is the Hispanic mortality advantage real or does it reflect artifacts stemming from data quality and the ways that the mortality estimates are calculated? Much of the focus has been on the reasons for what appears to be superior Hispanic immigrant health. Hence, the Hispanic paradox is often construed as the story of healthy immigrants or the story of return-migration among less healthy immigrants. Yet, the fact remains that most national studies show that some native-born Hispanic groups such as Mexican Americans have mortality risks comparable to native-born whites despite native-born Hispanics having much lower income, wealth, and education—factors typically thought to be at the root of race/ethnic differences in health (Hummer et al., 2000; Hummer et al., 1999; Rogers, Hummer and Nam, 2000). The Hispanic paradox encompasses both immigrant and native-born groups.

In the current study we focus on the Hispanic paradox among the native-born American population, assessing how mortality is associated with disability for non-Hispanic whites, blacks, and Mexican Americans. We make use of a life table model that integrates mortality and disability experiences over a lifetime to estimate how these groups differ in length of life, in life free of disability, and in life with disability. In the remainder of the paper we refer to the life table model as "active life expectancy" (ALE).

Our study makes a number of contributions to the scientific literature. First, we provide the first estimates of ALE for a nationally representative sample of native-born Mexican Americans. This allows us to assess whether the Hispanic paradox—i.e., low mortality among Mexican Americans despite low socioeconomic resources—also extends to the postponement of disability. In other words, is Mexican Americans' relatively long life accompanied by a "compression of disability" and an expansion of healthy life?

Second, we examine how education, an important marker of lifetime socioeconomic status, accounts for the race/ethnicity differences in active life expectancy. A number of studies have noted the importance of socioeconomic status (SES) as a fundamental cause of the race (black/white) gap in chronic conditions (Hayward et al., 2000; Schoenbaum and Waidmann, 1997; Smith and Kington, 1997), but much less is known about the ways in which SES contributes—or does not contribute—to an understanding the Hispanic paradox. At first glance, however, the Hispanic paradox points to the explanatory limits of the SES paradigm in understanding health disparities. At the very least, the Hispanic paradox suggests that SES and minority status potentially combine in complex ways to affect health disparities.

The only nationally representative study of Hispanic active life expectancy is that by Hayward and Heron (1999). Based on vital registration and Census data for 1990, they reported that Hispanics had a relatively short period of disability—especially compared to blacks. So, there is some evidence that the Hispanic paradox may extend to disability as well as to mortality.

A number of factors suggest caution in drawing strong conclusions from Hayward and Heron's analysis. First, Hispanics included both native-born and immigrant groups, and all Hispanic subgroups were combined into a single category. The subgroups are culturally and socioeconomically diverse and also appear to have very different health profiles (e.g., Cubans, Dominicans, Central Americans, Latin Americans, and Puerto Ricans). Moreover, recent research by Palloni and Arias (2004) supports the idea that the Hispanic paradox among immigrants is actually a story of older, less healthy Mexican immigrants returning home, which then biases (U.S.) Hispanic mortality rates (and presumably disability rates) downward. The paradox may be less of a paradox than it is often portrayed to be. By restricting our focus to native-born Mexican Americans we are able to gain a better analytical handle on the Hispanic paradox because the subgroup is "better defined" and "closed" compared to the highly heterogeneous Hispanic category.[1]

Second, Hayward and Heron used the traditional "occurrence/exposure" approach to calculate the rates used in constructing the ALE models. This approach can lead to biased estimates of life expectancy when a population is not closed—such as is true of the Hispanic immigrant population in the U.S., as noted above. The occurrence/exposure approach also introduces error into the calculation of mortality rates, because these rates are based on data from two independent sources—vital registration and the Census. Both sources have potential coverage errors (i.e., incomplete enumeration and death registration), both may have discrepancies in ethnic classification, and both may contain age misreporting. The disability rates in the Hayward and Heron study, while depending only on Census data, were based on persons' reports of the "inability to work," a basis that may lead to underreporting at older ages. Some researchers have suggested that elderly Mexican Americans may exaggerate their age, which could mean observed mortality rates are lower than they would be determined to be with accurate age reporting (Elo et al., 2004; Liao et al. 1998; Rosenberg et al., 1999). There also is evidence of the underreporting of deaths among Hispanics, which downwardly biases traditional occurrence/exposure mortality rates (Elo et al., 2004).

In the present study we rely on a prospective cohort approach to calculate ALE for the major race/ethnic groups. The prospective cohort approach minimizes many of the problems associated with traditional occurrence/exposure approach to calculating ALE: Population coverage problems are minimized because mortality and disability rates are calculated "within sample"; ages at which events occur can be associated with age-appropriate risk groups; and race/ethnicity self-identification minimizes problems of misclassification (Patel et al., 2004). In addition, the current study relies on more age-appropriate measures of disability (ADL) than the prior study by Hayward and Heron.

A prospective cohort approach nonetheless has limitations. Perhaps foremost is the fact that the sample sizes in cohort studies, though large *in toto*, do not have very large subsamples of minority groups such as Mexican Americans. As we will show later, this is also true for the present study. Second, a cohort approach

---

[1] Note that in well-defined epidemiologic studies such as the San Antonio Heart Study, Mexican Americans were at *greater* risk of all-cause, cardiovascular, and coronary heart disease mortality than non-Hispanic whites (Hunt et al., 2003).

may still suffer from the under ascertainment of deaths among older Hispanics—especially women after age 75—because of difficulties in accurately identifying respondents' deaths in the Vital Statistics registration data (Patel et al., 2004). Finally, possible cultural differences in responding to disability questions may introduce measurement error. We stress that our results are preliminary and should not be over interpreted.

7.2. Data and Measures

We use the Health and Retirement Survey (HRS) and the Assets and Health Dynamics Survey (AHEAD) to investigate changes in disability and mortality for a cohort of persons aged 51 years of age and older. The HRS is a panel survey that began in 1992; respondents have been reinterviewed approximately every two years thereafter. At baseline the HRS target population was the 1931–1941 birth cohort and their spouses. The AHEAD survey, a companion to the HRS, was started in 1993 and targeted those born before 1923 and their spouses. AHEAD respondents were reinterviewed in 1995. In 1998, the AHEAD and HRS were combined and two additional birth cohorts were added to the sampling frame: those born between 1942 and 1947 ("war babies") and those born between 1924 and 1941 ("children of the Depression"). The combined panel is known simply as the Health and Retirement Survey (HRS). We make use of five waves of data (1992–2000) in the present analysis. The data are weighted using the baseline sample weight from the wave when the respondent was first interviewed. The weights are normalized by race.

The HRS is an excellent source with which to investigate race/ethnicity disparities in age-related disability and mortality experiences because blacks and Hispanics were oversampled at baseline in both the original HRS and Ahead. Approximately 60% of the oversampled Hispanic population is Mexican American.

Race/ethnicity is identified in the HRS by a set of self-report measures. "Black" includes all respondents who self-identify as black, regardless of their report of Hispanic ethnicity. If a person is not black and reports that they are Hispanic, we use additional information about whether the respondent considers himself/herself to be a Mexican American or Chicano to classify Mexican Americans. All other Hispanics are excluded from our analysis. "Whites" refer to respondents who self-identify as white and non-Hispanic. Only native-born persons are included in the analysis.

Disability is measured by a series of ADL items common to both the original HRS and AHEAD surveys. The measures are:

- Walking several blocks
- Picking up a dime from a table
- Climbing stairs without resting
- Lifting or carrying weights over 10 pounds
- Pulling or pushing large objects like a living room chair.

Mobility limitations are referenced by the activities of walking several blocks and climbing stairs. Strength measures include lifting weights and pulling/pushing large objects. The measure of picking up a dime from a table reflects limitations in fine motor skills. For the present, we ignore the possible hierarchical nature of these items. We also have not systematically evaluated how the race/ethnicity groups differ on particular types of disability. Finally, we caution that these items do not necessarily characterize the full range of disability problems that can differentiate the health of the race/ethnicity groups. The lack of congruence in the HRS/AHEAD disability measures is a constraining factor.

Persons are defined as "inactive" or "disabled" if they report that *three or more* of these activities is:

- A little difficulty/very difficult/can't do (AHEAD survey)
- Some difficulty/a lot of difficulty (HRS survey)

The "inactive" or disabled state thus references the confluence of multiple forms of disability, tapping into the idea of severity. We chose this approach rather than focus on the severity levels defined by the specific item responses because of comparability problems across the surveys and over time. Although we have not conducted an exhaustive assessment of the shortfalls of our measurement approach, the results reported below are highly consistent with results from prior research using a variety of measurement approaches (Crimmins, Hayward, and Saito, 1994; Rogers, Rogers, and Belanger 1990; Rogers, Rogers, and Belanger 1989).

Education is measured in terms of years of completed schooling. Education is treated as a continuous measure in modeling disability and mortality, and we explore possible non-linearities in the association between education and disability/mortality incidence. In particular, we evaluate whether there are declining health returns to education.

7.3. Methods

We use Markov-based multistate life table models (MSLT) (Schoen, 1988) to calculate ALE for non-Hispanic whites, blacks, and Mexican Americans. Our multistate life table model is straightforward. It allows for bi-directional transitions between the active and inactive health states and mortality transitions from the two health states. In all, four transitions constitute our ALE model. Transitions were observed across four 2-year observation intervals spanning the 1992–2002 observation period.

We modeled the disability and mortality transition rates using a piecewise exponential hazard modeling approach (Hayward and Grady, 1990; Land, Guralnik, and Blazer, 1994). We tested for non-linearities in the age and education specifications, non-proportionality, race/sex interactions, and race/education interactions. Parameter estimates from the best-fitting models were used to calculate predicted transition rates for the race/ethnicity/sex/education subgroups.

TABLE 7.1. Observation intervals by race/Ethnicity and education group, HRS 1992–2000, persons aged 50 years and older.

	Race/ethnicity		
Education	Whites	Blacks	Mexican Americans
0–8	4,839	2,426	743
9–12	22,837	4,485	556
13 and more	17,181	1,994	225
Total	44,857	8,905	1,524

We then calculated multistate life tables for each race/ethnicity group broken down by education (8 years, 12 years, and 16 years) and sex—the exception being for Mexican Americans and blacks, for whom we only calculated a life table for persons with 8 and 12 years of education. We observed a highly skewed educational distribution among the two minority groups—especially the Mexican Americans— and the data for exposure and transitions were very thin for persons at upper levels of education (despite the number of observation waves) (see Table 7.1).

7.4. Results

The hazard model results used to calculate the transition rates for the MLSTs are shown in Table 7.2. As is evident in the first column, Mexican Americans have the highest rate of disability onset followed by blacks and then whites. There are no statistical differences for the recovery rate (column 2). Mexican Americans are the least likely to die, although the risks are not statistically different than whites.

TABLE 7.2. Hazard coefficients for transitions between active life, inactive life, and death among persons aged 50 years and older in the HRS (1992–2000)[a]

	Active to inactive	Inactive to active	Active to dead	Inactive to dead
Base model				
Intercept	−5.950***	0.844***	−9.722***	−6.448***
Age	0.045***	−0.0645***	0.083***	0.060***
Female	0.493***	−0.164***	−0.683***	−0.566***
Race and ethnicity				
Black	0.285***	−0.074	0.416**	−0.061
Mexican	0.317***	0.152	−0.158	−0.097
Education				
Education	0.042**	0.031*	0.073*	−0.001
Education squared	−0.006*	—	−0.005**	—
Log-likelihood	−12,637.89	−5515.63	−6616.17	−4029.35
Number of transitions	4,148	2,507	1,443	1,521

[a] 60,161 total person year intervals.
[b] Non-Hispanic white reference group.
*$p < 0.05$, **$p < 0.01$, ***$p < 0.001$ (two-tailed tests).

Blacks are the most likely to die among persons without disability, but there are no race/ethnicity differences in mortality among the disabled. This pattern suggests that as health deteriorates the race gap in mortality declines. In models not shown we observed that the black/white and Mexican American/white gaps in disability onset were reduced when controlling for education, although not eliminated, as one can see. Although we will show the implications of these rates for ALE, the overall pattern suggests that Mexican Americans and whites are likely to have roughly comparable life expectancies, but that Mexican Americans will spend more years disabled.

More years of schooling is associated with lower rates of disability (see by combining the education and education squared terms) and lower rates of death for persons who are not disabled. Although we have not shown the results for models that omit education, our analysis clearly demonstrates a persistent effect of race/ethnicity that is not accounted for by education. Race/ethnic disparities in the transitions making up ALE are not a simple reflection of differences in educational attainment.

7.5. Multistate Life Table Models

For each race/ethnicity/sex group, we show in Table 7.3 the educational gradient in active and inactive life. For Mexican Americans and blacks, we limit our results to two groups: persons with 8 and 12 years of education, doing so because of sparse data at higher levels of education (see Table 7.1).

The life table results for Mexican Americans differ substantially from the patterns shown for blacks and whites. Although both Mexican Americans and blacks in the group have low educational attainment, blacks' overall life expectancy falls far below that for Mexican Americans. For example, among men aged 50 years with 8 years of education, life expectancy is about 22 years for blacks compared to over 24 years for Mexican Americans. Mexican American life expectancy, however, is very similar to that for whites, despite the two groups' different locations on the socioeconomic ladder. At least for mortality, our results are consistent with the idea of an Hispanic paradox.

Although our results support the idea of an Hispanic paradox in mortality among native born Mexican Americans, our results also identify that Mexican Americans can expect to live more years with disability relative to whites—and blacks. Among men aged 50 years with 8 years of education, for example, Mexican Americans can expect to live over 6 years with disability compared to 5.4 years for blacks and 5 years for whites. Longer life for Mexican Americans is not associated with a compression of disability that is characteristic of whites; The Hispanic paradox in mortality does not appear to extend to disability among native-born Mexican Americans.

Blacks are clearly the most disadvantaged group in terms of living the fewest years and living the fewest years without disability. Although Mexican Americans live more years with disability than blacks, they also live more years without disability. This suggests that Mexican Americans occupy some sort of middle ground

TABLE 7.3. The educational gradient in active and inactive life expectancy (e_x) by race/ethnicity, native-born respondents in the HRS (1992–2000)

Panel A: Males

Black

Age	8 years			12 years			16 years		
	Active	Inactive	Total	Active	Inactive	Total	Active	Inactive	Total
50	16.4	5.4	21.8	18.2	4.7	22.8	—	—	—
65	7.3	4.3	11.6	8.5	3.8	12.2	—	—	—
80	2.5	1.7	4.2	2.8	1.6	4.9	—	—	—

Non-Hispanic White

Age	8 years			12 years			16 years		
	Active	Inactive	Total	Active	Inactive	Total	Active	Inactive	Total
50	19.0	5.0	24.0	21.2	4.2	25.4	25.2	3.2	28.5
65	9.1	4.0	13.1	10.7	3.5	14.2	13.9	2.7	16.6
80	3.3	1.7	5.0	4.0	1.5	5.4	5.5	1.2	6.7

Mexican American

Age	8 years			12 years			16 years		
	Active	Inactive	Total	Active	Inactive	Total	Active	Inactive	Total
50	18.1	6.2	24.3	20.2	5.4	25.7	—	—	—
65	8.6	4.8	13.4	10.0	4.4	14.3	—	—	—
80	3.3	2.0	5.3	4.9	2.4	7.3	—	—	—

Panel B: Females

Black

Age	8 years			12 years			16 years		
	Active	Inactive	Total	Active	Inactive	Total	Active	Inactive	Total
50	17.9	9.7	27.6	19.5	9.0	28.5	—	—	—
65	8.9	7.4	16.3	9.7	7.2	16.9	—	—	—
80	3.6	3.2	v6.8	3.9	3.2	7.1	—	—	—

Non-Hispanic White

Age	8 years			12 years			16 years		
	Active	Inactive	Total	Active	Inactive	Total	Active	Inactive	Total
50	19.9	9.3	29.2	21.9	8.4	30.3	26.0	27.0	33.0
65	10.1	7.3	17.4	11.3	6.9	18.2	14.3	6.0	20.3
80	3.3	4.3	7.6	4.7	3.2	7.9	6.0	3.0	9.0

Mexican American

Age	8 years			12 years			16 years		
	Active	Inactive	Total	Active	Inactive	Total	Active	Inactive	Total
50	19.2	10.7	29.9	21.9	10.0	31.9	—	—	—
65	10.0	8.2	18.2	10.9	7.9	18.8	—	—	—
80	4.4	3.7	8.1	4.8	3.7	8.5	—	—	—

between blacks and whites. Mexican Americans are clearly advantaged relative to blacks in terms of the length of life and the length of life free of disability— but Mexican Americans are disadvantaged in terms of disability relative to whites.

7.6. Conclusions

The Hispanic paradox is often construed as primarily an immigrant story. It is not, at least with respect to native-born Mexican Americans. As shown here and in a number of other studies, native-born Mexican Americans have mortality rates that are comparable to those for native-born whites, despite Mexican Americans' obvious socioeconomic disadvantages.

The question we asked here was whether the Hispanic paradox also extended to disability? Is the native-born Mexican American mortality advantage accompanied by an advantage in disability? Expressed in the parlance of the ALE model, is the relatively lengthy life expectancy of Mexican Americans associated with a compressed period of disability?

On the whole, we do not find much evidence to support the idea of an Hispanic paradox with respect to disability among native-born Mexican Americans. While Mexican Americans and whites share life chances, they have different disability experiences over their lifetimes. Mexican Americans are much more likely to be disabled, and to be disabled for longer periods of time compared to whites.

The current study builds on the earlier work by Hayward and Heron (1999) to improve ALE estimates, but this new analysis is clearly preliminary and circumscribed. We have focused only on native-born Mexican Americans as our Hispanic group of interest, so how our results hold up for other native-born Hispanic groups is not clear. Nonetheless restricting our attention to native-born Mexicans has some clear analytical advantages over examining Hispanics as a whole—or even just native-born Hispanics. We have also attempted to improve the quality of ALE estimates by using a cohort approach to minimize error of a variety of types arising from multiple data sources. On balance, we believe that we have improved the quality at the ALE estimates. At the same time, we reiterate that the sample sizes for blacks and Mexican Americans in the HRS are not large despite over sampling and the data become very thin at higher levels of education and at the highest ages.

A glaring gap in our nation's population health monitoring is the lack of information on ALE in the Hispanic population, now the largest minority group in the United States. Certainly, there are enormous challenges in collecting high quality data for a culturally diverse group. Still, without high quality data that accurately document the health problems of our nation's largest minority group, targeting health policy will continue to rest on best guesses and frequently fail to reach those persons most in need. At present, there seems little political support to address this problem.

References

Crimmins, E.M., Hayward, M.D., and Saito, Y. (1994). Changing mortality and morbidity rates and the health status and life expectancy of the older population. *Demography, 31*, 159–175.

Crimmins, E.M., Hayward, M.D., and Saito, Y. (1996). "Differentials in Active Life Expectancy in the Older Population." *Journal of Gerontology: Social Sciences, 51B*, S111–S120.

Crimmins, E.M., and Saito, Y. (2001). Trends in healthy life expectancy in the United States, 1970–1990: Gender, racial, and educational differences. *Social Science and Medicine, 52*, 1629–1641.

Elo, I.T., Turra, C.M., Kestenbaum, B., and Ferguson, B.R. (2004). Mortality among elderly Hispanics in the United States: Past evidence and new results. *Demography, 41(1)*, 109–128.

Geronimus, A.T., Bound, J., Waidmann, T.A., Colen, C.G., and Steffick, D. (2001). Inequality in life expectancy, functional status, and active life expectancy across selected black and white populations in the United States. *Demography, 38(2)*, 227–251.

Hayward, M.D., Crimmins, E.M., Miles, T.P., and Yu, Y. (2000). The significance of socioeconomic status in explaining the racial gap in chronic health conditions. *American Sociological Review, 65*, 910–930.

Hayward, M.D., and Heron, M. (1999). Racial inequality in active life among adult Americans. *Demography, 36*, 77–91.

Hummer, R.A., Rogers, R.G., Amir, S.H., Forbes, D., and Frisbie, W.P. (2000). Adult mortality differentials among Hispanic subgroups and non-Hispanic whites. *Social Science Quarterly, 81*, 459–476.

Hummer, R.A., Rogers, R.G., Nam, C.B., and LeClere, F.B. (1999). Race/ethnicity, nativity, and U.S. adult mortality. *Social Science Quarterly, 80*, 136–153.

Hunt, K.J., Resendez, R.G., Williams, K., Haffner, S.M., Stern, M.P., and Hazuda, H.P. (2003). All-cause and cardiovascular mortality among Mexican-American and non-Hispanic white older participants in the San Antonio Heart Study—Evidence against the "Hispanic paradox." *American Journal of Epidemiology, 158(11)*, 1048–1057.

Liao, Y., Cooper, R.S., Cao, G., Durazo-Arvizu, R., Kaufman, J.S., Luke, A., and McGee, D.L. (1998). Mortality patterns among adult Hispanics: Findings from the NHIS, 1986 to 1990. *American Journal of Public Health, 88*, 227–232.

National Center for Health Statistics. (2001). Health, United States, 2001, with urban and rural health chartbook. DHHD Publication No. (PHS) 01–1232. Hyattsville, MD.

Palloni, A. and Arias, E. (2004). Paradox lost: Explaining the Hispanic adult mortality advantage. *Demography, 41(3)*, 385–415.

Patel, K.V., Eschbach, K., Ray, L.A., and Markides, K.S. (2004). Evaluation of mortality data for older Mexican Americans: implications for the Hispanic paradox. *American Journal of Epidemiology, 159(7)*, 707–715.

Rogers, A., Rogers, R.G., and Belanger, A. (1990). Longer life but worse health? Measurement and dynamics. *The Gerontologist, 30*, 640–649.

Rogers, R.G., Hummer, R.A., and Nam, C.B. (2000). *Living and dying in the USA*. New York: Academic Press.

Rogers, R.G., Rogers, A. and Belanger, A. (1989). Active life among the elderly in the United States: Multistate life-table estimates and population projections. *The Milbank Quarterly, 67*, 370–411.

Rosenberg, H.M., Maurer, J.D., Sorlie, P.D., Johnson, N.J., MacDorman, M.F., Hoyert, D.L., Spitler, J.F., and Scott, C. (1999). Quality of death rates by race and Hispanic origin: A summary of current research, 1999. Hyattsville, MD: National Center for Health Statistics. *Vital Health Statistics, 2(128)*.

Schoen, R. (1988). *Modeling multigroup populations*. New York: Plenum Press.

Schoenbaum, M., and Waidmann, T. (1997). Race, socioeconomic status, and health: Accounting for race differences in health. *Journal of Gerontology: Social Sciences, 52B*, (Special Issue), 61–73.

Smith, J.P., and Kington, R. (1997). Race, socioeconomic status, and health in later life. In L.G. Martin and B. Soldo (Eds.), *Racial and ethnic differences in the health of older Americans* (pp. 106–162). Washington, DC: National Academy Press.

Section 2
Contextualizing Support and Mexican-Origin Health in Old Age: Issues of Family, Migration, and Income

8
Dynamics of Intergenerational Assistance in Middle- and Old-Age in Mexico

Rebeca Wong and Monica Higgins*

8.1. Introduction

Numerous studies have shown that transfers of privately held resources across generations of the vertically extended family are common but vary in type and intensity with age and the life-cycle stage of the donor and/or recipient. Although resources typically flow from parent to child regardless of age in the United States, middle-aged adults have a heightened and substantial risk of making transfers to their kin in both ascending and descending generations. The literature on *inter-vivo* transfers shows that the patterns of giving are responsive to a range of sociodemographic, economic, cultural, and health attributes of both potential donor and recipients. The magnitude and direction of these effects vary with respect to the type of transfer modeled, the treatment of economic covariates, and the characterization of potential donors and recipients (Henretta et al., 1997; Schoeni, 1997; Soldo and Hill, 1993).

Less is known about the patterns of *inter-vivo* transfers for developing countries, in particular for countries that lack the infrastructure for public transfers to support the elderly beyond the productive years of the population. In developing economies, characterized by scarce or non-existent institutional systems for old-age support, and financial markets that are largely not available for the general population, the laterally and vertically extended family is the central institution in which investments in human capital are secured. Intra-family transfers are expected to flow from parents to children as investments in their old age security, from children to parents as repayment for past human capital investments in the child, or from children to parents in exchange for future bequests. The literature suggests,

*Paper prepared for presentation at the Second Conference on Aging in the Americas, University of Texas, September 2005. A previous version of the paper was presented at the Population Association of America Meetings, Philadelphia, PA, April 2005.

The authors gratefully acknowledge excellent research assistance from Juan José Díaz of the Maryland Population Research Center, University of Maryland. Please address correspondence to: Rebeca Wong, wongr@umd.edu.

that over time, transfers may flow in both directions to smooth consumption as a strategy for survival, as an expression of altruism of family members toward each other, or as repayment for services received or prior gifts (Lillard and Willis, 1997; Biddlecom et al., 2002).

Mexico is a middle-income country undergoing an epidemiological transition, characterized by a scheme in which infectious diseases are still among the main causes of death but chronic conditions have started to gain importance. Health status varies widely among the population because of historical social and economic inequalities including unequal access to health care (Parker and Wong, 1997). An older adult has access to health care and to retirement income only if the individual or a relative that claims the elderly as a dependent is affiliated to one of the institutions of the social security system, or if the older adult is retired and was formerly affiliated to one of the systems. In other words, only those who participate or used to participate, or whose relatives are participating in the formal labor market have access to the coverage of institutions. According to 2000 figures, only about half of the population aged 60 or older has health insurance coverage in a public or private institution. In addition, only 30% of men and 15% of women aged 60 or older have a retirement pension in Mexico (Parker and Wong, 2001; Gomes, 2001). In this context of scarce or no institutional protection, the possible vulnerability of elderly adults and the forms for meeting their financial and care needs become of great research importance. In addition, the historical lack of functioning financial and credit markets overall implies that non-financial means of securing old-age consumption must play important roles, with two examples being accumulation of assets in non-financial forms and support from family members. In such contexts of generalized lack of institutional protection for all ages, however, it is expected that family support flow in various directions, that is, that elderly adults simultaneously receive and provide help of various types.

We know little, however, about how prevalent the support of family members is, whether the support is in financial or non-financial form, and we know even less about the dynamic nature of the support, about how stable, temporary, or permanent is the support observed. We aim to learn also about the socioeconomic, health, and demographic determinants of such support. With the advent of recent specialized survey data it has become possible to study intergenerational assistance for developing countries such as Mexico, and in particular with the availability of longitudinal data, to examine the dynamic nature of the flows of family help.

The purpose of this chapter is to contribute to the literature on intergenerational transfers in developing countries by describing the patterns of transfers observed between adults of middle-and-old age and their children in Mexico, using the two waves of the Mexican Health and Aging Study, 2001 and 2003. We focus on the dynamic nature of the behaviors, and describe the determinants of becoming active as well as those of turning inactive in the two directions of the flow of transfers (giving and receiving), using two types of *inter-vivo* informal transfers: economic and non-economic. We focus on the attributes of the older adult, and pay particular attention to the role that fluctuations in health can play in the change of transfer behavior (what we term a "switch"). We focus on the following questions:

- What determines a switch toward giving help? Toward receiving?
- What determines a switch toward stopping to give help? Stopping to receive help?
- Do health shocks of the index person prompt a switch in transfer behaviors?

Do the same determinants that trigger a switch in one direction (say, *toward receiving* help) also trigger the switch *away from* the same (say, *away from receiving*), but with an opposite-sign effect?

8.2. Data

The Mexican Health and Aging Study (MHAS) is a prospective panel study of health and aging in Mexico and nationally representative of populations aged 50 and older in Mexico at the baseline in 2001. MHAS was supported by a grant from the National Institutes of Health/National Institute on Aging[1]. Interviews were sought with spouse/partners of sampled persons regardless of their own age. States with high rates of out-migration to the U.S. were over sampled at a rate of 1.7 to 1. Anthropometric measures also were obtained from a 20% sub-sample of respondents. The survey instruments and protocols were designed to maximize comparisons with the U.S. Health and Retirement Study (HRS).

MHAS was fielded in Mexico by the Instituto Nacional de Estadística, Geografía e Informática (INEGI), the equivalent of the U.S. Census Bureau. The sampling frame for MHAS was the household listing of about 136,000 dwelling units from the 4th Quarter of the ENE-2000 (National Employment Survey), also fielded by INEGI. Hence, the weighted first wave MHAS data are both a baseline for the MHAS panel and a representative cross-section of older Mexicans. Baseline interviews were completed in 9,845 households with about 15,000 respondents. By U.S. standards, the individual non-response rate of 10.5% for a population-based survey is very low. These data are particularly useful for the purpose of this paper because the survey gathers information on health, current income, and wealth, exchanges of money and in kind, and non-monetary help (in time) between the index adults and their children, whether co-residents or not. Because of the importance of international migration from Mexico to the U.S. the survey also includes an emphasis on past migration of the index persons and current migration of their children.

---

[1] The Mexican Health and Aging Study was supported by NIA grant no. AG18016 (P.I. B Soldo). The MHAS data are public; data and documentation can be obtained from the study website, www.mhas.pop.upenn.edu

The second wave of MHAS was successfully fielded in 2003. Both waves of the survey were carried out through pencil-and-paper, face-to-face interviews. For the 2003 re-visit, if a baseline respondent had died, a special exit interview was sought that year with a next-of-kin or informed respondent. Sample attrition was small by U.S. standards. At the individual level, 93.4% response rate was obtained, for a total of about 12,000 follow-ups with surviving individuals who were age 50 or older at the baseline. Of these, about 10,500 completed direct interviews both in 2001 and 2003. The number of next-of-kin interviews obtained in the follow-up (about 540) was consistent with the a priori expected number of deaths in the MHAS population during a two-year interval, estimated using Mexican life tables.

The content of the survey instrument included:

- Health: On multiple domains such as self-reported health, symptoms, functionality, depression, cognition, pain, healthy behaviors (alcohol and tobacco), health insurance, use of and expenditures on health care services
- Background: Childhood health and socioeconomic conditions, education, literacy, migration history and marital history
- Family: Roster of all children including those already deceased, with their demographic characteristics, summary of current health, education, economic activity and migration
- Transfers: Financial and in-time[2] help provided to children and received from children, indexed by child; financial and in-time help to parents
- Economic data: Work history, sources, and amounts of income including wages, pensions, and government help; type and value of assets
- Dwelling: Type, location, and materials of construction, other indicators of quality of the dwelling, and availability of consumer durables

For the purposes of this paper, we use the baseline sample of individuals aged 50 years or older who completed a direct interview in 2001 and were successfully re-interviewed in 2003. The effective sample size is 11,263 observations.

8.3. Methods

We use the 2001 and 2003 reports on help given to and received from children and/or grandchildren[3], and construct outcome variables for each of two types of help (monetary and in-time) that capture the 2-year change in the direction of the

---

[2] We use the terms financial/monetary/economic interchangeably to mean the help received or provided in monetary terms. We use the terms non-financial/in-time/non-economic interchangeably to mean the help received or provided to assist in household chores, run errands, care of children, help for functionality problems, or providing transportation. This includes help received for functionality problems such as activities of daily living (ADL) or instrumental activities of daily living (IADL).

[3] In the rest of the paper we refer to help given/received to/from children, and assume that grandchildren are also included.

flow of help: switch to receiving, switch away from receiving, switch to giving, and switch away from giving. We provide further details on the construction of these variables later, in the descriptive results.

We estimate probit regression models for each type of outcome. We use the following explanatory variables, defined at the baseline (year 2001):

- Demographic (age, sex, marital status, education, urban/rural residence, resides in high-migration state, number of living children, children currently residing in the U.S.)
- Economic (individual income and net worth, employment, coverage by a health insurance)
- Living arrangements (resides with married children, unmarried children, others)

We include also as explanatory variables the health changes between 2001 and 2003 in the form of:

- Chronic conditions, if a chronic condition was reported in 2003 that had not been reported in 2001 (among cancer, stroke, heart and lung disease, diabetes, and high blood pressure)
- Functionality problems, if the number of functionality problems reported in 2001 are more or fewer than those in 2001
- Self-rated health, if the self-assessed global health reported in 2003 is better or worse than the self-report in 2001

The description of variables and their codes are provided in Appendix Table A.

8.4. Descriptive Results

Table 8.1 presents a descriptive summary of the baseline data, which is representative of the Mexican population aged 50 or older. Slightly more than half are women (55%), and about one-half are aged 50–59. Roughly one-quarter report zero years of formal education, and over 60% report less than six years. Close to one-fifth (18%) are widowed, while close to 70% are married or in consensual union. Among women, roughly one-quarter are widowed, compared to 9% of the men. The average number of living children is about 5, evenly distributed on average between sons and daughters. About one-half reside with unmarried children, 17% live with spouse only, and 17% live alone. Only about half (54%) have health insurance. Migration to the U.S. is part of the life experience for some of the population; about 14% of men and 3% of women aged 50 or older report that they have been migrants to the U.S. at some point during their lives, and 22% of those who have children have at least one child currently residing in the U.S. The distribution of income and wealth are quite skewed in particular across educational groups as has been documented elsewhere. About one-third of those aged 50 or older reports family help as a source of income. Of the total, roughly 10% report that they receive income only from labor and family help, whereas another 10% report family help as their only source of income (Wong and Espinoza, 2003).

TABLE 8.1. Demographic characteristics of Mexican population 50 and over in 2001.

	2001
Gender	
Male	45.2
Female	54.8
Age	
50–59	46.3
60 or older	53.7
Number of years of education	
0	25.1
1–5	35.8
6	18.3
7+	20.9
Currently employed	43.0
At least one child living in the U.S.[1]	22.0
Resides in high-migration states[2]	27.9
Urban residence[3]	65.8
No. of living sons and daughters (mean)	5.3
No. of living sons	2.7
No. of living daughters	2.7
Living arrangements	
Lives alone	17.4
Resides with at least one married child	8.0
Resides with spouse only	17.0
Resides with unmarried child(ren) but not married child(ren)	50.2
Resides with others	7.5
Mean income (annual pesos)	
Low income (<30th percentile)	−1,073.2
Medium income	1,313.3
High income (>70th percentile)	13,481.5
Mean net worth (pesos)	
Low assets (<30th percentile)	17,164.9
Medium assets	110,281.1
High assets (>70th percentile)	504,956.5

1/ Includes only those who have children.
2/ Point of origin for over 40% of migrants to the U.S.: Durango, Guanajuato, Jalisco, Michoacán, Nayarit, and Zacatecas.
3/ Communities of more than 100,00 population.
Source: MHAS, 2001–2003.

The two-year health transitions are described in Tables 8.2A, 8.2B, and 8.2C. While we downplay the levels of health and concentrate on the dynamics, we note that the proportion of cases reporting chronic conditions is low in absolute terms but the relative *increase* in the proportion of people reporting chronic conditions is significant. About 5% of individuals report having at least one chronic condition in 2003 when there was none in 2001 (among heart, lung disease, stroke, or cancer). With respect to self-rated global health, roughly 28% of the baseline sample report worse health in 2003 than in 2001, and 21% report better health. About 10% of the panel sample report having one or more ADL/IADL difficulties in 2003 compared to 2001, and about the same proportion report fewer functionality problems.

TABLE 8.2A Comparison of principal health conditions* (in percentages).

	2001	2003
Self-rated health[1]		
Excellent	1.8	1.4
Very	4.2	3.1
Good	30.6	27.6
Fair	47.4	49.8
Poor	15.9	18.2
Chronic conditions		
Cancer	1.8	2.3
Respiratory problems	6.1	8.8
Heart problems	3.2	4.8
Stroke	2.3	3.1
Diabetes	15.4	19.8
Hypertension	37.6	48.5
Any of four chronic conditions[2]	12.2	16.9
Number of chronic conditions[2]		
0	87.9	83.1
1	10.9	14.9
2	1.1	1.9
3	0.1	0.2
At least one ADL or IADL[3]	16.7	17.9

*Mexican population aged 50 and over in 2001.
[1] Includes single, married, divorced, and separated.
[2] Among the following four chronic illnesses: cancer, heart, respiratory, and stroke.
[3] ADLs include: difficulty walking taking a bath, eating, going to bed, and using the toilet.
IADLs include: difficulty preparing a hot meal, shopping for groceries, taking medications, and managing own money.
Source: MHAS 2001–2003.

TABLE 8.2B New reports of chronic conditions* (in percentages).

	2001 to 2003
Chronic conditions	
Cancer	0.5
Respiratory problems	2.6
Heart problems	1.6
Stroke	0.8
Diabetes	4.3
Hypertension	10.9
Any of four chronic conditions[1]	4.7
Number of chronic conditions (among four)[1]	5.3

*Mexican population aged 50 and over in 2001.
[1] Among the following four chronic illnesses: cancer, heart, respiratory, and stroke.
Source: MHAS 2001–2003.

TABLE 8.2C Change in self-reported health and functionality* (in percentages).

	2001 to 2003	
	Worse off	*Better off*
Self-rated health	28.3	21.4
At least one ADL or IADL[1/]	10.7	9.6

*Mexican population aged 50 and over in 2001.
[1/] ADLs include: difficulty walking, taking a bath, eating, going to bed, and using the toilet.
IADLs include: difficulty preparing a hot meal, shopping for groceries, taking medications, and managing own money.
Source: MHAS 2001–2003.

We begin the description of the dynamics of transfers in Table 8.3. We use the variables in the survey regarding assistance given to and received from children to construct four categories that capture the direction of transfers for each type (economic and non-economic). The four categories are: only donor, only recipient, both (donor and recipient), and neither.

TABLE 8.3 Distribution of the population 50 and older providing/receiving economic or non-economic help to/from children in 2001 and 2003* (in percentages).

	Economic help in 2003						
	Only donor	Only recipient	Both	Neither	Total	# of Obs.	%
Economic Help in 2001:							
Only Donor	28.7	13.1	7.9	50.3	100.0	1,385	13.0
Only Recipient	3.6	58.7	5.2	32.4	100.0	4,469	41.8
Both	9.8	41.3	12.8	36.1	100.0	1,124	10.5
Neither	10.8	28.0	5.3	55.9	100.0	3,704	34.7
# of Obs.	1,068	4,315	689	4,647			
%	10.0	40.3	6.4	43.4			
	Non-Economic help in 2003						
	Only Donor	Only Recipient	Both	Neither	Total	# of Obs.	%
Non-Economic Help in 2001:							
Only Donor	21.9	13.9	23.9	40.4	100.0	1,910	17.9
Only Recipient	9.9	29.0	20.9	40.2	100.0	2,098	18.4
Both	15.8	19.6	32.2	32.4	100.0	3,087	28.9
Neither	12.2	19.5	18.2	50.1	100.0	3,952	37.0
# of Obs.	1,539	2,179	2,527	4,473			
%	14.4	20.3	23.6	41.7			

* Mexican population aged 50 and over in 2001.
Source: Mexican Health & Aging Study MHAS 2001–2003.

We present the distribution of the cases in 2001 and 2003 in Table 8.3. The top panel shows the economic help while the bottom panel shows the non-economic help. First, we note that the analysis based on the two cross-sections is somewhat deceiving; they hide a large entry-exit activity in the two-year interval. Focusing on economic help, the last row and last column of the panel (labeled "%") imply little change in the distribution of the categories of transfers if we compare the 2001 and 2003 cross-sections. For example, 13% were only donors in 2001 compared to 10% in 2003, 42% were only recipients in 2001 and 40% in 2003, 10% were both recipients and donors in 2001 and 6% in 2003, while 35% were neither in 2001, compared to 43% in 2003. However, the rest of the data in the panel show that there is quite a bit of movement in the two-year period to giving and receiving, as well as away from receiving and giving economic help. For example, of those who were donors only in 2001, by 2003 about one-third (29%) remained as donor only, 50% turned to neither giving or receiving, and 13% switched to being only recipients of economic help.

This pattern seems characteristic of those who are donors only, however. The pattern of change for those who in 2001 were recipients only is quite different. Of these, more than half remain as only recipients, and one-third turns to neither receiving nor giving. In contrast, of those who start both giving and receiving help, about 40% convert to only receive help and another 40% to neither receive nor give. And finally, of those who start as neither, about 56% remain in the same category, and 28% turn to only receiving. In summary, this panel shows a vast array of switches in giving and receiving economic help to/from children (see Figure 8.1).

The lower panel of Table 8.3 shows quite an active set of behaviors for non-economic help as well. We note here that donors only stay as only donor in 22% of the cases, with 40% switching to neither, and 24% to both recipients and donors. Of those starting as only recipient, 40% convert to neither, 21% to both, and 29% stay as only recipient. In contrast, of those who start as donor and recipient, one-third stay as both, and one-third switch to neither giving nor receiving help. And of those starting doing neither, about one-half stay as neither, and the rest distributes across the other three categories.

For the multivariate models of the determinants we want to describe the changes in time in the *direction* of the flows between the target person and their children. Thus, we define a switch *toward* a direction of flow (say, receive) and a switch *away from* the same direction for each of economic and non-economic help. We define the outcome variables as follows:

1. Switch *toward receiving* if the target individual was donor only or neither in 2001 and became a recipient only or both (donor and recipient) in 2003.
2. Switch *away from receiving* if a person was recipient or both in 2001 and moved to donor or neither in 2003.

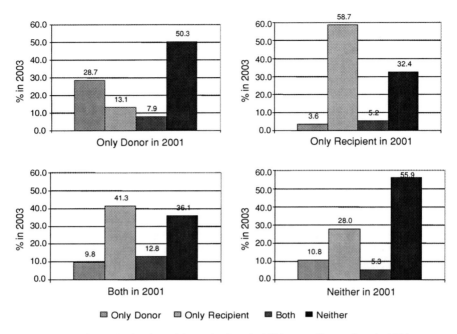

FIGURE 8.1. Distribution of financial flow in 2003 according to flow in 2001.

3. Switch *toward giving* if the individual was recipient or neither in 2001 and turned to donor only or both (donor/recipient) in 2003.
4. Switch *away from giving* if the person was donor or both in 2001 and converted to recipient or neither in 2003.

The resulting distributions are shown in Table 8.4 for economic and non-economic help. Note that since each switch outcome is defined conditioned on the initial (2001) state, the number of cases for each outcome varies.

Regarding economic help, proportionately more persons aged 50 or older who were actively exchanging help with their children at baseline switched away from giving (70%), about 38% moved away from receiving, 30% moved to receiving, and only 12% switched to giving. With respect to non-economic help the proportional changes are more homogeneous. Almost 53% moved away from giving, 49% away from receiving, 38% toward receiving, and 30% toward giving. Thus, on average, these unadjusted trends indicate a larger propensity to move *away from giving* than in any other direction, and this is true for both economic and non-economic exchanges with children (see Figure 8.2)

We next look at the determinants of these switches in behavior. We estimate separate probit regression models for each of the outcome variables. Table 8.5 presents the marginal effects of switches toward/away from categories of economic help, and Table 8.6 shows the equivalent results

TABLE 8.4 Distribution of switches in the flow of economic and non-economic assistance to/from children. Population aged 50 and older in 2001*.

	Economic Help		Non-economic Help	
	Total	%	Total	%
Donor or Neither in 2001				
Recipient or Both in 2003 (switch = 1)	1,516	29.8	2,137	37.6
Donor or Neither in 2003	3,516	70.2	3,546	62.4
Total	5,032	100.0	5,683	100.0
Recipient or Both in 2001				
Donor or Neither in 2003 (switch = 1)	2,120	37.9	2,433	48.7
Recipient or Both in 2003	3,473	62.1	2,564	51.3
Total	5,593	100.0	4,997	100.0
Recipient or Neither in 2001				
Donor or Both in 2003 (switch = 1)	988	12.1	1,781	30.4
Recipient or Neither in 2003	7,185	87.9	4,081	69.6
Total	8,173	100.0	5,862	100.0
Donor or Both in 2001				
Recipient or Neither in 2003 (switch = 1)	1,741	69.6	2,535	52.6
Donor or Both in 2001	768	30.6	2,283	47.4
Total	2,509	100.2	4,818	100.0

* Mexican population aged 50 and over in 2001.
Source: Mexican Health & Aging Study MHAS 2001–2003.

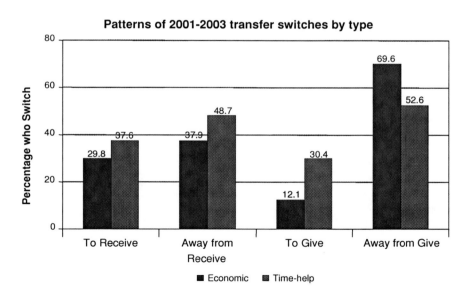

FIGURE 8.2. 2001–2003 switches in transfer of flows.

TABLE 8.5 PART A. Marginal effects of 2-year switches in economic transfer status.

Explanatory variables	(1) To receive	(2) Away from receive	(3) To give	(4) Away from give
At baseline (2001):				
Age 60–69 (ref: 50–59)	0.043***	−0.020	−0.046***	0.086***
	(0.008)	(0.211)	(0.000)	(0.000)
Age 70 or older	0.085***	−0.065***	−0.085***	0.151***
	(0.000)	(0.001)	(0.000)	(0.000)
Single woman (ref: married man)	0.075**	−0.123***	−0.046***	0.044
	(0.020)	(0.000)	(0.000)	(0.286)
Married woman	0.024	−0.013	−0.007	0.044*
	(0.166)	(0.482)	(0.419)	(0.076)
Widowed woman	0.118***	−0.127***	−0.036***	0.056
	(0.000)	(0.000)	(0.002)	(0.188)
Single man	−0.095**	0.076	−0.042*	−0.113*
	(0.015)	(0.207)	(0.075)	(0.053)
Widowed man	−0.036	−0.049	0.025	−0.174***
	(0.317)	(0.188)	(0.222)	(0.006)
Lives in high-migration state	0.032**	−0.006	0.005	0.016
	(0.039)	(0.672)	(0.564)	(0.457)
Lives in more-urban area	−0.090***	0.087***	−0.003	0.019
	(0.000)	(0.000)	(0.684)	(0.419)
Resides with married child	0.052	−0.020	0.022	−0.016
(ref: lives alone)	(0.146)	(0.531)	(0.275)	(0.786)
With spouse only	0.013	−0.070**	0.026	0.010
	(0.720)	(0.037)	(0.228)	(0.878)
With unmarried child	0.049	−0.000	0.041**	−0.019
	(0.138)	(0.989)	(0.026)	(0.730)
With others	0.037	−0.081**	0.038	0.004
	(0.394)	(0.028)	(0.123)	(0.951)
No. of living children	0.009***	−0.002	0.001	0.004
	(0.001)	(0.376)	(0.339)	(0.297)
Has 1+ child living in US	0.062***	−0.116***	−0.012	0.093***
	(0.001)	(0.000)	(0.171)	(0.001)
Has no children—DK/DR	−0.060***	0.073**	0.051***	−0.074***
	(0.003)	(0.023)	(0.000)	(0.008)
Schooling: 1–5 years	−0.040**	−0.030*	0.032***	0.034
(ref: 0 years)	(0.029)	(0.071)	(0.001)	(0.291)
6 years	−0.016	−0.010	0.013	−0.030
	(0.445)	(0.625)	(0.259)	(0.413)
7+ years	−0.100***	−0.018	0.063***	−0.078**
	(0.000)	(0.470)	(0.000)	(0.032)

Notes: Probit regressions. Dependent variables:
   To receive: if donor-only or neither donor nor recipient in 2001, and recipient-only or both (donor and recipient) in 2003.
   Away from receive: if recipient-only or both in 2001, and donor-only or neither in 2003.
   To give: if recipient-only or both in 2001, and donor-only or both in 2003.
   Away from give: if donor-only or both in 2001, and recipient-only or neither in 2003.
p-values in parentheses
* significant at 10%; ** significant at 5%; *** significant at 1%

(*Continued*)

TABLE 8.5 PART B. Marginal effects of 2-year switches in economic transfer status.

Explanatory variables	(1) To receive	(2) Away from receive	(3) To give	(4) Away from give
At baseline (2001):				
Low income–medium assets	−0.038	0.009	0.010	0.016
(ref: low–low)	(0.185)	(0.760)	(0.558)	(0.799)
Low income–high assets	−0.081***	0.036	0.033*	0.005
	(0.004)	(0.266)	(0.074)	(0.935)
Med income–low assets	−0.024	−0.041	0.017	0.058
	(0.396)	(0.139)	(0.305)	(0.313)
Med income–med assets	−0.026	−0.035	0.047***	0.027
	(0.334)	(0.199)	(0.005)	(0.626)
Med income–high assets	−0.067**	0.048	0.048***	0.030
	(0.020)	(0.105)	(0.007)	(0.601)
High income–low assets	−0.060*	0.022	0.050**	−0.018
	(0.075)	(0.499)	(0.014)	(0.774)
High income–med assets	−0.087***	−0.024	0.037**	−0.024
	(0.002)	(0.425)	(0.037)	(0.670)
High income–high assets	−0.122***	0.063**	0.071***	−0.063
	(0.000)	(0.030)	(0.000)	(0.248)
Employed	−0.036**	0.055***	0.027***	−0.029
	(0.022)	(0.001)	(0.001)	(0.206)
Has health insurance	−0.011	−0.006	0.008	−0.038*
	(0.478)	(0.698)	(0.277)	(0.093)
Health shocks 2001–2003:				
Cancer, new report	0.076	−0.051	−0.007	0.010
	(0.416)	(0.564)	(0.882)	(0.936)
Stroke, new report	−0.015	−0.047	0.047	−0.019
	(0.841)	(0.498)	(0.240)	(0.901)
Heart disease, new report	0.071	0.067	−0.007	−0.011
	(0.188)	(0.194)	(0.802)	(0.903)
Lung disease, new report	−0.003	−0.046	−0.025	−0.083
	(0.939)	(0.243)	(0.261)	(0.205)
Diabetes, new report	0.046	−0.041	0.005	−0.054
	(0.154)	(0.227)	(0.777)	(0.286)
Hypertension, new report	−0.005	−0.047**	0.019*	−0.009
	(0.806)	(0.031)	(0.093)	(0.774)
Worse ADL/IADL	0.024	0.002	0.019	−0.027
(ref: no change)	(0.291)	(0.942)	(0.114)	(0.449)
Better ADL/IADL	0.021	−0.013	−0.025**	0.048
	(0.401)	(0.561)	(0.042)	(0.138)
Worse self-rated health	−0.005	0.021	−0.004	0.023
(ref: no change)	(0.744)	(0.212)	(0.673)	(0.310)
Better health	−0.019	0.004	−0.011	0.010
	(0.286)	(0.808)	(0.225)	(0.701)
Became proxy interview	0.055**	−0.002	0.012	0.077**
	(0.037)	(0.926)	(0.406)	(0.031)
No. of observations	4,980	5,464	7,979	2,465

Notes: Probit regressions.
p-values in parentheses
* significant at 10%; ** significant at 5%; *** significant at 1%

TABLE 8.6 PART A. Marginal effects of 2-year switches in non-economic transfer status.

Explanatory variables	(1) To receive	(2) Away from receive	(3) To give	(4) Away from give
At baseline (2001):				
–Age 60–69 (ref: 50–59)	−0.054***	0.008	−0.065***	0.066***
	(0.001)	(0.671)	(0.000)	(0.000)
Age 70 or older	−0.039**	−0.028	−0.181***	0.141***
	(0.047)	(0.222)	(0.000)	(0.000)
Single woman	0.005	−0.062**	−0.034	−0.055*
(ref: married man)	(0.865)	(0.040)	(0.194)	(0.076)
–Married woman	0.008	−0.014	−0.019	0.017
	(0.672)	(0.482)	(0.249)	(0.386)
–Widowed woman	0.045*	−0.033	−0.022	−0.049*
	(0.063)	(0.186)	(0.322)	(0.054)
–Single man	−0.110**	0.017	−0.153***	0.168***
	(0.011)	(0.758)	(0.000)	(0.004)
–Widowed man	0.059	−0.017	−0.055*	0.123**
	(0.116)	(0.668)	(0.079)	(0.017)
Lives in high-migration state	−0.078***	0.086***	−0.049***	0.097***
	(0.000)	(0.000)	(0.000)	(0.000)
Lives in more-urban area	−0.056***	0.040**	−0.031**	0.013
	(0.000)	(0.020)	(0.030)	(0.488)
Resides with married child	0.196***	−0.071*	0.112***	−0.069
(ref: lives alone)	(0.000)	(0.074)	(0.000)	(0.110)
–With spouse only	0.058*	−0.031	−0.003	0.027
	(0.087)	(0.478)	(0.929)	(0.557)
–With unmarried child	0.143***	−0.034	0.054*	0.006
	(0.000)	(0.375)	(0.054)	(0.886)
–With others	0.074*	−0.050	0.062	−0.035
	(0.055)	(0.293)	(0.106)	(0.478)
No. of living children	0.006**	−0.003	0.003	0.004
	(0.035)	(0.254)	(0.205)	(0.217)
Has 1+ child living in US	−0.024	0.032	−0.052***	0.025
	(0.139)	(0.107)	(0.001)	(0.211)
Has no children–DK/DR	0.045*	−0.047*	−0.049**	0.078***
	(0.066)	(0.060)	(0.019)	(0.003)
Schooling: 1–5 years	−0.004	−0.047**	0.052***	−0.036*
(ref: 0 years)	(0.799)	(0.015)	(0.002)	(0.080)
6 years	0.009	−0.035	0.087***	−0.062**
	(0.665)	(0.150)	(0.000)	(0.010)
7+ years	−0.022	−0.027	0.033	−0.091***
	(0.359)	(0.293)	(0.151)	(0.001)

Notes: Probit regressions. Dependent variables:
  To receive: if donor-only or neither donor nor recipient in 2001, and recipient-only or both (donor and recipient) in 2003.
  Away from receive: if recipient-only or both in 2001, and donor-only or neither in 2003.
  To give: if recipient-only or both in 2001, and donor-only or both in 2003.
  Away from give: if donor-only or both in 2001, and recipient-only or neither in 2003.
p-values in parentheses
*significant at 10%; ** significant at 5%; *** significant at 1%

(Continued)

TABLE 8.6 PART B. Marginal effects of 2-year switches in non-economic transfer status.

Explanatory variables	(1) To receive	(2) Away from receive	(3) To give	(4) Away from give
At baseline (2001):				
Low income-medium assets	0.019	0.039	−0.028	−0.071*
(ref: low-low)	(0.521)	(0.250)	(0.297)	(0.053)
–Low income-high assets	−0.016	0.051	−0.023	−0.051
	(0.593)	(0.147)	(0.417)	(0.155)
–Med income-low assets	0.025	0.051	0.007	−0.028
	(0.393)	(0.107)	(0.801)	(0.394)
–Med income-med assets	0.049*	−0.008	−0.005	−0.056*
	(0.082)	(0.798)	(0.855)	(0.081)
–Med income-high assets	0.018	−0.002	−0.052*	0.017
	(0.544)	(0.955)	(0.058)	(0.624)
–High income-low assets	0.044	0.035	−0.043	−0.005
	(0.196)	(0.361)	(0.171)	(0.890)
–High income-med assets	0.022	−0.000	−0.001	−0.011
	(0.473)	(0.995)	(0.965)	(0.750)
–High income-high assets	0.009	0.041	−0.003	−0.010
	(0.758)	(0.187)	(0.895)	(0.755)
Employed	0.013	0.003	−0.006	−0.016
	(0.396)	(0.855)	(0.691)	(0.377)
Has health insurance	0.007	−0.012	0.047***	−0.024
	(0.652)	(0.461)	(0.001)	(0.161)
Health shocks 2001–2003:				
Cancer, new report	0.275***	−0.152	−0.034	−0.165
	(0.003)	(0.133)	(0.673)	(0.144)
Stroke, new report	0.027	−0.116	−0.074	0.006
	(0.731)	(0.130)	(0.250)	(0.947)
Heart disease, new report	0.001	−0.001	0.052	0.015
	(0.979)	(0.989)	(0.272)	(0.806)
Lung disease, new report	−0.065	−0.039	−0.007	0.050
	(0.124)	(0.385)	(0.850)	(0.303)
Diabetes, new report	0.066**	0.026	0.033	−0.030
	(0.034)	(0.497)	(0.276)	(0.414)
Hypertension, new report	0.011	−0.017	0.032	0.011
	(0.592)	(0.491)	(0.101)	(0.645)
Worse ADL/IADL	0.102***	−0.099***	0.025	−0.038
(ref: no change)	(0.000)	(0.000)	(0.226)	(0.140)
–Better ADL/IADL	−0.027	0.103***	−0.035	0.030
	(0.268)	(0.000)	(0.122)	(0.248)
Worse self-rated health	0.000	0.007	−0.029**	0.006
(ref: no change)	(0.983)	(0.709)	(0.049)	(0.747)
–Better health	−0.032*	0.032*	−0.036**	0.024
	(0.069)	(0.097)	(0.030)	(0.220)
Became proxy interview	0.114***	−0.079***	0.013	0.011
	(0.000)	(0.007)	(0.593)	(0.716)
No. of observations	5,535	4,907	5,700	4,742

p-values in parentheses
*significant at 10%; **significant at 5%; ***significant at 1%

for non-economic help[4]. For the purpose of presentation, we split the tables into two parts: part A, containing the socio-demographic and migration characteristics, and part B, containing the effect of economic attributes and health shocks.

In describing the results we focus on two aspects. First, the covariates and how they are associated differently with the changes in the four directions we study: toward receiving help, away from receiving, toward giving, and away from giving help. Second, we summarize the differences in the results across monetary and in-time help.

8.4.1. Dynamics of Transfers with Age

The results show that certain socio-demographic factors are consistently associated with the switches we study: age, gender/marital status, number of children, having children who currently reside in the U.S., and education. Regarding age effects, except for one direction, the switches are as expected; that is, older individuals in general would be expected to move toward receiving help and away from giving. We find that older individuals are more likely to switch toward receiving economic help, less likely to start giving, and more likely to stop giving. There is no significant effect of age on the likelihood of stop receiving help, however. These effects are similar for economic and non-economic help, but there is one particular age-effect that differs between financial and time help: Older persons are less likely to switch to receive non-economic help. That is, if an individual were *not receiving* help in 2001, holding all else constant, older age implied a lower propensity to start receiving in-time help but a higher propensity to start receiving economic help.

8.4.2. Dynamics of Transfers and Gender, Marital Status, Number of Children

We find that being a widowed woman consistently is associated with a switch toward receiving economic help; the same result applies to a start for receiving in-time help. Having more living children is associated also with switching toward receiving economic and non-economic help. Having at least one child who currently resides in the U.S. has several switching effects. It increases the likelihood of moving toward receiving economic help, and decreases the chances of moving away from receiving. On the other hand, it reduces the chances of giving non-economic help. Holding all else constant, the effects of education are different

---

[4] We also estimated alternate specifications (not shown). We tried models with continuous age and age-squared terms, income and wealth terms separately instead of interacted, and models with an aggregate variable capturing the change in the report of any chronic condition of the six included in the models. The results do not differ from the ones presented here, and we chose to include those that show the effect of each chronic condition to highlight the contribution of diabetes and cancer in triggering changes in the flow of transfers.

for economic and non-economic help. Being in a high-category of education (7 or more years) reduces the chances of switching to receiving economic help and increases the odds of becoming a donor of economic help to children compared to those with zero years of education. However, having middle-level education (not the highest) increases the chances of switching to giving in-time help and also reduces the chances of getting away from giving in-time help.

8.4.3. Dynamics of Transfers and Economic Factors: Income, Wealth, Health Insurance, and Employment

Holding all else constant, we find little effect on transfers of availability of health insurance: It reduces the odds of stopping gifts of economic help, and increases the odds of giving in-time help.

On the other hand, we find a strong and consistent effect of the income/wealth variables included in the model. Remarkably, as well, the income/wealth variables seem to be associated with switches in economic help, but not as consistently or strongly as switches in non-economic help. Compared to those in the lowest income and wealth group, having high wealth (upper 30th percentile), regardless of income level within that level, is associated with reduced odds of becoming a recipient and increased odds of becoming a donor of economic help. We find the same pattern for those in the highest income level regardless of wealth level: Having the highest level of income reduces the odds of receiving and increases the odds of giving economic help (see Figure 8.3).

In contrast to the effects for economic help, in which being in the high end of the distribution seems to matter, the few effects shown for the income/wealth variables on the switches of non-economic help are for those in the low/medium

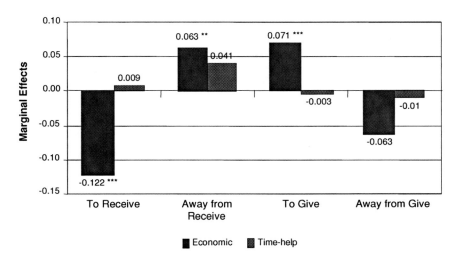

FIGURE 8.3. Effect of high income/assets on transfer switches by type.

end of the scale. Compared to those in the lowest income/wealth scale, having low income/medium wealth is associated with reduced odds of moving away from giving non-economic help; having medium/medium wealth increases the odds of receiving non-economic help and lowers the odds of moving away from giving non-economic help.

We also find that employment of the older adults has several switching effects in economic help only. Being employed decreases the likelihood of receiving economic help, and it increases the chances of moving away from receiving help and of giving economic help.

8.4.4. Dynamics of Transfers and Health Shocks

We find no effects of the two-year health shocks for the changes in flows of economic help. This is true for chronic conditions, functionality variables, as well as self-reported health. In contrast, the health shock variables are associated with switches in time-help. We find significant effect only for two of the chronic conditions that we included in the models and only for one of the directions. A person's new report of cancer or diabetes increases the odds of becoming a recipient of help. The effects are more consistent for functionality and self-reported health, however. A *worse* functionality report (ADL/IADL) is associated with higher chances of becoming a recipient and lower odds of moving away from being a recipient of in-time help. On the other hand, an *improvement* in functionality increases the odds of moving away from receiving help. Having a worse self-report of global health reduces the odds of becoming a donor of help, but an improved self-reported health has several effects: it reduces the odds of becoming a recipient, increases the chances of moving away from receiving, and reduces the odds of becoming a donor of non-economic help.

In summary, we find that socio-demographic as well as economic factors and health shocks are associated with the changes in transfer flows between older adults and their children in Mexico. We find that the same factors that increase the chances of starting a flow do not necessarily reduce the chances of stopping the same flow. We also conclude that economic factors captured by income and wealth have important effects on the changes in flows of economic help only; they seem to have no general effect on flows of in-time help. In contrast, health shocks seem to be associated with changes only in-time help. Of the health shocks we consider, functionality and self-reported health seem to be closely associated with changes in transfer flows; this is not seen so much in self-reports of chronic conditions.

8.5. Conclusions

We aimed to describe the patterns of flow of help exchanged between older adults and their children in Mexico, using the two waves of the Mexican Health and Aging Study (2001 and 2003). We focused on the dynamics of the flows and chose to define switches toward and away from giving and receiving help of two types: economic and non-economic. In our descriptive analyses, we find that although in cross-section the propensity to give and receive help appears quite stable (the unadjusted propensities are similar from one wave to the next), there is quite a bit of activity in and out of giving and receiving help. This finding could imply that transfers occur largely for other reasons, in addition to cultural norms of filial care. This is, of course, a tentative conclusion as our analyses were preliminary and basic, and did not seek to test such hypothesis explicitly. However, these patterns of entry/exit when we treat the index individual as the donor/recipient and the total set of children as the other end of the flow of help would seem to indicate that older adults enter into and exit from the help system, and do not necessarily remain in a state of continuously giving and receiving help, as a cultural norm would indicate. Still, active flows are prevalent between older adults and their children; in both waves about 40% of target individuals are only recipients of economic help from their children. But of those who started as recipients, about one-third move to receive no help two years later.

We examined the covariates of the switches in the flows of assistance: What determines that a person becomes a recipient, a donor? What determines that a person stops giving/receiving? In general we find that older age increases economic dependency, for in old age a person is more likely to be a receptor and less likely a donor of economic help. But, one particular finding deserves more attention in the future: Holding all else constant, older individuals are *less* likely to become recipients of non-economic help. This implies that old age per se may not necessarily imply higher demands of time-assistance from the younger generation. Other aspects, such as health and functionality may determine the ultimate in-time help demands. We confirm the importance of international remittances and U.S. migration from Mexico in our analyses of the flows of intergenerational assistance. Older adults with children residing in the U.S. are more likely to receive monetary help and less likely to give non-monetary help to children. This is perhaps because the distance between the older adult and the child in the U.S allows only for monetary transfers.

We find that the economic ability of older adults to support their financial needs determine strongly the flows of monetary transfers. What seems worth noting is that both wealth and income seem to have a protective effect, but only for those above the highest 30th percentile of the income and wealth distributions. On the other hand, the financial means of older adults do not seem to influence flows of in-time help to and from their children.

We found no association between economic assistance and changes in the reports of chronic conditions, functionality or self-reported global health. Two-year health changes seem to affect the flow of assistance to and from children only in non-monetary terms. Furthermore, there seems to be no relationship between new chronic conditions reported and intergenerational help, but changes in functionality and self-reported health affect the flow of time-help. Of notice is that not only a movement toward worse health seems to trigger help, but also *improvements* in these two dimensions of health trigger moves *away from receiving* help. This result implies "non-stickiness in help," that is, assistance triggered by a health shock seems reversible if gains in health or in functionality are experienced.

APPENDIX TABLE A. Definition of variables

Variable	Definition
Measured at 2001	
Marital status	Dichotomous variable for the categories: 1 = Single (*); 2 = Married/in union; and 3 = Widowed.
Age	Continous variable, range 50 to 105.
Education	Dichotomous variables for the levels: None (0 years) low (1 to 5 years), medium (6 years), high (7 or more).
Proxy interview	1 = proxy interview 0 = direct interview.
Currently employed 1/	1 = yes 0 = no
At least one child resides in the US 2/	1 = yes 0 = no
Resident of high-migration states 3/	1 = yes 0 = no
Urban residence 4/	1 = yes 0 = no
No. of living sons and daughters	Continuous variable, from 0 to 20.
Living arrangements	Dichotomous variable for the levels: 1 = live alone; 2 = resides with at least one married child; 3 = resides with spouse only; 4 = resides with unmarried child(ren) but not married child(ren); and 5 = resides with others.
Individual income	Dichotomous variables for the levels: Low, Medium, and High.
Household net worth	Dischotomous variables for the levels: Low, Medium and High.
Change from 2001 to 2003:	
Self-reported/rated health 5/	Dischotomous variables for the levels: no change, worse, better than in 2001.
Hypertension 6/	1 = worse than in 2001 0 = no change
Diabetes 6/	1 = worse than in 2001 0 = no change
Cancer 6/	1 = worse than in 2001 0 = no change
Lung disease 6/	1 = worse than in 2001 0 = no change
Heart disease 6/	1 = worse than in 2001 0 = no change
Stroke 6/	1 = worse than in 2001 0 = no change
Number of chromic conditions 7/	1 = more chromic conditions than in 2001, 0 = no change
Problems with one or more activity of daily living (ADL) or one or more instrumental activities (IADL) 8/	Dischotomous variables for the levels: no change, worse, better than in 2001.
Given Economic help 9/	1 = yes 0 = no
Received Economic help 10/	1 = yes 0 = no
Given Non-Economic help 11/	1 = yes 0 = no
Received Non-Economic help 12/	1 = yes 0 = no
Only Donor	1 = Gives help but does not receive any help; 0 = other
Only Recipient	1 = Receives help but do not give any help; 0 = other
Both	1 = Gives and receives help; 0 = other
Neither	1 = Does not receive/ give any help; 0 = other

(*) Includes single, married divorced, and separated.
1/ Currently employed in 2001. Not employed includes: those not working but had a job, looking for work, students, working around the house, not working.
2/ Includes only those who have children.
3/ Point of origin for over 40% of migrants to the U.S.: Durango, Guanajuato, Jalisco, Michoacán, Nayarit, and Zacatecas.
4/ Communities of more than 100,000 population.

(Continued)

Appendix Table A (*Continued*)

5/ Variable was classified as: 1 = excellent 2 = very good, 3 = good, 4 = fair, 5 = poor.

6/ If a doctor or nurse has ever told the person that s(he) has the illness or condition.

7/ Among four chronic illnesses: cancer, heart, respiratory, or stroke.

8/ ADLs include: difficulty walking, taking a bath, eating, going to bed, and using the toilet. IADLs include: difficulty preparing a hot meal, shopping for groceries, taking medications, and managing own money.

9/ If in the last two years, respondent (or spouse) gave financial or in-kind support (at least 5000 pesos) to any children and/or grandchildren.

10/ If in the last two years, respondent (or spouse) received financial or in-kind support (at least 5000 pesos) from any children and/or grandchildren.

11/ If in the last two years, respondent (or spouse) spent at least one hour a week helping any children and/or grandchildren.

12/ If in the last two years, respondent's (or spouse's) children and/or grandchildren spent at least one hour a week helping respondent (or spouse).

Source: Mexican Health and Aging Study MHAS 2001–2003.

Overall, the analyses we have performed seem to indicate adequate quality of the Mexican data, with inter-wave consistency and enough power to measure the dynamics of transfers and their covariates. The complexity of the movements that we observe is vast and our analyses preliminary, but it seems that the system of exchanges between older adults and their children is quite dynamic despite the fact that we observe only a two-year inter-wave interval, and transfers seem differentially responsive to economic, demographic, and health factors. We have limited our preliminary analyses to the propensity to provide and receive help, leaving aside the amounts and the specific child with whom the exchanges take place. We also omit the individual attributes of the children. Further analyses that consider these characteristics and the interaction between the two types of transfers are part of the future work to increase our understanding of the dynamics of intergenerational transfers. Cross-national research shall also further our understanding of the role of institutions (Hermalin, 1999; National Research Council, 2001) within the system of intergenerational exchanges aimed at securing the well-being of older adults and their families.

References

Biddlecom, A., Chayovan, N., and Ofstedal, M.B. (2002). Intergenerational support and transfers," In A.I. Hermalin (Ed.), *The well-being of the elderly in Asia: A four country comparative study* (pp. 185–230). The University of Michigan Press.

Chen, X., and Silverstein, M. (2000). Intergenerational social support and the psychological well-being of older parents in China. *Research on Aging, 22(1)*, 43–65.

Henretta, J.C., Hill, M.S., Li, W., Soldo, B.J., and Wolf, D.A. (1997). Selection of children to provide care: The effect of earlier parental transfers. *Journals of Gerontology, 52B*, 110–119.

Hermalin, A.I. (1999). Challenges to comparative research on intergenerational transfers. *Southeast Asian Journal of Social Science, 27(2)*, 9–20.

Lillard, L.A., and Willis, R.J. (1997). Motives for intergenerational transfers: Evidence from Malaysia. *Demography, 34(1)* 115–134.

National Research Council (2001). Intergenerational transfers. In *Preparing for an aging world: The case for cross-national research* (pp. 155–199). Washington, DC: National Academy Press.

Ofstedal, M.B., Knodel, J., and Chayovan, N. (1999). Intergenerational support and gender: A comparison of four Asian countries. *Southeast Asian Journal of Social Science, 27(2)*, 21–42.

Parker, S., and Wong, R. (2001). Welfare of male and female elderly in Mexico: A comparison. In E.G. Katz and M.C. Correia (Eds.) *The economics of gender in Mexico* (pp. 249–290). Washington, DC: The World Bank.

Palloni, A. (2001). Living arrangements of older persons. *Population Bulletin of the United Nations, Num, 42/43,* United Nations, New York.

Petrova, P. (2004). Does health status affect transfers from children? Evidence from Mexico. Dissertation paper, Boston College.

Quisumbing, A.R. (1994). Intergenerational transfers in Philippine rice villages: Gender differences in traditional inheritance customs. *Journal of Development Economics, 43*, 167–195.

Schoeni, R.F. (1997). Private interhousehold transfers of money and time: New empirical evidence. *Review of Income and Wealth, 43(4)*, 423–448.

Soldo, B. J., and Hill, M. (1993). Intergenerational transfers: Economic, demographic, and social perspectives. In G.L. Maddock and M.P. Lawton (Eds.), *Annual review of gerontology and geriatrics: Focus on kinship, aging, and social change, Vol. 13* (pp. 187–218). New York: Springer Publishing Co.

Wong, R., and Espinoza, M. (2003). Ingreso y bienes de la población en edad media y avanzada en México (Income and assets of middle- and old-age population in Mexico), *Papeles de Población, 9(37)*, 129–166.

Wong, R (2003). Relación entre salud y nivel socioeconómico entre adultos mayores: diferencias por género" (Relationship between health and socioeconomic level among older adults: gender differences). In N. Salgado y R. Wong (Eds.), *Envejeciendo en la pobreza: Género, salud y calidad de vida (Aging in poverty: Gender, health and quality of life)*, México: Instituto Nacional de Salud Pública.

9
Aging, Health and Migration: The Voices of the Elderly Poor in Mexico

V. Nelly Salgado de Snyder*

9.1. Introduction

The World Health Organization (WHO) defines the elderly as people aged 60 and older and reports that there are about 600 million elderly people in the world (WHO, 1999). However, according to demographic projections, by the middle of the current century this number will increase to 2 billion. The situation in Latin American countries is similar; in the year 2025 there will be almost 40 million elderly people (WHO, 1999; PAHO, 2002a,b). In Mexico, statistics from the national census (INEGI, 2000) revealed that in the year 2000 there were almost seven million people 60 and older; and according to demographic projections, in the year 2050 one in every four Mexicans will be in this age group (CONAPO, 2001; Tuiran, 1999). Falling fertility and mortality rates and emigration to the Unites States account for Mexico's rapidly aging population. Increasing numbers of elderly people represent one of the most important challenges for Mexico, particularly with respect to health.

The acknowledgement that social factors play a key role in the health status of people goes back to the 19the century. Furthermore, WHO acknowledges social roots of health problems in their definition of health as a state of complete physical, mental, and social well-being. Therefore, health status is a complex variable that goes beyond physiological functioning. Social determinants seem to be the strongest indicators of health status and quality of life for people (PAHO, 2002a). Furthermore, poverty has been identified as the single most important social determinant of health. Chronic poverty (understood as the poverty that affects more than one generation) and old age is a combination that in the developing world: is both, a threat and a challenge for the health system.

The elderly in Mexico are over represented among people with lower incomes. Currently, 28% of the households with elderly people are located in rural villages of 2,500 inhabitants or less, where houses are built with fragile materials, have dirt

* Project partially funded by The fulbright new century scholars program and indesoc.

floors, and lack running water, electricity, drainage system, and other basic services (Solis, 1999). Although urban poverty is growing, the rural poor still account for over 80% of the total number of poor people in the world. They face problems such as weak infrastructure, unemployment, lack of education and training, and geographical isolation. Additionally, they are vulnerable to exploitation, disease, and natural disasters because they lack productive assets and suffer from physical weakness, illness, and powerlessness (Desjarlais, Eisenberg, Good, and Kleinman, 1995). In Mexico, 50% of the population lives in moderate poverty, of these, 30% live in extreme poverty (Hernandez and Velazquez, 2002). The elderly poor are one of the most vulnerable groups because their poverty is compounded by mental and physical disadvantages associated with aging.

Poverty and old age are two related concepts in the developing world. The relationship between these two variables is determined by the fact that older people have reduced capacity in their physical functioning to work and produce goods and benefits for themselves and their families. With the epidemiological transition affecting the world, the number of elderly people in developed countries is rapidly increasing and so is the number of benefits they receive, unlike their counterparts in less developed countries where people tend to age in poverty. The care of older people is now falling on fewer younger people, making both generational groups vulnerable and in need of social and financial support structures that help them to carry out their tasks (Solis, 1999; Tuiran, 1999).

Health in old age is greatly determined by lifestyles, exposure, and opportunities for health protection over the life course (Kalache and Sen, 1998). Among other outcomes of this is a consistent underestimate of the numbers of older people in the developing world, since the threshold of old age is reached much sooner by those who have experienced arduous earlier lives (marked by hard physical labor).

Poverty is one of the most important reasons why Mexicans seek work in the United States. Currently, Mexico is a country of origin, transit, and final destination for international migratory fluxes. Annually, about 500 thousand people, mostly men (57%), cross the border between Mexico and the U.S. Many of them come from states with a high migratory tradition such as Jalisco, Zacatecas, Guanajuato, and Michoacán (CONAPO, 2005; Kocchar, 2005). Other migrants originate from geographical regions new to this phenomenon, such as the states of Puebla, Guerrero, Oaxaca, Veracruz, and Morelos, whose residents have incorporated themselves to the migratory flux to the U.S. only in the last couple of decades (CONAPO, 2003; 2002; Kochhar, 2005).

Almost one half of Mexican labor migrants to the U.S. are of rural origin and going north is considered a survival strategy for the family (Leite, 2003). Many of them return to their communities of origin in Mexico; however, an unknown number of Mexicans remain and settle in the U.S. with or without legal residency papers (CONAPO, 2003; 2002). It has been consistently reported that family reunion is the main reason for returning to their home country (Acevedo, 2004; Espinosa, 1998). Immigrant men and women return to their localities of origin with the clear purpose of enjoying their family, friends, and money earned (Espinosa, 1998; Salgado de Snyder, 2001).

For instance, in a recent publication conducted with elderly rural men in Mexico, their perception of well being was found to be associated with having migrated to the U.S. and having children living in that country (Gonzalez, Jauregui, Bonilla, Yamanis, and Salgado de Snyder, in press). Findings revealed that the money saved and remittances sent by the offspring gave the elderly men a sense of security. Many older people, however, go back to Mexico and wait for their death in their own land (Espinosa, 1998). Among the many issues ignored regarding Mexican immigrants in the U.S. is that which deals with those who return with health problems associated with psychological stress, physical efforts, and accidents during the time these migratory workers remained in the U.S. The return of the immigrants modifies the socio-cultural and demographic characteristics of the town of origin and determines the distribution of goods and services offered to the inhabitants (CONAPO, 2001).

The purpose of this study was to analyze how a group of elderly men with migratory experience perceive aging, poverty, migration, and social support. This research project discloses new knowledge about a topic seldom addressed in the research literature: the well being of elders in the context of poverty and migration from Mexico to the United States.

9.2. Materials and Methods

This chapter draws from a larger descriptive, cross-sectional study that used qualitative and quantitative approaches for data collection carried out from February to October of 2002. The quantitative part was carried out with a purposive sample of 604 elders residing in rural localities (defined in this study as having less than 15,000 inhabitants) of the Mexican states of Guerrero, Morelos and Jalisco. The findings of the quantitative analyses have been reported in previous publications (Salgado de Snyder, 2003; Salgado de Snyder et al., 2005; González et al, in press). The qualitative data presented in this chapter were generated through semi-structured interviews with a subsample of 18 elderly male migrants.

9.2.1. Participants

The elderly men who participated in this study were all return migrants who had worked for at least one year in the U.S. and that at the time of the interview lived in their communities of origin in the Mexican states of Guerrero, Morelos, and Jalisco. Prior to the interview they responded to a mini-mental test to assure their understanding of the questions; none of them had hearing or speech problems.

This publication was based on findings obtained from the quantitative analyses of data from the same study as this chapter.

9.2.2. Procedure

The participants were selected using a snowball technique. Interviewers explained the purpose of the study as well as the confidential treatment of the data collected and asked participants to sign an informed consent form. Interviewers, who were the researchers involved in the study, had masters and doctorate degrees; additionally, graduate students who received training for this particular project also participated as interviewers.

9.2.3. Instrument

The interview guide consisted of a series of semi-structured questions that asked about the participant's perceptions of the process of getting old in the context of poverty. For the purpose of the present chapter we analyzed the following sections of the interviews: perception of poverty, aging, migration, the U.S. and Mexico, return, health, sickness, and social support. The semi-structured interviews were transcribed and later analyzed using the program ATLAS.Ti 4.2._

9.3. Results

9.3.1. Socio-Demographics

A total of 18 men who had a mean age of 74.3 years participated in the interviews. All of them had lived most of their lives in small rural villages of the states mentioned above and they all knew how to read and write. Twelve men were married; the rest were widowed or separated. All reported having children (average 8 children). Eleven men continued working independently, four men did not work because of illness and three did not work because they received retirement money or a pension.

9.3.2. Poverty

All men interviewed for this project lived in poor rural communities. Some of these communities were very small, geographically isolated, and lacked basic services. In these communities all members shared the same environment of scarcity, and very often they also shared a background of chronic poverty in their families. Many made reference to having been born and raised in extreme poverty, and not wanting such difficult lives for their offspring. They also said that they aged faster than the non-poor because they had to work harder even in their old age. The following are some of the statements by the interviewed men in this study:

- "I have been very poor since I was a child. I don't think this is a defect, but when you are poor you suffer every day. All I can do is ask God to give me license to stay alive some days with nothing, and others with very little. (. . .) When you are poor and have many children to support, you work harder, you get tired faster,

and this makes you older (...) as time goes by and things do not change much, poverty becomes natural. I did not want my children to grow up being poor, like myself, this is why I worked so hard" (80 years old).
- "Uh, poverty is one of man's worst enemies, we could be rich in faith; but that is something else. Economic poverty is the worst enemy of the family" (64 years old).

In spite of living in evident poverty, men who had worked in the U.S. for a longer time did not perceive themselves "very" poor. They talked about different types and levels of poverty. Those who were receiving a pension or owned at least one piece of land did not feel poor. In fact, this group acknowledged that they were better off than other men who did not work in the U.S. and who had not food on the table, as well as others who were living in extreme poverty.

•"Well, there are many types of poor people, some are poor because they are lazy, others because they have vices, or did not look for work in other places, or they did not work hard enough, or for many other reasons. I thank God because at least I have enough money to go on with my life. I receive a pension and I have [agricultural] lands that I no longer work, but they are mine" (83 years old).
•"It is very sad to get old and to be poor. I'm not, because I worked hard in the U.S. (...), many others in this town are poor, but as long as they have at least something to eat they are fine. You see, here in this town, there are old men who are very poor and even crazy, they are good for nothing (...). I feel sorry for them" (65 years old).

9.3.3. Aging

Another section of the interview was about experiencing being old. Three men said that they were happy in their old age and felt satisfied because they had a long life and they had accumulated much experience. Some mentioned that the only way to face aging was to become aware of the experience they had gained in their long life. Such experiences made them feel useful to others; they enjoyed counseling the younger and felt proud of not having any vices.

- "I feel happy because I got to be this age, not everyone gets here (...) Happiness for me is a feeling of satisfaction and feeling proud of being alive" (78 years old).
- "With age I have understood life itself. I have gained knowledge and experiences, I understand what is going on with my old age and I have accepted it (...). It would be sad if at my 80 years I wished I was 20" (80 years old).
- "Aging means experience, but it also means greater needs. Depending on your financial situation and your health it could also mean only worries and concerns" (78 years old).

Other factors were mentioned as important in having aged successfully (understood as having a long life) such as religion, education, satisfaction throughout their life, good health, hard work, and staying active. Many men mentioned that aging had been an easy process because they were at peace with themselves.

- "For me getting old has been easy, I only went to school until 6th grade and the math that I learned there helped me a lot to solve many problems" (83 years old).
- "I have lived my life peacefully, I feel happy because I have no problems with anyone. At my age I can say that I am a happy man" (80 years old).
- "You have to fight daily to stay alive, you have to go ahead. If I just stay lying in bed, nothing happens. To live a long life it is necessary to stay active, to keep moving" (78 years old).

Some elderly men reported that they had aged with many problems. For instance, some had to continue working, others felt useless and tired. Despite the physical wear and tear that came naturally with age some men still had to keep struggling. Only three men associated aging with death; they said they were just waiting for the end and knew it was close.

9.3.4. Migration and Perception of the U.S.

Most of the elderly men who went to the U.S. to work left their family behind and had to travel back and forth. In the interviews, the concept "migration" was associated with a higher income and a better quality of life for the family. They said that going north was not an easy decision to make, and that they went out of need, not out of pleasure. They mentioned that the main reason for migrating was because they were unable to find work or did not earn sufficient money to support the family. Some suggested that it was better to risk their lives than to live in poverty.

- "Many people are confronted with the need to leave their country because there are no jobs, they pay very little here and it is impossible to support a family. People do not care even if they die trying to cross the border as long as they are given an opportunity to work and earn money. They suffer, they sacrifice for the good of the family. It is very sad not to be able to give your children what they need, if you cannot find it here, you go look for it in the U.S." (65 years old).

In some cases the interviewed men had children that followed their footsteps and went to the United States. For instance, one man mentioned that when he went to work he took his eleven children and his wife. Most elderly men indicated that working in the U.S. was a positive experience because they were able to improve the quality of life of the family they left behind. Some of the improvements were financial, such as the purchase of goods and land for the family.

A positive aspect about migration was receiving social security retirement money from the U.S. government. Having these benefits made them feel at ease and tranquil, they reported that they could enjoy their old age and continue contributing for the support of the family. Two men who worked for the Bracero Program were

---

The Bracero Program, carried out from 1942 to 1964, was a guest worker program in which Mexican laborers went to work in the U.S.

receiving such benefits. The men considered that having this money helped them to feel secure and have better quality medical care.

- "I lived and worked there [U.S.] and I still receive [here in Mexico] my social security check; this money is of great help to me" (80 years old).

9.3.5. Returning to Mexico

In spite of the fact that most of the men interviewed indicated that living in the U.S. meant having a better income that allowed them to purchase many goods, they mentioned that they preferred living in their own country, Mexico. One man said he preferred living in poverty in his own country to having more money and staying in the U.S. His reasons were related to having to leave the family behind, feeling lonely and sad, and having no family members or close friends in the U.S. One mentioned that the U.S. was only good to work, but not to be happy; he emphasized that in order to reach happiness one had to live in his place of birth.

- "There is nothing like living in your own country, even if you are very poor" (83 years old).
- "I have only good things to say about the U.S. That country gave me work and put food on my table until I got old. I worked there in two jobs for 37 years and with God blessings I returned to Mexico ten years after my father passed away" (65 years old).

About one half of the men interviewed perceived migrating to the U.S. as a negative experience, because they "lost their freedom," particularly regarding their social life. For instance, one mentioned that in the U.S. people worked very hard and many hours, and there was no time to meet new people or to make friends. Some mentioned they lived with more freedom in Mexico because in their hometowns they could walk the streets freely and they knew people and socialize with them.

- "I was there, of course many years ago. I know how life is like in the north, and I know I don't want that kind of life for me, never. No, I will never live there again. For me there is only one place to live and that is right here, in my Mexico" (78 years old).
- "The U.S. is the richest country in the world but I would not like to live there. I have many friends there and they tell me: 'Why don't you come to live in the U.S.?' I tell them, no, even if they give me retirement money I do not like that country to live. In the U.S. people do not have the freedom we have here. For instance, in Mexico, we can sit in the park and talk for a long time, or have the freedom to go anywhere you want, you could not do these things in the U.S. because everybody is working. People are never home, the houses are empty, it's like being in prison. I do not see life in the U.S." (65 years old).

9.3.6. Health and Illness

Most of the elderly men indicated that they suffered from illnesses, although most health-related problems were associated to old age (chronic conditions) and to the type of work they performed when they were younger (physical efforts and work-related accidents). They associated illness with dysfunction and sadness, although they also mentioned that the process of aging had to be faced with dignity.

- "When you get to be this old, there is always some type of problem. For instance right now I have a pain right here in my arm; also I have problems with my knees and my left foot. I feel crucified!" (88 years old).

As for medical attention for their health problems, most of them mentioned they were affiliated to Mexican Social Security through their offspring and had access to medical care in Mexico and/or in the U.S. They considered the medical attention received in the U.S. of better quality than in Mexico. They also acknowledged that heath care was related to income.

- "Those who do not have money cannot afford to go to the doctor, unlike me. I have been with many doctors. In fact, people can even die because of lack of money. I have a problem with my leg and I have seen many doctors. Nobody seems to know what my problem is. It hurts and all they do is give me painkillers" (83 years old).
- "I have blood circulation problems (. . .). Here in Mexico the doctors cut the toes of my left foot (. . .) I went to the U.S. because I have health insurance and it covered everything. The doctor there told me he was going to cut only half of my foot, and not the complete leg, like the doctor in Mexico said. Doctors there seem to know better what they are doing" (83 years old).

It is clear that suffering in old age is a fact in both countries, but this suffering could be faced differently with sufficient financial resources. Aging in the U.S. was perceived as positive because of the governmental support (economic and medical care) that elderly people received. However, they also made reference to negative perceptions of aging in the United States.

It is interesting that two elderly men made reference to misconceptions about the treatment of the elderly in the U.S. A 65-year old man said that old people were not allowed to be left alone, that government agents followed them to make sure they do not fall or have an accident because this would be very costly for the government. He also mentioned that if older people were found working the U.S. government would issue them a fine they had to pay. Another 80-year old man indicated that the U.S. Social Security Administration had personnel working in health clinics and hospitals that gave elderly people lethal injections without their consent because "older people are no longer productive an their health care is expensive" (sic).

As for their perception of aging in Mexico, it was clear from the interviews that there was no government support for the elderly, but it appeared that those interviewed felt the compensation for this was the greater liberty and happiness that resulted from being in their own country. Respondents also mentioned that people in Mexico are more "humanitarian." Taking into account these differences between aging in Mexico and in the U.S., for the majority of the elderly interviewed it is better to get old in Mexico.

- "I think it is the same here and there, but I believe that when we come to Mexico we become weakened. I cannot work. I just spend my days sitting down" (76 years old).
- "In the U.S. elderly people receive more help from the government than (. . .) here in Mexico. They get everything they need, like doctors, medicine, a check, everything. They even pay for a nursing home if you need it. In Mexico, people turn poor as they become old. But people in Mexico are very humanitarian; they always bring food and help others to get by in tough times" (65 years old).

9.3.7. Social Support

The majority of the elderly men interviewed indicated that they received support from their family, particularly members of the nuclear family, like wife and children. The help provided by the nuclear family, mostly from offspring, was instrumental and financial. However the economic help provided by some of their children was erratic in terms of periodicity and amount of money. It is interesting to note that most of the men mentioned that the most efficient support was received from their children living in the U.S.

- "The money my children send me from the U.S. helps me to meet my expenses. When I ask for it they help, they respond, sometimes one, sometime another. Because they are 11 children and they don't mind helping a little. One day one, other days the other children, this is why I can live free of problems. I do not have any extra money, but I don't lack anything" (80 years old).
- "It is difficult to go here and there, and not having a place of your own. However, I do not need anything because my children pay all my expenses, so I really do not have any problems, I do not suffer like others, I am not complaining" (76 years old).
- "My son is there [in the U.S.] and he sends us money every year. Sometimes we go to visit him, we go there because we like it, everything is very nice in the U.S." (88 years old).

A small proportion of the men mentioned problems with nuclear family members or extended family, who, in some cases, neglected and even refused to help their elderly relative. Some older men reported having had lost contact completely with their children. In one case the elderly man lived alone because his wife and children were living in the U.S. and refused to return to Mexico.

- "My son (by a previous marriage) who has money, kicked us (father and wife) out of his house and told us to get out of his way. I don't think this right" (80 years old).

- "To be happy, I don't need anything. I only need my family, but I lost them because they don't want to come back, my wife and children they are all very happy in the U.S." (70 years old).

It should be noted that the elderly person oftentimes was not only a recipient of support, but also a provider. They reported helping around the house with domestic chores, contributing money if they had it, and caring for children, older people, and sick members of the family.

- "The very first thing I do every day is to go and take care of my brother who is older than me and is mentally ill (...) I have to take care of him because he cannot be alone. When I go I take him food and feed him. I leave more food for the evening. I go back the following day" (88 years old).

The help received by family members was supplemented by other types of help such as retirement and pension money (some of this income came from the U.S.), as well as additional income generated through rental of property they had purchased years ago.

- "God has been good to me. I get along with what He lets me have and then I have some extra because I receive the pension from the Seguro. I have other income from business and rents. Thanks to God I don't need any more. If I had more [money] I would give it away to needy people." (65 years old).
- "I don't need any more, with God blessings I receive my pension's money from the U.S. and my children also send me money. I also have my discount card. When I go to visit my children I take the bus and I only pay one- half, if I go by plane they give me a discount. Thank God I have not suffered for lack of money." (70 years old).

9.4. Conclusions

The results presented in this contribution throw new light on the relationship between migration and well being in old age. Results suggest that elderly men who had worked in the U.S. felt satisfied with their life accomplishments. In spite of living in poverty, and some in extreme poverty, these men reported that they worked hard their entire life because they did not want their children to suffer the precarious life they themselves had suffered when they were young. For many, this was the principal motive for going north and working with or without legal papers. Many men viewed migration as the providing the only possibility for survival.

As for the process of aging, most of the men interviewed felt good about becoming old. They referred to the many experiences that they had accumulated throughout the years, which gave them satisfaction, particularly when they were given the opportunity to share these experiences with other people or to provide advice to younger men. Their satisfaction with old age was found to be associated with a feeling of inner peace and contentment derived from staying active, working hard, and enjoying overall good health. Others reported feeling tired, but only a few men associated getting old with death.

Most men perceived migration to the U.S. as an opportunity through which they were able to improve the living conditions of their families in Mexico. In fact, many elderly men had children living in the U.S. and were proud of this because they considered migration a positive experience, one that had many benefits for their families and for themselves in old age, such as being of the age to receive retirement money. However, returning to Mexico was a cherished goal for all of them. Some men manifested how they felt like they had been in prison while living in the U.S. They attributed this to its "lack of freedom," absence of family members and friends, and the feeling it created in them of being lonely and sad. It is interesting to note the men's conflicting perceptions about migration to the U.S. On the one hand, migration had provided a better quality of life, a means for getting ahead and a way to cover basic necessities; on the other hand, being in the U.S. was described as being in "a prison" because they only worked, which provoked loneliness and a lack of social and emotional freedom.

Most men reported suffering from diverse health problems, which made them feel dysfunctional and sad. A few of them had access to health services in the Mexican Social Security system because their children had enrolled them. They considered the medical attention they received in the U.S. to be much better than the attention they receive in Mexico.

Some of the men who participated in the study received benefits from the U.S. government (Social Security retirement) and other received income from remittances sent by their offspring. The men who had children in the U.S. felt they had income security in case of an emergency. Unfortunately, this was only the case of those elderly men with migratory experience to the U.S. and did not reflect the thoughts of men who did not go north and who did not have children in the U.S. who could respond in case of an emergency. It is evident that elderly men, particularly those living in poverty, need adequate social protection and at least a minimum income to help them meet their basic needs.

The act of migrating to the U.S. has put these men in a better economic position due to the fact that they possess money, land, housing, and food. They no longer consider themselves poor, despite having been very poor earlier in their lives. Finally, our results suggest that migratory experience, assets, and family support in old age had a positive impact on the elderly men's perception of their own quality of life.

References

Acevedo, L., Leite, P., and Felipe Ramos, L. (2004). Los adultos mayores en la migración internacional [The elderly in international migration]. In CONAPO *La situación demográfica de México, 2004* [*The demographic situation of Mexico, 2004*]. Mexico City, Mexico: CONAPO.

Consejo Nacional de Población. (2003). *Migración a Estados Unidos por entidad federativa, según migrantes de retorno y migrantes a Estados Unidos durante el quinquenio,*

1997–2002 [*Migration to the United States by federal entity, according to return migrants and migrants to the United States between the years 1997 and 2002*]. Retrieved February 15, 2006, from: http://www.conapo.gob.mx/mig_int/series/ 070402.xls

Consejo Nacional de Población. (2001). Migrantes mexicanos en Estados Unidos [Mexican migrants in the U.S.]. *Boletín de Migración Internacional,* 5(15), 1–12.

Consejo Nacional de Población. (2002). *Índices de intensidad migratoria Mexico–Estados Unidos, 2000* [Index of Mexico-United Status migratory intensity].Mexico City, Mexico: CONAPO.

Consejo Nacional de Población. (2005). *Migración México–Estados Unidos: Temas de Salud* [Mexico–U.S. Migration: Health Issues] (Report). Mexico City, Mexico: CONAPO.

Desjarlais, R., Einsberg, L., Good, B., and Kleinman, A. (1995). *World Mental Health. Problems and priorities in low income countries.* New York: Oxford University Press.

Espinosa, V.M. (1998). *El dilema del retorno: Migración, genero y pertenencia en un contexto transnacional* [The dilemma of return: migration, gender and belonging in a transnational context]. Zamora, Mexico: El Colegio de Michoacán.

Gonzalez, T., Jáuregui, B., Bonilla, Yamanis, T., Salgado de Snyder, V.N. (in press). Well-being and family support among elder rural Mexicans in the context of migration to the United States. *Journal of Aging and Health.*

Hernandez Laos, E., and Velazquez Roa, J. (2002). *Globalización, desigualdad y pobreza: Lecciones de la experiencia mexicana* [Globalization, inequality and poverty: Lessons from the Mexican experience]. México: UAM-Plaza y Valdez.

Instituto Nacional de Estadística Geográfica e Informática. (2000). *Tabulados básicos nacionales y por entidad federativa, bases de datos y tabulados de la Muestra Censal* [Basic national and federal entity tabulations, databases and tabulations from the Census Survey]. From the XII General Census of Population and Housing, 2000. Mexico City, Mexico: INEGI.

Kalache, A. and Sen, K. (1998) Aging in developing countries. In M.S.J. Pathy (Ed.), *Principles and practice of geriatric medicine.* Hoboken (NY): John Wiley and Sons.

Kochhar, R. (2005). *Survey of Mexican migrants, part three: The economic transition to America.* Retrieved on March 3, 2006, from the Pew Hispanic Center. Website: http://pewhispanic.org/files/reports/58.pdf

Leite P., Ramos F., and Gaspar S., (2003). Tendencias recientes de la migración Mexico–Estados Unidos [Recent trends in United States-Mexico migration]. In *La situación demográfica de México 2003* [*The demographic situation of Mexico, 2003*] (pp. 97–115). Consejo Nacional de Población. Mexico City, Mexico: CONAPO.

Pan American Health Organization. (2002a). *La salud como característica estratégica del proceso de envejecimiento. Políticas para el envejecimiento saludable en America Latina.* [*Health as a characteristic of aging. Policies for healthy aging in Latin America*]. Washington, DC: PAHO.

Pan American Health Organization. (2002b) Envejecimiento, ¿qué nos depara el futuro? [Aging, what does the future hold?]. *Noticias e Información para el Centenario.* Retrieved March 4, 2006 from: www.paho.org/Spanish/DPI/100/100feature02.htm

Salgado de Snyder, V.N. (2001). *The psychosocial functioning of return migrants from rural origin—executive summary.* Mexico City, Mexico: UCMEXUS-CONACYT.

Salgado de Snyder, V.N. (2003). Envejecimiento, genero y pobreza en México rural [Aging, gender and poverty in rural Mexico]. In V.N. Salgado de Snyder and R. Wong (Eds.), *Envejeciendo en la pobreza: genero, salud y calidad de vida* [*Aging in poverty: gender, health, and quality of life*] (pp. 37–56). Cuernavaca, México: Instituto Nacional de Salud Pública.

Salgado de Snyder V. N., Gonzalez, T., Jáuregui, B. and Bonilla, P. (2005). No hacen viejos los años sino los daños: Envejecimiento y salud en varones rurales [Injuries, not years, make people old: Aging and health in rural men]. *Salud Pública de México, 47,* 294–302.

Solis, P. (1999). La población en edades avanzadas [The population in advanced age]. In Gomez de Leon

Cruces, J. and Rabell Romero, C. (Coords.), *La Población de México: Tendencias y Perspectivas Sociodemograficas hacia el siglo XXI* (pp. 835–869). Mexico City, Mexico: CONAPO: FCE.

Tuiran, R. (1999). Desafíos del envejecimiento demográfico en México [Challenges of aging in Mexico]. In *Consejo nacional de población envejecimiento demográfico de México: Retos y perspectivas [Demographic aging of Mexico: Challenges and perspectives].* Mexico: CONAPO, Cámara de Senadores y Cámara de Diputados.

World Health Organization. (1999, December). *Report by the Director General on Poverty and Health* (Executive Board EB105/5 meeting). Geneva: WHO.

10
Aging and Health Interrelations at the United States-Mexico Border

Roberto Ham-Chande

10.1. A Demographic Definition of the Border Region

Between Mexico and the United States there is a border more than 3,200 kilometers long. It is not only a political division between two countries, it also marks the encounter between development and underdevelopment, where two different cultures and lifestyles meet. In an apparent paradox, strong interrelations across the border are generated not in spite of differences but because of such disparities. Relationships go from north to south and from south to north. The intensity and regularity of these short-distance international relations characterize the life of the region in all aspects, from big business to family economy, permeating social perceptions and creating a new culture, driving demographics, and affecting health outcomes. In fact, the border as a division is at the same time a shared space that imposes special conditions on very concrete situations and problems that deserve bi-national attention and policies.

In the study of the border there are always questions like, "What is the border region and how it is physically delimited?" In fact, these questions arise every time a border issue is undertaken because one must consider which key elements are to be addressed and what are the tools available to address them. For the *border population* the United States–Mexico border region is defined by the set of counties in the U.S. that border Mexico and the set of *municipios* in Mexico that are adjacent to the United States. There is a practical reason for this definition that is due to the availability of statistics, in which the smaller administration units are counties and municipios.

10.2. Border Population Growth

Table 10.1 compares the development of population sizes every ten years between 1930 and 2000 in the U.S., its border counties, Mexico, and its border municipios. From 1930 to 2000 the population of the U.S. grew 2.3 times, going from 122.8 million to 281.4 million, whereas the border population went from 836,000 to 6.3 million, thus it was multiplied by 7.5. In that same time period the population of

TABLE 10.1. Population growth in the U.S., border counties, Mexico and border municipios. 1930–2000

		U.S.	Border counties	Mexico	Municipios fronterizos
1930	Population	122,775	836	16,553	276
	% growth	0.68	1.30	1.67	3.58
1940	Population	131,410	952	19,564	395
	% growth	1.50	4.59	2.76	7.50
1950	Population	152,671	1,507	25,791	836
	% growth	1.68	4.50	3.03	5.58
1960	Population	180,671	2,364	34,923	1,461
	% growth	1.26	1.91	3.73	4.05
1970	Population	204,878	2,862	50,695	2,190
	% growth	1.01	3.37	3.13	2.83
1980	Population	226,546	4,009	69,347	2,906
	% growth	0.93	2.60	1.57	2.71
1990	Population	248,710	5,197	81,141	3,810
	% growth	1.24	1.92	1.83	3.79
2000		281,422	6,296	97,483	5,564

Mexico was multiplied by 5.9, from 16.6 million to 97.5 million. But the greatest growth occurred in the Mexican border population where the inhabitants went from a small 276,000 to 5.6 million, a growth of as much as 20.2 times.

Such numbers indicate clearly the remarkable increase of the border population. With regard to the natural increase due to the difference between births and deaths, the part mortality plays in population change is not very different from what is observed in other urban areas of each country. But the crude birth rate is higher because people in reproductive age migrate toward the border region. And it is exactly the migratory issue the main determinant high growth of the border region. Migration toward the border and across the border has been quite heterogeneous and like all population movements in times of peace, the main driving forces have to do with economic opportunities and family reunion. The Mexican border receives migrants from all over Mexico; the U.S. border has a large input from Mexico, mainly from the Mexican border.

10.3. Border Urban Pairs

The U.S.–Mexico border spans geographic zones that to a large extent are desert or semi desert, with little annual precipitation. Such a dry environment requires human settlements to concentrate in a few urban zones. With regard to transborder relations these areas always contain twin-cities or even urban systems with an international character. Table 10.2 lists the most populated eight counties and municipios and their population sizes. On the U.S. side those eight counties have 91.1% of the total border population and the eight municipios make up 85.5% of the entire Mexican part. In the U.S. most of that population is settled in the main cities, as indicated in parentheses. In the Mexican part the name of the main city is always the same as the municipio.

TABLE 10.2. Population in major U.S. border countries and Mexico border municipios: 2000

	U.S. population	Mexican population		Both populations
San Diego (San Diego)	2,813,833	1,210,820	Tijuana	4,024,653
Imperial (El Centro)	142,361	764,602	Mexicali	906,963
Yuma (Yuma)	160,026	145,006	SL Río Colorado	305,032
Pima (Nogales)	843,746	159,787	Nogales	1,003,533
El Paso (El Paso)	679,039	1,218,817	Ciudad Juárez	2,072,538
Webb (Laredo)	193,117	418,141	Nuevo Laredo	611,258
Hidalgo (McAllen)	569,463	420,463	Reynosa	989,926
Cameron (Brownsville)	335,782	418,141	Matamoros	753,923
	5,737,367	4,755,777		
% from total	91.1	85.5		

10.3.1. The Mexicanization of the U.S. Border

One of the demographic phenomena recently publicized in the U.S. is the growth, and presence of the Hispanic population. Hispanics are not only the largest minority but they are also gaining in cultural impact, economic status, and political partic-ipation in a block cemented by linguistic and historical backgrounds that together provides a solid basis for a shared identity. A main feature of the Hispanic presence in the U.S. is that Mexico is the main source not only of people but also of ideas and attitudes. We should consider how the border participates in this acculturation process, not just along the Mexican side of the order, but along both sides.

Table 10.3 shows for the eight larger counties of U.S. border the distribution of populations that are Hispanic, non-Hispanic, and also Mexican in origin. Figures are from the last three population Censuses, 1980, 1990, and 2000. These statistics show that present and impacting 14.0% Hispanic in the U.S. was exceeded along the southern border of the country several decades ago it. In fact, it has always been like that. In any case, the remarkable circumstances border the presence of the Hispanic and Mexican-origin population keeps growing. In San Diego the least

TABLE 10.3. Distribution of Non-Hispanic, Hispanic, and Mexican origin population in U.S. border countries: 2000

	% Non-Hispanics			% Hispanics			% Mexican-origin		
	1980	1990	2000	1980	1990	2000	1980	1990	2000
San Diego	85.2	79.6	73.3	14.8	20.4	26.7	12.2	17.6	22.6
Imperial	43.9	34.2	27.6	56.1	65.8	72.4	53.8	63.1	66.5
Yuma	70.6	59.4	49.5	29.4	40.6	50.5	27.4	38.7	44.8
Pima	79.0	75.3	70.6	21.0	24.7	29.4	18.8	22.1	24.7
El Paso	38.1	30.4	21.7	61.9	69.6	78.3	58.8	66.2	67.3
Webb	8.5	6.1	5.6	91.5	93.9	94.4	87.2	89.3	78.0
Hidalgo	18.7	14.8	11.6	81.3	85.2	88.4	78.4	81.2	77.3
Cameron	22.9	18.1	15.5	77.1	81.9	84.5	66.0	77.2	69.6

Hispanic of the counties, the 14.8% in 1980 almost doubled to 26.7% in 2000, with the corresponding figures of 12.2% and 22.6% of Mexican-origin population. In Webb County, site of the city of Laredo, all past and present numbers of the Hispanic population exceed 90.0% and those of Mexican origin are seen to be always close to that number. Trends for future scenarios are generally considered to be leading upward.

10.4. Differences in Demographic Structures

Although the border of the U.S. is "Mexicanizing," the region will never be exactly like Mexico nor will things be Mexican in all respects. From one side of the border to the other there are either little or great social and economic variations, as seen in the demographic characteristics. Table 10.4 depicts some of the age structure differences in three border pairs: San Diego and Tijuana; El Paso and Ciudad Juárez; Cameron and Matamoros. The percent distribution of population in three broad age groups is shown, including median age. For statistical discussions, the ethnicity categories at the U.S. border include "non-Hispanic," "Hispanic," and "Mexican-origin."

Population structures are younger along the Mexican part of the border than along the U.S. part of the border; the Mexican-origin population along the U.S.-side border is less young. From young to aging Hispanics on the U.S. side, and finally the most aged are the non-Hispanics. In large part these differences are due to migration toward the borders of Mexico and the U.S. of people of working age. In the three sites of the Mexico –border, populations structures are similar. Similarities are also present in San Diego and El Paso, but there is a significant

TABLE 10.4. Distribution of population by ethnicity and groups of age in selected U.S. border counties and in selected border municipios in Mexico. 2000

	Total	Non-Hispanic	Hispanic	Mexican-origin	Total
		San Diego			Tijuana
0–14	21.7%	18.2%	31.5%	32.2%	33.4%
15–64	67.1%	68.2%	64.1%	63.6%	63.4%
65+	11.2%	13.6%	4.4%	4.2%	3.2%
Median age	32.4	36.1	23.8	23.4	22.8
		El Paso			Juárez
0–14	26.7%	18.9%	31.2%	28.8%	32.9%
15–64	63.6%	67.7%	62.0%	62.9%	63.6%
65+	9.7%	13.4%	6.8%	8.3%	3.5%
Median age	29.2	37.9	26.9	27.2	22.5
		Cameron			Matamoros
0–14	28.3%	14.2%	30.8%	30.6%	33.3%
15–64	60.5%	55.0%	61.6%	61.6%	62.7%
65+	11.2%	30.8%	7.6%	7.8%	4.0%
Median age	28.2	44.1	25.2	25.5	22.4

difference in Cameron. In this last county there is a high degree of aging in the non-Hispanic population. Apart from a lower fertility the reason for this is most probably the white non-Hispanic young population in places like Cameron is migrating away from the border.

10.5. Defining Health-Related Interrelations

Household structures and kinship networks have a main roll in lifestyles, health outcomes, and how one manages and cares for ill family members. Possibilities, behaviors, and consequences are shaped by economic, social, and cultural factors that determine family and individual practices. In Mexico there is also a critical socioeconomic heterogeneity that is seeing significant dynamics. To begin, there are not adequate health conditions for a large part of the population, and public institutions are inadequate. This insufficiency assigns to the family vital responsibilities, resulting in great stresses due to poverty, demographic change, and the narrowing of public programs and institutions resulting from trends towards a market economy. An important growing factor is seen in the remittances and help received in Mexico from the Mexican-origin population in the U.S. A question arises as to how much of these resources are used on health care and what are their effects. With all these things considered, the U.S.-Mexico border is worthy of study. Interrelatedness of Mexico with the U.S. on health outcomes must be analyzed, including how good outcomes are prevented and how a family copes with health issues according to transborder possibilities from family and other ties. The approach of looking at this interrelatedness provides grounds for identifying the participation and help that kinship in Mexico is supplying to the Mexican-origin population on the U.S. side of the border.

From other studies about the border and its characteritics, transborder relations that have an effect on health issues and support can be studied through theoretical frameworks that address the following issues:

- A first main frame of reference is surely the demographic dynamics at the border region. The general characteristics of border populations have already been described in the previous sessions.
- Another category includes social and economic factors that may act on one or the other side of the border or on both sides, but that always show bi-national elements.
- An important category are family relationships and social networks that extent across the border.

Interrelations across the border that have a direct effect on health and well-being can be classified in three main categories.

- Economic transferences
- Use of services of health
- Family care provided in the border side that is not the habitual residence

10.6. A Pilot Study in Tijuana

A methodologic aspect of importance is the recognition that the kind of relations required, offered, or received are regarded differently depending upon the side of the border on which one observes them. In this article a reference is made to a pilot study about identification and weighting of border interrelations related to health. It is the perspective from the population in Tijuana. The objective is to extend the study to a representative sample of Tijuana, to the whole Mexican border, and to propose a similar research program on the U.S. border.

The transborder relationship indicators touch the following concepts.

- Place of birth, nationality
- Border crossing: frequency, purpose, crossing documents (or lacking)
- Schooling in the U.S.
- Knowledge and use of English
- Job in the U.S./income in the U.S.
- Family and social networks in the U.S.
- Use of health care services in the U.S.

The pilot survey included 168 households with 653 members, including 38 households with at least one person 60 or over, for a total of 62 persons in that age break. The sections of the survey were:

a. Demographics of household members
b. Economic characteristics of household and family
c. Household support
d. Household and family strategies for health care
e. Access and use of health services
f. Health status
g. Dwelling conditions

10.6.1. Preliminary Findings: All Participants in the Pilot Survey

The preliminary findings included the following data:

- 2% born in the U.S.
- 41% born out of Baja, California.
- 2% U.S. citizens; 1% held both citizenships.
- 65% had crossed the border; from these 83% have crossing documents
- 3% had studied in the U.S.
- 33% spoke English

10.6.2. Preliminary Findings: Persons 60+

The preliminary findings included the following data:

- Everybody was a Mexican citizen
- 93% of men headed their household; 35% of women were heads of household and 50% were spouses
- 84% have crossed the border; 75% have visa; 18% are green card holders
- Crossing purposes were mainly shopping and <u>family visiting</u>
- 35% crossed at leas once a week, 46% about once a month. Most of the stays lasted a few hours.
- 17% did not have medical insurance (in any place)
- 82% went to primary school but less than half completed studies
- 16% claimed to speak English
- 67% of men had worked only in Mexico; 20% only in the US; 13% in both
- 41% of men had a pension, 12% of women did, mainly from IMSS

References

Alegría Tito (1989). "La ciudad y los procesos transfronterizas en México y Extados Unidos". *Frontera Norte, Vol.1, N°2*. El Colegio de la Frontera Norte. Tijuana.

CPS (2006). Current Population Survey. www.bls.census.gov/cps/

Durand, Jorge, Edward Tellez, Jennifer Flashman (2006). "The Demographic Foundatoins of the Latino Population". In Tienda, Martha abd Faith Mitchell (editors) *Hispanics and the future of American*. National Research Council. The national Academies Press. Washington DC.

Inegi (2006). Censos Generales de Población y Vivienda. www. inegi.gob.mx

Ham-Chande, Roberto & John R. Weeks (1992). "A demographic perspective of the US-Mexico border". In Weeks & ham-Chande *Demogrpahic Dynamics of the US-Mexico Border*. Texas University Press.

Partida, Virgilio (2006). "Impacto demográfico de la migración de México a Estados Unidos". En Zúñiga et. al. (coordinadores), *Migración México-Estados Unidos*. Consejo Nacional de Población. México, DF.

US Bureau of the Census. Population censuses. www.census.gov

Wallace, Steve Elmyra Ibáñwz, Verónica Gutiérrez (2005). *Relaciones transfronterizas y etención a la salud en Tijuana*. Paper presented at the Bi-national Health Week. UC-Mexus and the University of California Los Angeles. Chicago, October, 2005.

Zenteno, René & Rodolfo Cruz, (1989). "Un contexto geográfico para la investigación demográfica en la frontera norte". *Estudios demográficos y urbanos*. Vol. 3, N° 3. El Colegio de México. México, DF.

11
Colonias, Informal Homestead Subdivisions, and Self-Help Care for the Elderly Among Mexican Populations in the United States

Peter M. Ward

11.1. Introduction: Intersections Between Households, Housing, and Care for Elderly Parents Among Mexicans

In this chapter the aim is to explore briefly how Mexican low-income households traditionally care for their elderly parent populations and to examine the intersection between changing demography, household structure, tenure, and dwelling arrangements. This chapter will inform how we start thinking about housing and parental care dynamics and the way they may be changing with the demographic changes of the last few decades. The demographic transition that began in the 1970s shifted the population age structure from one of heavy dependency of the young from the 1960s through 1980s to one in which dependency ratios are shared more equally between young and the older cohorts. Who will care for the elderly given the low level of formal care provision offered by the formal sector for the elderly in Mexico? In particular, how will low-income households earning less than the equivalent of $10 a day support and provide for their parents in old age?

In order to begin to think about these questions we need an understanding of self-help issues and of the informal housing production that provides a residential context—housing in which aging parents may come to spend their latter years, which would also be real property that their children aspire to inherit. For many households in Mexican cities, informal housing arrangements and multiple lot sharing allow extended and kin sibling households to live cheek by jowl with an elderly parent or parents.

After reviewing housing and household arrangements in Mexico, I will look to the United States. In the context of the rising Hispanization of many regions of the country, outlined by the editors in their introduction, this chapter aims to explore how poor Mexican-origin families in the U.S. today will cope with looking after their aging parents (whether here or in Mexico), and, concomitantly, to ask how that same generation will be cared for over the next 20–30 years as they, themselves, enter old age. Here, I will draw upon my research, which begins to point to the intersection of household arrangements and care for the elderly, and informally produced homesteading opportunities afforded to some low-waged Hispanic families living in low land-cost areas in the United States. I outline an argument that such areas are environments suitable for non-institutionalized and culturally congruent low cost care for elderly parents. In the U.S., as in Mexico, joint living arrangements and co-residence among pre-retirement parents and their children often satisfy the need to support the offspring rather than the older generation's needs (Flippen and Tienda, 1998).

That said, relatively little is known about such baseline arrangements and how they work—even in Mexico; but there are a number of dynamics that we can expect to impact upon care for elderly parents both in Mexico and in the U.S. Many Mexican families are now transnational, with household members living in the U.S. either temporarily (albeit often long term) or, for a growing proportion, permanently. Mexicans in the U.S. will increasingly have to provide for their elderly parents, either by remitting resources to them directly (as they often do at present), or through indirect supports to their siblings, either to share the costs or to take on the lion's share of costs of care and medical treatment. Alternatively, as elderly parents have fewer children who are willing or able to care for them at home (i.e., in Mexico), pressure is likely to grow for family unification into (now) citizen or permanent-resident, son-and-daughter households in the United States. This raises the question of the receiving family's capacity to accommodate its parents in the new country and city of residence. Thus there is an urgent need to better understand the formal and informal housing arrangements that exist among cohorts of first generation Mexican migrants, for whom the pace of immigration has quickened and intensified since the 1980. And, by extension—and perhaps even more urgently, second and third generation Mexican American households now have to cope with greater longevity of their parents and grandparents. So, to paraphrase closely the Angels' (1997) book title: Who will care for them?

Subjects of much research and debate, as widely touched upon in other chapters in this volume, are the issue of care for the Hispanic elderly; their particular epidemiological status among the low income population; and the apparent paradoxical effects that the migration-selective process appears to have had (lowering morbidity rates compared to other low income groups). However, although the infrastructure for formal care for the elderly is much better developed in the U.S. than it is in Mexico, it is still far from adequate, and many Mexican-born (especially) as well as Mexican Americans eschew or are ineligible for such care. The answer to "who will care for them?" is largely a matter of informal arrangements and high dependence upon kin networks. Moreover, as well as being poor with limited formal insurance support and Medicare coverage (Angel, Angel, and Lein, 2006), the data suggest that U.S.-born Mexican American households do have particularly high rates of morbidity in certain debilitating and chronic degenerative diseases such as type-2 diabetes, and Hayward et al. (in this volume) show that Mexican Americans are more likely than non-Hispanic Whites to suffer from protracted disability. For those living in the border region, and even some living in the interior of the U.S., so-called "*maquila*" nursing homes and residential care facilities may offer a viable economic alternative. These already constitute a fast growing industry in Mexican border cities, where proprietors take advantage of cheaper nursing and medical services and lower levels of health care regulation.

However, many of these elderly will not have access to formal institutional and even home-based support, whether in Mexico or the U.S., and instead, perforce, will need to be cared for informally *in situ*, in their own or their children's homes. The extent to which this occurs makes salient the need for social scientists to gain a better understanding of the dwelling places of these elderly, including intra-lot residential arrangements in the United States. This brings me to the principal purpose of this chapter. I will compare informal housing and residential dynamics in the two countries, and, in particular, outline how the to-date little known universe of *colonias* and informal homesteading arrangements that are widespread in border states and metropolitan peripheries of the interior are likely to provide home and succor for elderly parents of Mexican and Mexican origin populations.

11.2. Irregular Settlement and Informal Housing in Mexico

Upon marrying, a couple in rural Mexico will, traditionally, move in with the man's parents, making the daughter-in-law subservient to her husband's mother. In such cases, ill treatment of the daughter-in-law may ultimately lead to ill treatment of the elderly mother when in later years the fortunes are reversed and the latter is dependent upon her son's family and his wife's good graces for care and support (Varley and Blasco, 2000; Varley, 2000). Of course, until relatively recently, life expectancy was considerably foreshortened so that even for women (who tended to live longer than men), such a predicament was less likely to be extended over many years. But in 2000 the life expectancy in Mexico for men and women was 69.5 and 75.5 years, respectively—a dramatic increase of the 39.5 and 41.5 years of 1940, and of the 59.5 and 63.6 years even as late as 1970. The annual population growth rate in Mexico has declined from approximately 3% in the 1960s to 1.3% today, with an expectancy that it will be negative by 2050. Fertility rates have declined dramatically from 6.9 children per woman in 1955 to 2.5 today. And as people live longer and expect to live out their days with their sons and daughters, family size has decreased (from an average of 5.1 in 1960, 4.9 in 1990, to 4.0 in 2005), and so, concomitantly, the dependency ratio on a smaller number of adult children will increase in future. As Angel and Whitfield point out in the introduction to this book, in Mexico the absolute number of older Mexicans is expected to increase by 227% over the next 30 years, and the current high ratio of women to men aged 60 or over (1.63 females to every male) is likely to be exacerbated, since women live longer. This is significant since women are more vulnerable economically, and given that labor market participation in formal employment for women is much lower than it is for men, so their access to pension and other health care resources is more marginalized.

Indeed, few of the elderly have a formal pension to help them—period. According to the National Survey on Aging, in 1994 only 26.8% of men and 12.5% of women over 60 years of age received a pension, and no less than two-fifths therefore depended financially upon a co-resident relative and a further 10% on other relatives (Varley and Blasco, 2000; Citeroni 1998). Relatively few aged 60 or older live alone (8.8% of women and 5.8% of men—the lower number of men being explained by the fact that they marry younger women and are more likely to remarry). In Mexico families are expected to care for their elderly and receive scant help from health or welfare agencies. Indeed, such an idealized construction of the caring family and of culturally embedded responsibilities and expectancies may prolong public policy neglect. However, the so-called "new poverty" in Mexico, which makes extended support of one's parents increasingly difficult (Gonzalez et al., 2004), longer life expectancy together with the declining number of siblings to share the responsibility of care, and maybe the windfall expectancy that "rich" brothers and in-laws living in the U.S. will provide directly or indirectly, are facts that may combine to change popular responses and capacities in less than a single generation. In many respects the elderly in Mexico today are more vulnerable than ever before.

But what of the status quo ante? In Mexico today, most low-income elderly formed part of the labor force engaged in the import-substituting industrialization boom period of the 1950s through the 1970s. Rapid urbanization with growth rates of 3 to 5% per annum, as in many developing countries, was a product of this phase of industrialization that drew labor into the cities from rural and provincial areas—labor that would man the factories (Roberts, 1994). Although low paid, these migrants did achieve some socioeconomic mobility that accompanied their physical displacement to cities. Accommodated at first in inner city tenements, these populations later moved out to occupy unserviced lots in the periphery upon which the squatted or that they purchased at relatively low cost lots (Gilbert and Ward, 1985). Thereafter, through mutual aid and self-help, these shantytown settlements were gradually consolidated into working class settlements that belied their humble and illegal origins. At the household level, dwelling structures were consolidated through the process of self-build, with young migrant households adding rooms to accommodate their fast growing families, and sometimes renting out rooms to more recent arrivals, or allowing siblings to share the lot and raise their own families. Thus, dwelling arrangements varied greatly, with some homes being two or even three stories, while others remained provisional single-story shacks. Individual lot arrangements also varied, comprising a mixture of nuclear families, extended households, single female-headed households, and sometimes "compound" patterns where kin-related households shared the lot. Another frequent feature was for now grown-up kids and their own young families to share the lot with their parents—the successful migrants of yesteryear.

This pattern of self-help informal urbanization was the norm until the middle-to-late 1980s, since which time a tightening land market and more concerted and effective public intervention to prevent the unfettered growth of spontaneous settlements has reduced their scope for development. In turn, this has led to densification of these former irregular settlements, within which new rental tenements are also being created, offering some opportunity for second generation households to move out of their parents' home and to set up on their own. But few of these barrio-raised young adults are willing to emulate their parents' endeavors by moving several miles out to the current periphery and engage in the cardships of self-building. Instead, they prefer to live nearby in one or two rented rooms, or to live "close-up" (*arrimados*) with kin—usually the man's parents.

This scenario portrays the contemporary scene for many low-income households today in Mexico's cities: namely, two or three siblings and their nuclear families share the lot with their parents or more usually with their widowed mother. As she (the aging parent) is also the owner of the house and lot, she has some leverage over her sons and daughters, and her status ensures that she will be cared for. However, it may also be "pay back" time if she has ill-treated her daughter-in-law earlier in the son's marriage. But, under Mexican law the lot does not go to the eldest son upon her (intestate) death; instead all will probably continue to share the lot, albeit under some heightened level of intra-familial conflict and insecurity once the matriarch has passed on. The point is that this is a *functional arrangement* that has evolved in Mexico in response to the nature of informal housing production, the demand for housing among young adult families, declining wages and employment opportunities, and a general cultural expectancy that the immediate family "step up to the plate" and care for elderly parents.

Such successful consolidation and integration also slowed dramatically during the "lost decade" of the 1980s and later in the 1990s, when neo-liberal policies and export-oriented industrial growth reduced the opportunities for formal employment and the capacity to invest in home improvements. This outcome was not least because that same orthodoxy also demanded fiscal sustainability, and that services and consumption be paid for. The pace of home improvement declined, therefore, and remittances (rather than salary) became an important determinant in house purchase and self-build extensions and improvements. Moreover, the nature of migration and transnational family structures has changed dramatically in the past two decades. No longer is out-migration to the U.S. largely a preserve of the poorest agricultural regions in which remittances are integral to continued subsistence of the populations at the place of origin, but today migration to the U.S. also increasingly involves migrants from larger urban areas, and from a wider range of states. According to a 2003 survey by the Inter-American Dialogue it is estimated that 41 percent of families today have a close relative living in the U.S. and one in five adults in Mexico receive some level of regular remittances. Moreover, the absolute amount sent back to Mexico in the form of remittances has increased dramatically from a range of \$2.5−5.0 billion between 1990 and 1998 to \$16+ billion in 2005. The remittances are also changing in nature, in that they are no longer almost exclusively for subsistence, but also go toward productive activities, medication, home improvement, etc. One might expect also, that increasingly they will provide support to care for elderly parents and kin back in Mexico.

The point here is that the realm of informal housing production provides an important support context and set of informal opportunities for household extension, intra-lot sharing, and, ultimately, care for the elderly who were, in many cases, the young adult successful pioneer self-builders of the 1960s and 1970s. Through irregular settlement and self-help they have successfully acquired a home and created a *"patrimonio para los hijos,"*—a patrimony of home ownership that has served them well, creating a modest investment as well as allowing for the use value of providing a place to raise a family (albeit, at considerable social cost). Germane to the argument here is that this residential arrangement stands to serve young adults well in their old age, if only as a lever to ensure that resident sons and daughters (in-law) care for them and treat them reasonably.[1]

11.3. Caring for the Elderly Among Hispanic Populations and Communities in the United States

11.3.1. Migration Patterns and the Hispanic Challenge

I move now to examine the situation pertaining in the U.S., asking what insights, if any, may be drawn from these findings and dynamics? The data show a rapidly rising Hispanic population in the U.S., which, for some writers, has become a major cause for concern, vis-à-vis the absolute size of the flow; its provenance (from adjacent Mexico), its overarching low-income nature; high rates of natural increase; and alleged low propensity for assimilation (Huntington, 2004). For Huntington and others the threat is one of undermining the "traditional" U.S. (read "Anglo" and "Protestant") culture and the traditional patterns of integration associated with "melting pot" immigration of the late nineteenth and first half of the twentieth centuries. Here is not the place to address those debates and the conclusions drawn by Huntington, although many of the facts he cites are not in dispute. Hispanics do indeed now constitute the second minority, displacing African Americans, and in a rising number of locations they are already, or will shortly become, the "majority minority." Whether or not this is seen as a problem or as an opportunity is a matter for individual interpretation. Regardless of this interpretation, there are an estimated 41 million Hispanics in the U.S. today (14% of the U.S. total), and of these, the Mexican-origin population is far and away the largest group, with 59% of all Latin American immigrants (23–24 million, which includes between 3–5 million estimated undocumented migrants).

While we know much about the quickening of Mexican immigration since the 1980s and about the new gateway cities to which many are arriving, there are very few studies that have begun to examine the housing and intra-urban mobility

---

[1] The Mexican Health and Aging Study (MHAS) is a major panel study of a representative population aged 50 and above at the baseline in 2001. Although the survey instruments include data and housing variables, these have not yet been fully analyzed, and are correlative in nature, rather than dynamic, related as they are to household structure and extension, lot sharing, etc.

patterns of the Mexican-origin population in U.S. cities.[2] Moreover, instead of the "circular migration" of yesteryear, the hardening of the border has made crossings more costly and dangerous, so it is increasingly difficult for undocumented workers to move back and forth, and so also, by default, they become permanent (albeit illegal) residents. Even if efforts to regularize guestworker status are successful in the future, transnational families are already commonplace, with family networks spread across the two countries; and even if individuals cannot move back and forth with ease, telecommunications and improved banking and other financial services allow for ready communication and remittance transfers. Thus, many temporary migrations become de facto permanent, but most migrants will remain in close contact with kin back home. Some will marry or enter consensual unions in the U.S. and begin to raise their own families. And even if more enlightened policies prevail in the future, allowing for greater ease of movement between the two countries and for eventual naturalization of formerly undocumented workers, many of these Mexican-origin families will live out their lives in the U.S. According to the Pew Hispanic Center the share of first generation immigrants within the total Hispanic population will drop from around 40% in 2000 to around one-third by 2020. Increasingly, the Mexican American-born will dominate, and many adults today, along with the earlier first generation cohort arrivals from Mexico during the 1970s, are those who are hoping to break into the housing market as owners, even though for many their low overall household income precludes formal means of buying a home.

Below I explore the extent to which we might we look in the U.S. for similar sorts of housing dynamics as exist in Mexico, ones that allow for household extension, lot or accommodation sharing, and opportunities to care for elderly parents, whether they are U.S.-born or are first generation arrivals. I also consider the situation of elderly parents of the first generation cohorts, persons who come to live out their twilight years in the U.S. as part of family unification.

11.3.2. Variations in Housing Tenure Patterns by Region and Among Hispanics

The United States is a nation of homeowners with more that 66% of households owning their dwelling units. However, as one might imagine, there are significant variations by city and housing market. High land cost cities such as New York, Los Angeles, etc. have lower rates of ownership and more renting, which in turn correlates with ethnic and racial variation with respect to home ownership. Forty-six percent of African Americans own their home—a figure not greatly different from all Hispanics (45.57%), although a slightly higher proportion of Mexicans own their homes (48.3%). Among other Hispanic and Latino populations there

---

[2] "Intra-city" (within the city) as opposed to "interurban" (between cities). Elsewhere (Ward 1999) I have begun to formulate comparisons of intra-urban trajectories in Mexico cities with those migrant-receiving metropolitan areas of the U.S.

are sharp variations: only 34.6% of Puerto Ricans are owners, largely an effect of the high value housing market costs in New York, where a large Puerto Rican population lives. Cubans have a high level of ownership (57.6%), reflecting their clustering in Miami; their earlier immigration stream to the U.S., and their higher skill (middle class) backgrounds as flight refugees from Castro's Cuba.

Overall, 52% of migrants rent and the Census data reveal important variations in age and household structure of Mexican households compared with other groups. Relatively few (only 14%) live in one-unit households compared with 36.6% nationally (32.5% African American; 24.6% Puerto Ricans, and 33.7% Cubans). Moreover, Mexican renters are younger and are more likely to be recently arrived cohorts, or represent the so-called "1.5" or second generation migrants.[3] Fifty-three percent of Mexican-origin rental householders are less than 35 years old, compared to a national average of 40.6 years (with 39.5% Puerto Rican and 24.2% among Cubans). And taking just the 15–24-year age cohort, 16% of Mexican renter householders are below 24 years, contrasting with 12.5% nationally (and 5.8% for Cubans and 11.5% Puerto Rican). This suggests that Mexican-origin householders are young and primarily renters.

Yet, somewhat counter intuitively, the household size among Mexicans is much larger than for other groups for both owners and renters. Just in the case of renters, some 18% of renter Mexican households have six or more members, (compared to 4.5% nationally; 6.1% for Puerto Ricans, etc.). Indeed, 50% of Mexican renter households in the U.S. have 4 members or more (21% nationally, and 31% for Puerto Ricans). These data reveal important variations among Mexican populations. Specifically, Mexicans—especially recent migrants—are likely to start their housing trajectories as young renters, but instead of living as singletons or in small household units, they appear to be sharing domestic and dwelling spaces. What this suggests is that Mexican-origin intra-city migration trajectories, tenure patterns, and paths to home ownership are substantially different from other Hispanic populations and ethnicities. Indeed, this tallies with what we are slowly beginning to discover about housing arrangements in many cities: the multiple sharing of two- or three-bedroom apartments (between two or three often kin-related members), thereby reducing the per family rental costs, but at significantly increased overall dwelling unit densities.

As those younger migrant households perceive their futures as permanent in the U.S., and as they embark upon their own family building cycles, so they also look for affordable home ownership housing options. For example, among cohort streams arriving at Texas gateway cities in the 1970s, 1980s, and 1990s, the proportion of migrant owner-households increases the longer these migrants reside in the U.S. (some 46.6 % for 1970s arrivals versus 23% for the period 1990–94, and 15.7% for 1995–2000). Protracted residence—and age generally— is associated with increasing rates of home ownership (Rogers, 2006). The question

---

[3] The 1.5 generation refers to those who arrive with their parents as infants or as young children. Second generation refers to those who are born in the U.S. to at least one Mexican migrant parent.

is, where and in what sort of residence and, how does this all tie in with care for elderly?

11.4. Poverty, "The American Dream," and the Rise of Informal Homestead Subdivisions

According to the U.S. Census Bureau (Proctor and Dalaker, 2002), there were 32.9 million people living in poverty in the United States in 2001, which represents 11.7% of the total population. Twenty-three (23.2) million people living in poverty are in families (equal to 9.9% of all people living as families), while 9.2 million unrelated individuals live in poverty (equal to 19.9% of individuals). After the 1990s, a decade of the largest economic expansion in history, the poverty rate improved only slightly compared with previous decades. The U.S. Census Bureau uses a set of monetary income thresholds that vary by family size and composition to determine who is poor. In 2001 the "poverty line," which is a weighted average of these thresholds, was $18,104 for a family of four.

There are over 23 million people living as "families" (as opposed to as "individuals") who are deemed to be living in poverty. With regard to ethnicity and race, Blacks and Hispanics are more than twice as likely to be living in poverty as Whites, although in terms of absolute population numbers, there are many more White households living in poverty, who therefore constitute an important part of the demand side for future informal homesteading and other housing options. Moreover, as we have observed already, the United States is a nation of homeowners, with two-thirds of all American households being classified as "owner-occupiers." Not surprisingly, poorer people are less likely to own and more likely to rent or share accommodation: Of the 12.5 million households living below the poverty line, 65% are renters but this also means that over are third as poor households are, in fact, owners. However, although we cannot control by race and poverty, there are indications that Hispanics are more likely to own than are African American households, especially disaggregating for those who are older and not recent migrants. For Hispanics generally, and for Mexicans in particular, the culture of home ownership is especially important, even among the poorest households, but clearly the capacity of the poor to break into the property market is severely limited.

Indeed, the evidence suggests that homeownership became less affordable during the 1990s, not only for the poor but also for low- and moderate-income households. In 1995 the U.S. Census Bureau estimated that only 56% of families (owners and renters) could afford to purchase a modestly priced house in the area where they lived, and the proportion was declining (Savage, 1999).[4] Affordability also differs by household type: more families than unrelated individuals could afford a

---

[4] A home is considered "modestly priced" if 25% of all homes in the area are below its value and 75% are above. Estimates assume a conventional fixed-rate 30-year mortgage at the market interest rate in that year with a 5% down payment.

modestly priced house; regionally, the Midwest was the most affordable, and the West least. Spatially, too, it was found that few people could actually afford to buy in the area in which they currently lived: In 1995 only 10% of renters could have afforded to purchase a modestly priced home in their current neighborhood.

These data, combined with the poverty structures discussed above, underscore the difficulties that many people experience in entering the housing market as owners. Nor have the multiple federal government programs and subsidies aimed at improving housing affordability in general, and at promoting home ownership specifically, had a great impact in improving access to home ownership for the poor and very poor. Thus, many households in the U.S. find themselves outside the formal market mechanisms and routes to home ownership. They must, therefore, explore other ways to reduce their housing acquisition costs and to find more flexible methods of financing—ways that can serve their economic realities. Thus we come to the discussion of informal housing production, described next.

11.5. *Colonias* and Informal Homestead Subdivisions

Relatively little systematic research exists about how low-income urban dwellers in the United States gain access to residential land and participate in the American Dream of homeownership. Since the 1990s an exception has been the growing concern and analysis of so-called *colonias* in Texas and other border states (Davies and Holz, 1992; Office of Attorney General, 1993; Ward, 1999, 2003; Larson, 1995, 2002; Donelson and Holguin, 2001). The growth of *colonias* is seen as primarily a rural border-housing phenomenon for, almost exclusively, Mexican-origin populations. In fact, the majority of these colonias house urban populations, even though their actual locations are often buried in the rural hinterlands of cities from which these working poor commute to engage in low-paid service activities. And although colonias are indeed concentrated around U.S.–Mexico border cities, where they are also characterized by some of the worst housing conditions, they are not exclusive to that region. In Texas alone there are estimated to be over 400,000 people living in some 1,600 or more *colonias* (Ward, 1999; Ward et al., 2003; see also http://twdb.state.tx.us/*colonias*/index/htm), and if one extends the definition to areas outside of the border, the numbers rise still further. In New Mexico and Arizona the numbers are also substantial. In Arizona the 1990 Census suggested that approximately 162,000 people lived in 77 so-called "*colonias* designated areas," while in New Mexico it indicated that 70,000 lived in 141 such settlements.

Despite the hazards and difficulties associated with acquiring *colonia*-type housing, this supply system remains the only mechanism of entering home ownership for low-income households earning less than $25,000 a year (and many, in fact, earn only half this amount). This is especially the case in the very poor counties of the border regions, such as Starr County, where the median household income in 1999 was $16,504 (compared to $39,927 for Texas as a whole), and 45% of all households earned below $15,000 a year, with 20% of households earning below $10,000. Data for the other border counties are also indicative of widespread poverty, albeit not quite as extreme as in Starr County. The data for median household income and proportion of households earning less than $15,000 total per year (figures in parenthesis) follows: Cameron, $26,155 (29.6%); Hidalgo, $24,635 (32%); Webb, $28,100 (26.9%); Zapata, $24,635 (32%). Upper Valley El Paso County, although also poor, has a more bifurcated income distribution, with $31,051 median, but 20% of households receiving less than $15,000 per year. As one moves away from the border, the median incomes usually rise, but there remains a significant proportion of the population in the "very poor" category, so that *colonia*-type subdivisions are also widespread (Ward and Koerner, 2007). Travis County in central Texas, where the relatively well-off capital of Austin is located, has a median household income of $46,761 (yet 12.2% receive less than $15,000), while Lubbock in the north has incomes of $32,198 (22.5% below $15,000). Both cities have *colonia*-type subdivisions in their peripheral urban (peri-urban) hinterlands, and although they are developed informally and are poor, the housing conditions and infrastructure deprivation are rarely as extreme as one finds along the border.

Preliminary research suggests that *colonias* and similar types of low-income (homestead) subdivisions are widespread in the peri-urban areas outside of a wide range of cities such as:

- Austin and Lubbock in central and north Texas;
- Albuquerque and Santa Fe in New Mexico;
- Tucson and Phoenix in Arizona;
- And in so-called "gateway" cities such as Charlotte and Greensboro in North Carolina
- Dalton and Atlanta in Georgia.

With the use of integrated remote sensing and GIS data techniques (Ward and Peters, 2005) it is becoming much easier to identify such *colonias* and what I am now calling informal homestead subdivisions (IFHS), and to see them as fairly ubiquitous throughout the United States (see Figures 11.1 & 11.2). They are likely to be found wherever relatively low cost land markets exist and where there are low-income populations wishing to embrace home ownership. For this group, informal homestead subdivisions (and *colonias*) are often the only viable option given their low absolute household incomes and/or irregularity of workers' earnings, and their subsequent ineligibility for formal finance (mortgage) assistance.

Moreover, for many of these households, "manufactured" housing represents one lower cost alternative to home ownership (Ward, 2003). Manufactured housing is built entirely in a factory under a federal building code administered by the U.S. Department of Housing and Urban Development and homes may be single or multi-section, and are transported to the site for installation. Manufactured homes do not include travel trailers, motor homes, or modular housing. The latter "modular" homes, while also manufactured either in units or as prefabricated parts, are built to the state or local building codes and are also transported to the site and installed,

FIGURE 11.1. Aerial photograph of an informal homestead subdivision (IFHS) outside of Lubbock, north Texas.

but they do not possess integral transportation gear. Both types of structure are common in *colonia*-type subdivisions examined here.

Elsewhere (Ward and Koerner, 2006) we offer a typology of different types of *colonia* and homestead subdivisions, and drawing upon that framework it is possible to consider that affordable housing opportunities are being created today, primarily as places in which to raise a family and which directly parallel the irregular settlement urbanization of Mexico described earlier (Ward, 1999). But just as in Mexico, where lot space provided opportunities for internal subdivision and shared residence, so, too, *colonias* in Texas and elsewhere offer similar housing and residential sharing opportunities for Mexican-origin populations. In fact they are even more conducive to multiple family occupancy or to large extended households, being much larger in lot size than their Mexican counterparts (1/4 to a full acre, usually). And, in particular, they are likely to provide safe and low-cost

FIGURE 11.2. Oblique photograph of an IFHS outside Austin, central Texas.

care environments in which contemporary homesteaders may themselves grow old with their children living *in situ* or in adjoining lots, the elders depending upon their children for care and comfort in old age. Alternatively, recent migrant cohorts who now live as homesteaders in these settlements may want to bring their elderly parents from Mexico to live with them.

Several types of low-income settlement can be identified, each offering different possibilities for household extension and for accommodating elderly parents.

1. *Classic colonias* in the border region, as mentioned, comprise very low-income, and almost exclusively Mexican or Mexican-origin populations. Just as these settlements offer a low cost mode of housing ownership, subsequent out-mobility and sale of one's homestead is unlikely, the market being stunted with little effective demand (Ward et al., 2004). To the extent that a "home is forever" (Gilbert, 2000), these dwellings will become not only the homes in which contemporary households will raise their families, but also are the games in which they—the parents—will grow old, whether gracefully or not. But the strong presence of close kin living nearby is an important source of social capital that will be important as the population age.[5] Average household incomes vary between $10–14K; lot size varies, but is usually ¼-acre minimum; homes are a mixture of self-built modular (but

---

[5] For example, in one random survey of 227 households across 12 *colonias* along the border, no less than 49 percent of households had kin living nearby in the same *colonia* (Ward et al., 2000: 52). It is not unusual to find close kin living in neighboring lots.

FIGURE 11.3. Sparks, El Paso. Mixture of dwelling types on 1/2 and 1/3 acre lots.

unfinished on the interior) and trailers or campers. Figure 11.3 for Sparks's colonia in El Paso is typical of the motley housing arrangements that exist (see also Figure 11.4). On the border incomes are so low that few have the wherewithal to use their lot space to full effect, but most households have kin or parents living in the same settlement or nearby. Adjacent lots holding parents and sons/daughters are commonplace. Family members (recent arrivals) living *arrimados* with kin is also quite frequent in such settlements.

2. *Non-border peri-urban informal subdivisions* are very similar to *colonias*, although they are usually not quite as poor (annual incomes of $20–30K household), and being further from the border they are less dominated by Hispanic populations and are more mixed, or even dominantly Anglo. Servicing levels, while austere, are much less likely to be entirely absent. These subdivisions are buried in the peri-urban rural areas and are low density. The homes are located on large, individual lots and streets that are often unpaved (see Figures 11.5 and 11.6). This less well-recognized housing alternative is the primary focus of this study. From the air, these are visible as low-density settlements with trailer or modular homes (Figures 11.1 and 11.2).

3. *Semi-urban or rural housing subdivisions* are usually extensive low-density settlements with similar physical dwelling structures and serious servicing deficiencies; they are often much older (nineteenth century or early to mid-twentieth century), and they are more likely to be the homes of elderly vestige populations whose children have moved away.

FIGURE 11.4. El Cenizo *colonia*, Webb County (outside Laredo).

4. *Recreational colonias and subdivisions* come in various shapes, sizes, and types. While they share the remote rural locations, low level of servicing, and trailer-type dwellings, they provide housing for better off working-class populations whose hobbies or preferences are for outdoor life or those wishing to have an affordable second residence for weekends and vacations. These are not usually established with the elderly in mind, although for first residence homes they do provide ample room for "granny annex" extensions or separate units. But at this stage of family building it is more likely that a family will have a pony in the paddock than an older family member living with them (Figure 11.6).

5. *Retirement colonias* are often physically similar to recreational ones, but provide relatively low cost options to so-called *downsizers*—parents whose children have left home and who are now living on modest or limited savings and pensions (Huntoon and Becker, 2002). They are not common, although many mobile home communities (see below) may be a secure and comfortable way to downsize. This arrangement is more common among Anglo populations.

The next two categories comprise manufactured homes that are located in *formal* subdivisions, usually within city jurisdictions rather than in the peri-urban (rural) area; we mention them here because they form an important mode of manufactured housing for low-income groups and are readily differentiated from the housing that we examine in this paper. They are:

FIGURE 11.5. Manufactured home in a *colonia*, Tucson, Arizona.

6. *Mobile home communities*, which offer an option for the moderately poor who can afford to buy a modular home or a new trailer home and lease or purchase the fully serviced lot site. Developed within code, they usually occupy low cost peripheral locations of cities and enjoy full services (Figure 11.7).

7. *Trailer parks* are also located within the city limits or its Extra Territorial Jurisdiction (ETJ), and homes (dwellings) that are owned or rented, are placed on small sites with full services, the site or lot being rented (Figure 11.8). They usually target small recently formed households, Anglo in particular, and are less commonly associated with Hispanic or elderly households.

In particular it is the homestead subdivisions that offer opportunities for accommodation of elderly parents with a reasonable or good quality of life. This is because lots are relatively large and separate; dwelling structures can be set aside (a separate trailer or unit, for example) so that the elderly parent has some degree of independence as they live within a "compound" arrangement. Garden/yard areas offer ample space for people to sit safely outside. For those showing symptoms of senility or Alzheimer's, as well as other chronic and mobility impaired and debilitating illnesses, *colonias* and subdivisions are relatively safe environments. This is because densities are low; lots may be fenced and made secure, and there are often

FIGURE 11.6. Recreational or amenity (trailer) home on one-acre lot with adjacent pad-dock/yard.

FIGURE 11.7. Mobile home community–Austin.

FIGURE 11.8. Aerial view of trailer park, Houston.

high levels of social control and observation from neighbors. If a family member goes walkabout, they are unlikely to go far, and will be noticed. Such dwelling environments also allow for cross-generational interaction of grandparents and grandchildren, with some opportunities for aging parents to engage in child care and support—at least until they become too old or too infirm. A major advantage, of course, is that all this can be achieved at low or minimal cost and reduces the need to place elderly parents in an institution, even if they could afford to do so, or had full Medicare support. For cultural reasons also, Institutional care would become a last resort in such cases rather than the only resort. And for many there is no alternative.

The downside is that these settlements are isolated and invariably unserved by public transport, so that during much of the day adults and children are at work or at school. But again, where lots are shared, or where kin live nearby as neighbors, there may be greater opportunity for one member of the (extended) family to serve as caregiver thereby releasing others to work. This is a common feature of household arrangements in Mexico. Another negative feature is that their location affords relatively little or no access to local social services, but this may be moot to the extent that these are limited or unavailable to uninsured or uncovered populations.

11.6. Conclusion: Looking to Informal Housing in the Future

It is not yet possible to assess with any accuracy how far settlements and informal housing systems outlined above will make a significant contribution to the care of the elderly among Mexican populations in the future. This is because most homesteaders are not yet of an age where they have to adjust to their own declining years, although many are giving serious consideration and concern to the position of their aged parents. But the combination of changing demographic patterns; the broadening of transnational families, and permanence of Mexican-born migrants living in the U.S., and the intersection of all these factors with their legitimate desire for home ownership that can only be accommodated through informal and self-managed means, makes likely an expansion of homestead settlements that parallel the traditional self-help housing production of previous generations in Mexico. And just as those settlements offered flexibility and opportunities for household extension and care for elderly parents within a cultural context in which family was expected to cater to the old, so we may expect *colonias* and informal homestead subdivisions to offer similar opportunities in many areas of the U.S., especially where there is a supply of low cost unserviced rural land in the peri-urban periphery close to metropolitan centers that are the new immigration gateways. Home-based caring for aged parents will probably become commonplace, but here I talked about how this might be achieved among low-income Mexican-origin households, either by adult children bringing their parents up from Mexico to live with them, and/or as younger adults themselves advance to retirement and old age, they look to their own kids.

Other non-Hispanic low-income populations are also likely to see the advantages of homestead subdivisions for precisely the same reason. Indeed, it appears that they are already moving in this direction. Once recent newspaper article cited an Anglo women living in Florida who had purchased a lot and trailer home for her elderly parents in a settlement outside Austin, precisely since she regarded it as a safe (and affordable) haven for her father who had advancing Alzheimer's.

Whether in Mexico or the U.S., informal housing arrangements and systems of caregiving to elderly parents will continue to be important. Mexico may develop greater institutional breadth of coverage for elderly low- and middle-income populations as it adjusts to the new demographic reality born of its successful demographic transition in the 1970s, and as options for informal housing development and lot acquisition and sharing recede under more tightly regulated land markets (Ward, 2005). Also, it is to be expected that as the proportion of Mexican-born declines in the U.S., so the level of remittances flowing back to family members is likely to decline. Increasingly, remittances are likely to be earmarked for care for elderly parents in Mexico itself. Alternatively, under sibling pressure from less "fortunate" households in Mexico, who perceive the opportunities for an aging mother or father to be accommodated with "successful" migrant brothers or sisters now resident in the United States in housing arrangements as those described above, family unification (documented or undocumented) may provide a viable and best-option alternative. And, for first and second generation Mexicans whose working lives are already being forged in the United States, informal homestead and lot sharing arrangements are likely to offer a modest and safe residential space for their own retirement, particularly where such an arrangement can build upon social capital and some institutional and pension assistance. Further research is urgently required to assess the social insurance scenarios for what are often already highly segmented populations living in the United States, and while it will not be appropriate for many households who are renting or living in cramped accommodations, informal homestead subdivisions and *colonias* are likely to become an important residential context in which to care for the elderly.

We now know more or less where and what to look for in such settlements, but much of the science about household adaptations, the blending of formal and informal care in each of these housing settings, and the cultural expectancies and sensitivities of one generation upon another has barely begun.

References

Angel, R., and Angel J. (1997). *Who will care for us? Aging and long-term care in multicultural America.* New York: University of New York Press.

Angel, R., Angel, J.L., Lee, G.-Y., and Markides, K.S. (1999). Age at migration and family dependency among older Mexican immigrants: Recent evidence from the Mexican American EPESE. *The Gerontologist, 39,* 1, 59–65.

Blank, S. (1998). Hearth and home: The living arrangements of Mexican immigrants and U.S.-born Mexican Americans. *Sociological Forum, 13,* 1, 35–57.

Blank, S., and Torrecilha, R.S. (1998). Understanding the living arrangements of Latino immigrants: A life course approach. *International Migration Review, 32,* 1, 3–19.

Citeroni, T.B. (1998). Neither sage nor servant: A sociological interpretation of older women's narrative perceptions of self, family, and social support (Cuernavaca, Mexico), Ph.D. Dissertation, University of Texas at Austin.

Contreras de Lehr, E. (1992). Aging and family support in Mexico. In H.L. Kendig, A. Hashimoto, and L.C. Coppard (Eds.), *Family support for the elderly: The international experience* (pp. 215–23). New York: Oxford University Press.

Davies, C.S., and Holz, R. (1992). Settlement evolution of *'colonias'* along the U.S.–Mexico border: The case of the Lower Rio Grande Valley of Texas. *Habitat International, 16(4),* 119–142.

De Vos, S. (1990). Extended family living among older people in six Latin American countries. *Journal of Gerontology: Social Sciences,* 45(3), 87–94.

Donelson, A., and Holguin, E. (2002). Homestead subdivision/*colonias* and land market dynamics in Arizona and New Mexico. *Memoria of a research Workshop: Irregular settlement and self-help housing in the United States.* pp. 39–41. Lincoln Institute of Land Policy, September 21–22, Cambridge, MA.

Fields, J., and Casper, L.M. (2001). U.S. Census Bureau, Current Population Reports, America's Families and Living Arrangements: 2000. pp. 20–537. Washington, DC: U.S. Government Printing Office.

Flippen, C., and Tienda, M. (1998). *Family structure and economic well-being of Black, Hispanic, and White pre-retirement adults.* Office of Population Research Working Paper No. 98-2, Princeton, NJ.

Gilbert A., and Ward, P.M. (1985). *Housing, the state and the poor: Policy and practice in Latin American cities*, Cambridge: CUP.

Huntington, S.P. (2004). *Who Are We? The challenges to America's National identition*, New york, Siman & Schuster.

Huntoon, L., and Becker, B. (2001). *Colonias* in Arizona: A changing definition with changing location. In *Memoria of a research workshop: Irregular settlement and self-help housing in the United States.* pp. 3438–3441, Lincoln Institute of Land Policy, September 21–22, Cambridge, MA.

Larson, J. (1995). Free markets in the heart of Texas. *Georgetown Law Journal, 84 (December),* 179–260.

Larson, J. (2002). Informality, illegality, and inequality. *Yale Law and Policy Review, Vol. 20,* 137–182.

OAG Office of the Attorney General, Texas. 1993. Socio-economic characteristics of *colonia* areas. White Paper. Austin, TX: Office of the Attorney General.

OAG (1996). *Forgotten Americans: Life in Texas* colonias. Austin, TX: Office of the Attorney General.

Proctor, B.D., and Dalaker, J. (2002). U.S. Census Bureau Current Population Reports, Poverty in the United States: 2001. pp. 60–219. Washington, DC: U.S. Government Printing Office.

Roberts, B. (1995). *The Making of Citizens*, London Edward Arnold.

Savage, H.A. (1999). U.S Census Bureau Current Housing Reports, h121/99-1, Who Could Afford to Buy a House in 1995? Washington, DC: U.S. Government Printing Office.

Varley, A., and Blasco, M. (1999 forthcoming). 'Reaping what you sow'? Older women, housing and family dynamics in urban Mexico. In United Nations International Research and Training Institute for the Advancement of Women (INSTRAW) (Ed.), *Ageing in a gendered world: Issues and identity for women.* Santo Domingo/New York: IN-STRAW/UN Publications.

Ward, P.M. (1999). Colonias *and Public Policy in Texas and Mexico: Urbanization by Stealth.* Austin: University of Texas Press.

Ward, P.M. (2003). Informality of housing production at the urban-rural interface: The not-so-strange case of *colonias* in the US: Texas, the border and beyond. In A. Roy and N. AlSayyad (Eds.), *Urban informality* (pp. 243–70). Lexington/Center for Middle Eastern Studies, UC Berkeley.

Ward, P.M. (2004). "From Marginality of the 1960s to the New Poverty of Today" (with Mercedes Gonzalez de la Rocha, Elizabeth Jelín, Helen Safa, Janice Perlman, and Bryan Roberts), *Latin American Research Review, 39,* 1, 183–203.

Ward, P.M. (2005). The lack of 'cursive thinking' with social theory and public policy: Four decades of marginality and rationality in the so-called 'slum.' In B. Roberts and C. Wood (Eds.), *Rethinking development in Latin America* (pp. 270–296). Pennsylvania State University Press.

Ward, P., with Guiisti, C., and de Souza, F. (2004). *Colonia* land and housing market performance and the impact of lot title regularization in Texas. *Urban Studies, 41,* 13. PP

Ward, P.M., and Peters, Paul. (2005). Integrating remote sensing/GIS methods in housing analysis. *Proceedings of the 2nd International Congress on Ciudad y Territorio Virtual.* 78–83.

Ward, P., and Koerner, M. (2006). Informal housing options for the urban poor in the U.S.: A typology of *colonias* and other homestead subdivisions. Unpublished paper.

Wong, R, Díaz, J.J., and Espinoza, M. (2004). Health care use among elderly Mexicans in the U.S. and Mexico: The role of health insurance. Paper presented at the Population Association of America Meetings, Philadelphia, April 2005.

Section 3
Access to Health Care Services Among Elderly Hispanics with Special Reference to Mexican Americans

12
Disparities and Access Barriers to Health Care Among Mexican American Elders

Fernando Treviño and Alberto Coustasse

12.1. Introduction

According to the year 2000 United States Census data, individuals from diverse racial and ethnic backgrounds are increasing, both in their absolute numbers and in their proportion to the general population. Among the elder population, the Latino population is projected to grow the fastest (Administration on Aging, 2000). In absolute numbers, it is expected that there will be over 13 million Latino elders by the year 2050. This would constitute or a change from 6% in 2003 of the population 65 and older, to 18% by 2050 (Federal Interagency Forum on Aging-Related Statistics, 2004). From this group, it is estimated that at least 4.5 million Latino elders will require Long-Term Care (LTC) (Markides, Rudkin, Angel, and Espino, 1997). Many of the same disparities and barriers that are present in younger populations continue into old age, when health care can be even more vital to a person's quality of life. Older minorities tend to be in poorer health than the general population, have more functional impairments, more limited educations and lower incomes, and in turn bear more out-of-pocket costs, which can be more than 31% for those at the lowest income levels (Krisberg, 2005).

This study has two objectives. First, we examine access barriers to acute and long-term health care for Mexican American elderly as perceived by sociologists/gerontologists and public health policy experts (PHPs). Second, we assess the relative influence of chronic poverty (CP) as compared to other identified factors in patients' accessing quality medical care services as perceived by the above professional groups.

In this chapter *acute care* is defined as medical treatment given to individuals whose illnesses or health problems are short term (usually under 30 days), and long-term care is characterized as a continuum of broad-ranged maintenance and health services delivered to the chronically ill, disabled, and others (U.S. Department of Health and Human Services, 1997).

Much research has been conducted in social gerontology and public health policy on Mexican Americans elders over the past two decades. Yet, systematic knowledge and agreement about health care disparities in access and barriers to acute and long-term care services varies *within and between* different professional, academic, and health policymakers groups. For more than a decade, numerous panels, committees, and commissions have been appointed to address gaps in Latino health research. Nonetheless, the problem of inadequate data on Latinos persists. This is explained in great part by the lack of a concerted or continuous effort in the past and by the lack of influence of bodies commissioned to address these needs to monitor health policy implementation (Aguirre-Molina and Pond, 2003).

Racial and ethnic disparities in health outcomes have been observed among persons with similar health insurance, within the same system of care, and within the same managed care plan, (Bierman, Lurie, Collins, and Eisenberg, 2002). Although targeted interventions have narrowed and even closed some of these gaps, others persist. Research into the underlying factors contributing to health disparities and the design, implementation, and evaluation of interventions to eliminate them is needed. These efforts have been hindered by the general lack of standardized data on race and ethnicity in health care settings, because without these data disparities cannot be assessed (Bierman, Lurie, Collins and Eisenberg, 2002). On the other hand, in part, there is consensus on some core factors or barriers of LTC services use. Research has revealed an underutilization of long-term services by Mexican American elderly and holds that this is due in part to their lack of knowledge, lack of health insurance, use of informal networks, socio-institutional and socio-cultural barriers (Parra and Espino, 1992; Angel, Angel, Aranda, and Miles, 2004).

Poverty has many faces; it changes from place to place and across time, and has been described in many ways. To know what helps to reduce poverty—what works and what does not, what changes over time—poverty has to be defined, measured, and studied. Various definitions and concepts exist for poverty, one being whether households or individuals possess enough resources or abilities to meet their current needs. This definition is based on a comparison of individuals' income, consumption, education, or other attributes with some defined threshold below which individuals are considered poor in that particular attribute. As poverty has many dimensions, it has to be looked at through a variety of indicators—levels of income and consumption, social indicators, indicators of vulnerability to risks, and of socio-political access (World Bank, 2006).

There is significant disagreement about poverty in the United States, particularly over how poverty ought to be defined. There are two versions of the federal poverty measure: the poverty thresholds (which are the primary version) and the poverty guidelines. The U.S. Census Bureau defines poverty as economic deprivation. A way of expressing this concept is that it pertains to people's lack of economic resources (e.g., money or near- money income) needed for consumption of economic goods and services (e.g., food, housing, clothing, transportation). Thus, a *poverty standard* is based on a level of family resources (or, alternatively, of a family's actual consumption) deemed necessary to obtain a minimally adequate standard of living, defined appropriately for the United States (U.S. Census Bureau, 2005a). The U.S. Census Bureau issues the poverty thresholds that are generally used for statistical purposes—for example, to estimate the number of persons in poverty nationwide each year and classify them by type of residence, race, and other social, economic, and demographic characteristics (Fisher, 1997). The Department of Health and Human Services issues the poverty guidelines for administrative purposes—for instance, to determine whether a person or family is eligible for assistance through various federal programs (US Department of Health and Human Services, 2006). The measurement and analysis of poverty, inequality, and vulnerability are crucial for cognitive purposes (to know what the situation is), for analytical purposes (to understand the factors determining this situation), for policy-making purposes (to design interventions best adapted to the issues), and for monitoring and evaluation purposes (to assess the effectiveness of current policies and to determine whether the situation is changing) (World Bank, 2006).

Financial strain is a term referred as "the specific stressor associated with the subjective sense that one's income is inadequate" (Angel, Frisco, Angel, and Chiriboga, p. 537, 2003) and is associated with actual income and poverty. They have also found strong association in the long term, with cognitive capacity, depression, self-esteem, and a weaker association with more objective measures, such as performance-based mobility and mortality, among Mexican American elders who experienced financial strain. As it appears to these scholars, financial strain is a part of a package of cognitions and emotions of low morale that has adverse effects on subjective health (Angel, Frisco, Angel, and Chiriboga, 2003).

Chronic poverty refers to the situation of people who remain poor for much or all of their lives, many of whom will pass on their poverty to their children and all too often die easily preventable deaths. People in chronic poverty are those who have benefited least from economic growth and development. Opportunity is not enough for chronically poor people to escape poverty. They need targeted support, social assistance, social protection, and political action that confronts exclusion (Chronic Poverty Research Center, 2005).

In 2004 in the United States, 37 million people were in poverty, up 1.1 million from 2003. Poverty rates remained unchanged for non-Hispanic Blacks (24.7%) and Hispanics (21.9 %), rates rose for non-Hispanic Whites (8.6% in 2004, up from 8.2% in 2003), and decreased for Asians (9.8% in 2004, down from 11.8% in 2003) (U.S. Census Bureau, 2005b). Among persons aged 65 and over the poverty rate decreased to 9.8% in 2004, down from 10.2% in 2003, while the raw number in poverty in 2004 (3.5 million) was unchanged (U.S. Census Bureau, 2005b). The poverty rate in 2004 for Hispanic older persons was 18.7 percent. This was almost twice the percent for the total older population (Global Action on Aging, 2005).

12.2. Methods

In order to address the research questions comprehensively, the study utilized primary data collection, data from the U.S. Census Bureau and other elder-related national databases, and literature review, combining quantitative and qualitative methods.

Primary data were obtained from interviews with key informants using a semi-structured, open-ended questionnaire that allowed for individual variations. The researchers used purposive sampling and snowball recruitment. The inclusion criteria for being a key informant were: 1) extensive knowledge of barriers to access acute and long-term health care services of Mexican American elders, and/or 2) work experience at an academic institution, in aging societies, or in a related governmental agency, and or 3) participation in the second conference on "*Aging in the Americas*" at the LBJ School of Public Affairs, University of Texas at Austin.

For literature review, three major publications were examined: The Institute of Medicine (IOM) Report "Unequal Treatment" (2002), The National Healthcare Disparities Report (NHDR) from the Agency for Healthcare Research and Quality for the U.S. Department of Health and Human Services (2004), and the Sullivan Commission Report on Diversity in the Healthcare Workforce: "Missing Persons: Minorities in the Health Professions" (2004). In addition, published material from some of the key informants was also used.

12.2.1. Data Collection and Procedures

Potential key informants were initially contacted by email or telephone and informed of the nature of the project and the contribution the researchers felt the participants could make to this study. All interviews were conducted either person-to-person or by telephone. The majority of interviews were conducted at the interviewee's workplace face-to-face; nevertheless, the practicalities of the situation in three instances required telephone interviews with experts from Los Angeles, New York, and Austin.

Twelve key informants were interviewed; two-thirds of them participants in the conference and one-third non-participants who were selected following snowball recruitment: two experts from Texas and one from each coast. Almost 42% were sociologists/gerontologists (5 participants), 50% were public health policy experts (PHPs) (6 participants) and one was a geriatrician (8.3%). One-third (4) were females; 5 participants were Hispanic descent. Results that follow summarize the respondents' answers to each question in the order presented at the interview. Six questions were used and two of those questions were excluded from this manuscript; the results of those responses will be reported elsewhere, as the researchers found them to be unproductive for the goals of this study. The operative questions are described as follows:

1. What are the main barriers to access to acute and long-term services for Mexican Americans (MA) elders?
2. Do you believe there is consensus regarding barriers to access to acute and long-term services for Mexican American elders?
3. Does Chronic Poverty (CP) fully explain differential access to acute and long-term care services for Mexican American elders?

4. Does chronic poverty fully explain the differential quality of acute and chronic care for Mexican American elders?

Key informants were asked these questions in relation to Mexican-origin immigrants and Mexican Americans. For the third question three major groups of professionals were pre-identified: 1) sociologists, social health scientists, and PHPs; 2) health care providers; 3) policymakers and state or federal legislators. For all questions, a preconceived spreadsheet with a list of possible barriers was used for responses, based on four pilot interviews performed prior to the study.

Most questions were asked directly, although some were covered in the respondents' elaboration on earlier questions and not asked directly. The interviews explored the topics in depth, bringing out relationships between topics and temporal sequencing of issues as the relative importance of topics to the interviewees. All interviews were audio taped and transcribed for textual analysis by the investigators. This material was then reconciled with field notes and comments of interviewers. Using template analysis (King, 2004), the transcripts were coded into broad themes based on the interview questions to create an initial template. A data sheet listed the major topics and subtopics of the interview guide in order to record responses in a logical manner. The interview transcripts were reviewed and summarized in a codebook identifying domains and subgroups within domains. The perceptions of acute and long-term barriers for Mexican American elders from the respondents followed the analytical framework for health care services barriers (primary, secondary, and tertiary barriers) for Hispanics at the organizational level, published elsewhere (Carrillo, Treviño, Betancourt, and Coustasse, 2001).

Triangulation of the quantitative and qualitative information through convergence or divergence of multiple data sources was employed as well as methodological triangulation, which involved the convergence of data from multiple data collection sources (Denzin and Lincoln, 1994). The qualitative data were retrieved from the semi-structured interviews and from analysis of documents that included official records, national reports, and published literature from many of the same participants in the interviews. This information was combined as supporting evidence and non-supportive evidence for the first three questions, the main theme being access and barriers to health care, and, for the fourth question, the main theme being quality of received health care. Additionally, frequencies of answers from the interviewees by each question were converged with quantitative data from the major national report cited above, the U.S. Census data, and other demographic related databases.

12.3. Results

First, we examined access barriers to acute and long-term health care faced by Mexican American elderly as perceived by sociologists/gerontologists and PHPs. Table 12.1 describes the barriers identified by the key informants, following the framework developed by Carrillo et al., (2001) for categorizing critical barriers levels.

TABLE 12.1. Primary, Secondary and Tertiary Access Barriers to Health Care Services for Mexican Origin Elders

Barrier identified for accessing acute and LTC (following Carrillo, Treviño, Betancourt and Coustasse framework)	Supportive evidence	Non-supportive evidence	Not mentioned/don't know
Primary Financial barriers, i.e. lack of insurance or ability to pay for care or treatment.	**For primary** 1. Lack of insurance: 8 respondents thought it was a major barrier to access (67%). 2. Financial factors: 10 respondents thought inability to afford co-payments was a barrier to access (83%). 3. Chronic poverty: 4 informants thought CP did fully explain access to health care for Mexican Americans (33%).	**For primary** 1. Lack of insurance: 4 interviewees thought was only a problem for Mexican immigrants and not for MA, as they are entitled to Medicare and/ or state programs (33%). 2. Chronic poverty: 7 of the informants thought CP did not fully explain access to health care for MA (58.3%).	**For primary** 1. Lack of insurance: Not mentioned by 0 individuals. 2. Financial factors: Not mentioned by 2 individuals. 3. Chronic poverty: 1 individual did not know.
Secondary Extramural: From door to HCF, i.e., transportation, proximity, availability of care, etc. Intramural: From HCF to provider's office, i.e., intake process, waiting times, interpreters, referrals, lack of Hispanic professional and leadership workforce, and lack of continuity with same provider.	**For secondary** 4. Availability of services: 6 individuals mentioned it as barrier to access for acute and LTC (50%). 5. Transportation: 7 respondents thought it was a major barrier to the access (58.3%). 6. Patients don't understand the system: 5 respondents thought this issue was also a factor/ barrier (41.6%).	**For secondary** 3. Transportation: 2 interviewees thought was only a barrier if they lived in rural areas and 2 said it was not a problem (33.3%). 4. Patients don't understand the system: 3 respondents didn't think it was a problem (25 %).	**For secondary** 4. Availability of services: Not mentioned by 6 individuals. 5. Transportation: Not mentioned by 1 individual. 6. Patients don't understand the system: Not mentioned by 4 individuals.
Tertiary Language and/or culture hinder the patient-provider communication, i.e., lack of English proficiency by the patient and lack of culturally competent workforce between providers.	**For tertiary** 7. Culture: 12 respondents thought was a barrier (100%). 8. Language: 9 respondents thought it was a major barrier to the access health care services (75%). 9. Discrimination: 11 thought some elements of discrimination were present as a barrier (91.7%). 10. Ageism: 7 respondents mentioned it (58.3 %).	**For tertiary** 5. Language: 2 interviewees thought it was a problem only for Mexican immigrants, but not MA (17%). 6. Discrimination: 1 respondent didn't think it was a problem (8.3%). 7. Ageism: 1 respondent didn't think it was a problem (8.3%).	**For tertiary** 7. Language: Not mentioned by 2 individuals. 8. Discrimination: Not mentioned by 0 individuals. 9. Ageism: Not mentioned by 4 individuals.

12.3.1. Primary Access Barriers

12.3.1.1. Supportive Evidence

The lack of insurance was the most frequently cited barrier among respondents. Eight participants recognized lack of insurance barrier, among them, 5 sociologists or gerontologists (100%) and 3 PHPs (50%) who agreed as well. However, 2 PHPs argued it was not a problem for Mexican Americans but rather for those elders of Mexican origin (see Table 12.1). This can be also supported by literature review: "Mexican Americans, in particular, are seriously underinsured for medical care at earlier ages, and after the age of 65 they are less likely than other groups to have supplemental Medigap coverage" (Angel and Angel p. 345 1997). Furthermore, a key interviewee stated, "Medicare is not a free program and this population, in general, cannot afford [the] several thousands of dollars per year that Medicare will not cover"

Financial and economical factors/barriers were mentioned by 10 out of 12 participants—80% and 83.3% of them sociologists and PHPs, respectively, and no one was disagreement (see Table 12.1). Another pointed out, "The major barrier to access of any type of services is lack of income, which is associated with lack of insurance, which is associated with poor employment opportunities. There are several factors combined together that hinder older Mexican American access to high-quality acute and long-term services" Another responded, "If you are a Mexican immigrant elder you are caught in between everything; here the financial factor comes to play significantly. If you don't have Medicare/Medicaid it will exacerbate the CP". Another answered, "I think CP explains a large part of the variance. CP explains acculturation, immigration, assimilation, and lack of economic opportunities. CP means no jobs, no assets, no housing assets, and no cash, [and being] unable to pass on to the next generation" Finally, the poverty rate in 2004 for Hispanic older persons was 18.7 percent. This was almost twice the percent for the total older population (Global Action on Aging, 2005).

12.3.1.2. Non-Supportive Evidence

One interviewee answered, "The Mexican American elders have better access than the middle age group for acute care". Another mentioned, "Medicare has eliminated for the most part disparities in access due to lack of health insurance for Mexican American elders, but the next question is: Is everyone getting the same quality of care in the same type of services?" Another responded, "Among lawmakers, if you qualify for Medicare and you need other services, you have to go to the emergency room or local senior clinic paid by private non-profit organizations, so how they define the problem is they don't view it as a disparity issue in access." A different individual pointed out, "I have one thing to agree on, poor people across the board have to share common challenges and barriers to care and it has to do with financial access and structural factors." Finally, one interviewee stated, "I think in the border [it] is different because you can use the services "both sides, and the services" the Mexican side are fairly inexpensive, and in some ways it is a benefit for them because they can pick and choose services on both sides."

12.3.2. Secondary Access Barriers

Availability of services was mentioned by 60% of sociologists (three participants), and by 50% of PHPs (three key informants). In addition, one interviewee replied, "Probably for LTC, the biggest access barrier we know is long-term facilities for Hispanics are not available. I don't think there are many nursing homes available in the Southwest of the U.S. for the very poor." Another respondent pointed out, "One thing that is clear is that there are fewer providers for this population along the border and that applies across the board to physicians, pharmacists, dentists, nursing, and other allied professions." Finally, from the Sullivan Commission Report: "The fact that the nation's health professions have not kept pace with changing demographics may be an even greater cause of disparities in health access and outcomes than the persistent lack of health insurance for tens of millions of Americans. Today's physicians, nurses, and dentists have too little resemblance to the diverse populations they serve, leaving many Americans feeling excluded by a system that seems distant and uncaring" (Sullivan Commission, p. 8, 2004).

Transportation was mentioned as a barrier by three sociologists, while of the rest, (40%) thought it depended on whether a person lived in a rural or an urban area; 50% of PHPs agreed it was a main barrier; however two of them responded it was not a problem for this population (see Table 12.1), while the geriatrician answered he saw "this problem every day in his practice".

12.3.3. Tertiary Access Barriers

Difficulty concerning proficiency in English was mentioned by 75% of all respondents and by 80% of sociologists, 67% of PHPs and by the geriatrician; however one sociologist and one PHP agreed that it was a problem for the first generation and not Mexican American elders (see Table 12.1).

On cultural competency there was agreement among all 12 key informants, who thought it was a major barrier to access to care. The geriatrician answered, "Without trust the family doesn't provide information at the medical encounter, they refuse to take the medications or, for example, refuse to take the patient out from the ICU. [T]hey were punishing the physician because he lacked respect and cultural competency".

On discrimination, 11 participants thought some elements of discrimination were present as a barrier (91.7%). All five sociologists interviewed and the geriatrician agreed on this social factor, while it was mentioned by five of six PHPs (83.3%). In addition, one respondent stated: "I think for Hispanics in the political environment outside of the health arena, it is amazing the anti-Latino sentiment and bias I see; if you follow the work of Sam Huntington, who has been an advisor of the White House, he has described Latinos, and the Mexican in particular, as one of the biggest threats for the U.S., by refusing to learn English, refusing to become acculturated, or refusing to limit the family size, and they are perpetuating [the] stereotype. [O]n the other hand African Americans are the best example of how the U.S. has not been able to move beyond race in this country."

12.4. Consensus Intra and Inter-Groups of Respondents

Based on the barrier factors presented above, we can state there was full consensus between sociologists and PHPs in cultural barriers and wide consensus on financial/economic factors, language or proficiency in English, elements of discrimination, and some consensus in availability of services and transportation. However, there was no consensus on the principal access barrier: lack of insurance. While 100% of sociologists supported it was a barrier, only 50%, or three, PHPs agreed and 33%, or two, disagreed, meaning the problem was seen as existing for immigrants' elders and not Mexican American elders (Table 12.1). Also, in lesser degree and relevance, 60% of sociologists thought patients did not understand the health system, while only one-third of the PHPs thought the same. Each group had one participant who thought they understood the system enough to know how to navigate it.

On the other hand, there was a direct question addressing this "perception" of lack of consensus for which individual answers are presented in Table 12.2. In this question, three major camps or groups of professionals were pre-identified: 1) sociologists, social health scientists, and PHPs; 2) health care providers; and 3) policymakers and state or federal legislators. Seven or 58.3% of respondents did not think there was consensus between sociologist and PHPs, health care providers and policymakers. Three or 60%, of the sociologists and four, or 66.7%, of PHPs did not think there was consensus between the different groups (Table 12.2). This notion is expanded with the comment from the geriatrician: "I think there is the wrong consensus, the only one thing these three groups can agree on, is financial issues; everyone says it is a matter of economics and there is nothing we can do about it." In additional a sociologist pointed out: "No, I don't think they speak to each other; the physicians are focused on compliance and they may have some concern with cultural competency, but they don't know what it is."

The 2004 National Health Disparity Report states: "Three key themes are highlighted for policymakers, clinicians, health system administrators, and community leaders who seek to use this information to improve health care services for all Americans (disparities are pervasive, improvement is possible and gaps in information exist)." (NHDR, p. 7, 2004). This report excluded all professionals from the first group (sociologists, social health scientists, and PHPs); however these professionals often trains future policymakers, administrators and clinicians, and are driving force for health service research and publication about health access and disparities. Finally, one respondent mentioned: "It is premature to say that consensus exists; I think there is a better understanding from leadership and emerging

TABLE 12.2. Consensus among Sociologists and Public Health Policy experts; Policy Makers and Health Care Providers Regarding Differential Quality of Acute and Chronic Care for Mexican American Elders.

Consensus between professional groups: Yes	Consensus between professional groups: No	Consensus between professional groups: Don't know
Comments	*Comments*	*Comments*
4 individuals thought there was agreement between groups (33%).	7 individuals thought there wasn't agreement between groups (58.3%).	1 individual did not know (8.6%).
1. "I think there is agreement, in the same factors that is what everyone talks about it."	1. "Probably not, because they start with different values and preconceptions."	1. "I really have no idea."
2. "There are numerous studies and agreement in these barriers to access to care."	2. "No agreement and I don't think they speak to each other."	
3. "There is agreement in the economic issues."	3. "No, among lawmakers many think there is no access problem if you have at least Medicare."	
4. "Yes, there is agreement, but the prism they look through is very different."	4. "If I would have to generalize I would have to say no."	
	5. "There [are] very different opinions because they have different patterns and mode of observations."	
	6. "It is premature to say there is consensus."	
	7. "No consensus between providers and even between sociologists."	

consensus between policymakers, but I don't see it in practice yet; In Texas there is a federal office of minority health and a state office for eliminating health disparities, so there is an effort to address the problem of health disparities, however, for Mexican American elders I am not sure I have seen a specific strategy targeting them as categorical funding"

To address the second objective we assessed the relative influence of chronic poverty as opposed to other identified factors on access to medical care services.

12.4.1. Primary Barriers

12.4.1.1. Supportive Evidence

Regarding the first level of barriers, four key informants thought quality services for MA elders was not fully explained by CP (33%). Only one sociologist, or

20%, and three PHPs supported it. The PHP responded, "I think it explains a large part of the variance. CP explains acculturation, immigration, assimilation and lack of economic opportunities." She added, "Many physicians do not accept Medicaid and Medicare payments and if they do, many Mexican American elders don't have outside money to make the co- payments which these programs don't cover for them, they are now being denied from high quality services and also the care they receive will be inferior." From literature review, the average cost of prescription drugs for any Medicare enrollee age 65 and older, was $1,191. Medicare covered only 4%, Medicaid 9% and 41%, at least, was out-of-pocket payments (Federal Interagency Forum on Aging-Related Statistics, 2004). Finally, another respondent stated, "Within CP, old Mexican women and immigrant elders are the most disadvantageous groups of all because they lack insurance."

12.4.1.2. Non-supportive Evidence

Seven key informants thought quality services for MA elders were not fully explained by CP (58.3%). A significant difference of opinion was found between camps. While four, or 80%, of sociologists and the geriatrician thought CP did not fully explain quality of services for MA elders, three, or 50%, of the PHPs thought it did and one did not know. One key informant responded, "Just because you have Medicare/Medicaid it doesn't mean you will get quality care. Just because you can walk into a physician's office it doesn't mean the care you will get will be of high quality."

12.4.2. Secondary Barriers

Regarding intramural barriers, such as lack of Hispanic professionals and leadership workforce, there was rich supportive evidence from literature review. The Sullivan Commission Report 2004 stated: "The work of the Commission comes at a time when enrollment of racial and ethnic minorities in nursing, medicine, and dentistry has stagnated despite America's growing diversity. While African Americans, Hispanic Americans, and American Indians, as a group, constitute nearly 25 percent of the U.S. population, these three groups account for less than 9 percent of nurses, 6 percent of physicians, and only 5 percent of dentists. The lack of minority health professionals is compounding the nation's persistent racial and ethnic health disparities. From cancer, heart disease, and HIV/AIDS to diabetes and mental health, African Americans, Hispanic Americans, and American Indians tend to receive less and lower quality health care than Whites, resulting in higher mortality rates." (Sullivan Commission, p. 3, 2004). In addition, The National Alliance for Hispanic Health states: "More than two-thirds (86%) of Hispanic community leadership recommendations were not fully incorporated into the final objectives under Healthy People 2010. Over one-third (40%) of Healthy People 2010 population-based objectives and sub-objectives do not have Hispanic baseline data and therefore progress on these objectives cannot be measured for Hispanic communities" (The National Alliance for Hispanic Health, p. 1, 2006).

Other factors besides CP explaining quality of services received by Mexican American elders included lack of available services, mentioned by 60% of sociologists and two-third of PHPs; lack of incentives for physicians under Medicaid, mentioned by five participants (two sociologists and three PHPs). Regarding the fact of physicians dropping from Medicaid due to low reimbursement, the geriatrician added, "Yes, that's true, but we are happy because those providers in the first place shouldn't have seen those patients... [We] are interested in physicians caring for old persons... [If they do] not, probably that will decrease the quality of care [elders] have been receiving".

12.4.3. Tertiary Barriers

There was plenty of supportive evidence: The New York Times summarized the IOM conclusions from the Unequal Treatment" report as follows: "even when members of minority groups have the same incomes, insurance coverage and medical conditions as Whites, they receive notably poorer care. Biases, prejudices and negative racial stereotypes, the panel concludes, may be misleading doctors and other health professionals" (New York Times, March 22, 2002). Five participants mentioned ageism explained the differential quality of acute and chronic care for Mexican American elders. The geriatrician, two sociologists, and two PHPs supported this view, and the rest did not mention it. Also from literature review: "Ethnic elders generally seek health services less frequently, and later in the course of an illness than White elders. One of the important components of access is the degree to which ethnicity influences an individual's perception of a given illness and the decision to seek health care. Not to be overlooked are the roles racism and discrimination play in both access and quality of care received. Current information seems to indicate that ethnic elders are generally poorer, less well educated, and their greater degree of chronic illness and disability significantly impacts their functional status and quality of life" (American Geriatrics Society, p. 4, 2001).

Other factors besides CP explaining quality of services received by Mexican American elders were: language, mentioned by all sociologists, the geriatrician, and 50% of PHPs; however, one PHP stated language was included in chronic poverty. Culture as a barrier was mentioned by 80% of sociologists, the geriatrician and two-thirds of PHPs; however the same PHP stated culture was as well included in chronic poverty. Three, or 60%, of sociologists, the geriatrician, and 50% of PHPs brought up trust. Finally, lack of culturally competent physicians was acknowledged by one-third of PHPs, the geriatrician, and two sociologists, or 40%; however, one sociologist disagreed and thought providers are sufficiently trained for this. On the other hand, the geriatrician stated, "Medical schools are not addressing the issue in a meaningful way because we are not prioritizing in this country; everything is important, AIDS, national security, etc; so how do you prioritize your priorities? There are enough culturally competent providers?

No. Will [there] be in the future? No. Will [there] be ever be in this country? No".

12.5. Conclusion

We found that the perceptions of acute and long-term barriers for Mexican American elders followed the analytical framework for health care services barriers for Hispanics at the organizational level published by Carrillo, Treviño, Betancourt, and Coustasse (2001). The financial barrier (primary) is not as severe as in middle age, if the elders have Medicare/Medicaid, but if the elderly lack Medigap, or have not worked enough to comply with Medicare requirements (or worked in jobs not covered by Medicare) the reality is quite different and becomes severe. Moreover, recent elderly immigrants or older women, who have stayed home and taken care of the family, will face many of same barriers as middle-aged Mexican-origin groups, which will lead to financial strain on an older and often sick or disabled vulnerable population.

At the level of structural or institutional barriers (secondary barriers), they will face access barriers such as transportation, availability of services, lack of information of health services, and lack of Hispanic professionals and leadership workforce. And they will face socio-cultural barriers (tertiary barriers) such as poor communication at the encounter with the physician, which will be even worse than for middle-aged Hispanics groups due to the likehood that cultural and language barriers are exacerbated at older ages. However, we need to add to this category ageism, discrimination, and stereotyping as recognized by the "Unequal Treatment" report.

We found among sociologists/gerontologists and PHPs consensus in most of the barriers identified; however, this was not the case for lack of insurance. This difference may be explained by different training and research experiences, which predispose each group toward a different set of values and perspectives. Also, this could be explained by different barriers that working differently for subgroups of Mexican-origin elderly and over different time periods, so answers by respondents could have been given with emphasis in different sub-groups of Mexican-origin elders (i.e. Mexican immigrants or Mexican American elders), which could have been a limitation of this study.

On the other hand, regarding their views of consensus between different professional groups such as providers and policymakers, on health care barriers experienced by Mexican American elders, most sociologists/PHPs were in agreement that there was not consensus between groups.

We found as well that the majority of our respondents did not believe that CP fully explained differential quality of acute and chronic care for Mexican American elders, and many respondents identified availability of services, ageism, lack of a culturally competent workforce, language, trust, and culture as critical components of the complex issues regarding the quality of the U.S. health care system for Mexican-origin elders.

12.6. Discussion

If not addressed, the combined impact of lack of financial resources and the limited availability of health care services and providers for the growing population of older Mexican Americans will exert a significant toll on the health system, Medicare, Medicaid, and on society in general in future years.

Although there is no national consensus with respect to the collection of data on race/ethnicity for performance measurement in health care settings and quality improvement in the services provided to Mexican American elders, the weight of prior research, the goals and needs of Healthy People 2010, and the demographic changes in the U.S. population are likely to increase the demand for such data in the years to come.

Finally, there are a number of factors affecting older Mexican Americans' access to acute and long-term care, factors that are precipitated by the limited Hispanic representation of health care providers, administrators, leadership, policy analysts and decision makers. Compounding the shortage of Mexican American providers is the shortage of Mexican American political leaders, senior agency directors, or advisors to such policymakers, whose presence influences the structure and availability of health services. This has consequences for the few Hispanic leaders in their fields, making it more difficult to develop a collective and coherent understanding of how to address access barriers that disproportionately affect the Mexican elder's community. The shortage decreases the community's capacity to ensure that institutions and decision makers become responsive to particular Mexican Americans health access needs.

References

Agency for Health Care Research and Quality (AHRQ). (2004). National health disparities report, 2004. Retrieved March 1, 2005, from http://www.qualitytools.ahrq.gov/ disparitiesreport/browse/browse.aspx?id=4483.

Aguirre-Molina, M. and Pond, A.N. (2003). *Latino access to primary and preventive health services. Barriers, needs and policy implications.* New York: Columbia University.

American Geriatrics Society (2001). Position statement on Ethnogeriatrics. Retrieved on January 28, 2006, from http://www.americangeriatrics.org/products/positionpapers/ ethno_committee.shtml

Angel, R.J, and Angel, J.L. (Eds). (1996). *Who will care for us? Aging and long-term care in multicultural America?* New York: New York University.

Angel, R.J, and Angel, J.L. (1997). Health service use and long term care among Hispanics. In K.S. Markides and M.R. Miranda (Eds.) *Minorities, aging and health.* Thousand Oaks: Sage.

Angel, R.J, Frisco, M., Angel, J.L., and Chiriboga, D. (2003). Financial strain and health among elderly Mexican-origin individuals. *Journal of Health and Social Behavior (Dec) 4(4)*, 536–51.

Bierman, A.S. Lurie, N., Collins K.S., and Eisenberg, J.M. (2002). Addressing racial and ethnic barriers to effective health care: the need for better data. *Health Affairs, 21(3 May-Jun)*, 91–102.

Carrillo, E. Treviño, F.M., Betancourt J.R, and Coustasse, A. (2001). The role of insurance, managed care, and institutional barriers. *Health issues in the Latino community.* In Molina, Molina, and Zambrana (Eds.), San Francisco; Jossey Bass.

Chronic Poverty Research Center. (2005). Chronic Poverty Report 2004–05. Retrieved on January 27, 2006 from http://www.chronicpoverty.org/resources/cprc_report_2004-2005.html

Denzin, N.K., and Lincoln, Y.S. (Eds.). (1994). *Handbook of qualitative research.* Newbury Park, CA: Sage.

Federal Interagency Forum on Aging-Related Statistics. (2004). Older Americans 2004: Key indicators of well-being. Federal Interagency Forum on Aging-Related Statistics. Washington, DC: US Government Printing Office. Retrieved February 9, 2006, from http://www.agingstats.gov/chartbook2004/pr2004.html

Fisher, G.M. (1997). The development and history of the U.S. poverty thresholds—A brief overview. *GSS/SSS Newsletter* [Newsletter of the Government Statistics Section and the Social Statistics Section of the American Statistical Association]. Winter, pp. 6–7.

Global Action on Aging, (2005). Statistical Profile of Hispanic Senior Citizens. Senior Journal, September 22, 2005. Retrieved on January 27, 2006, from http://globalag.igc.org/elderrights/us/2005/REMhispanic.htm

Institute of Medicine (2002). Unequal treatment: Confronting racial and ethnic disparities in health care. In B.D. Smedley, A.Y. Stirth, A.R. Nelson (Eds.). Washington, DC: National Academy Press.

Institute of Medicine (2002). *Unequal treatment: Confronting racial and ethnic disparities in health care.* Washington, DC: National Academy Press.

King, N. (2004). Using templates in the thematic analysis of text. *Essential guide to qualitative methods in organizational research.* In C. Cassell and E. Symon (Eds.), London: Sage.

Krisberg, K. (2005). Cultural competencies needed to serve all older Americans: Cultural skills will help bridge health gaps. *The Nation's Health, June–July* (pp. 29–30).

Markides, K.S., Rudkin, L., Angel, R.J. Espino, D.V. (1997). Health status of Hispanic elderly in the United States. *Racial and ethnic differences in late life in the health of older Americans.* In Martin, Soldo, L. G. Martin and B. J. Soldo, (Eds.). Washington, DC: National Academy of Sciences.

National Alliance for Hispanic Health, (2006). *Policy brief: Healthy People 2010: Hispanic concerns go unanswered.* Retrieved on February 10, 2006, from http://www.hispanichealth.org/hpzolo.lasso

New York Times, (2002, March 22). Subtle Racism in Medicine, Section A, p. 24)

Parra, E.O., and Espino, D.V. (1992). Barriers to health care access faced by elderly Mexican Americans. *Clinical Gerontologist*, 11, 171–177.

Sullivan Commission on Diversity in the Healthcare Workforce. (2004). Missing persons: Minorities in the health professions 2004 Report. Retrieved March 1, 2005, from http://admissions.duhs.duke.edu/sullivancommission/documents/Sullivan_Final_Report_000.pdf

U.S. Census Bureau(2005a). *Measuring poverty.* Retrieved on January 28, 2006 from http://www.census.gov/hhes/www/img/povmeas/intro.pdf

U.S. Census Bureau, Housing and Household Economic Statistics Division (2005b). *Poverty: 2004 highlights.* Retrieved on January 28, 2006 from http://www.census.gov/hhes/www/poverty/poverty04/pov04hi.html

U.S. Department of Health and Human Services. (2006). *The 2006 HHS poverty guidelines: One version of the U.S. federal poverty measure.* Retrieved on January 28, 2006 from http://aspe.hhs.gov/poverty/06poverty.shtml

World Bank. (2006). *Understanding poverty.* Retrieved on January 27, 2006 from http:// web.worldbank.org/WBSITE/EXTERNAL/TOPICS/EXTPOVERTY/EXTPA/0,, contentMDK:20153855~menuPK:435040~pagePK:148956~piPK:216618~theSite PK:430367,00.html

13
Lack of Health Insurance Coverage and Mortality Among Latino Elderly in the United States*

Rogelio Saenz and Mercedes Rubio

13.1. Introduction

The Latino population in the United States represents one of the most rapidly growing and dynamic racial and ethnic groups in the country. For example, the Latino population in the U.S. increased by 58% between 1990 and 2000 compared to a growth of only 13% in the nation's overall population (Saenz, 2004). However, the Latino population, especially the Mexican and Puerto Rican populations, has consistently been close to the bottom of the socioeconomic hierarchy of the United States. Indeed, Mexicans have the lowest educational level among the major racial and ethnic groups in the country, while also, along with Puerto Ricans, having relatively high levels of poverty. Yet, Latinos have high levels of labor force participation. Nonetheless, they tend to occupy jobs that are disproportionately located in the secondary labor market—jobs that offer low wages, limited opportunities for advancement, and hardly any benefits. Because employment represents the major route for securing health insurance coverage, Latinos, especially Mexicans, have the lowest level of health insurance coverage in the country (Amey, Seccombe, and Duncan, 1995; Angel and Angel, 1996; Angel, Angel, and Markides, 2002; Treviño et al., 1991).

The lack of insurance has immediate and long-term consequences, especially among the poor and the working poor. For example, it is quite expensive to receive health care without health insurance coverage. Certainly, for those without resources, health care is a luxury, with the result being that people do not seek health care until it is no longer possible to postpone it. Research has demonstrated that Mexicans have the lowest rates of health care utilization largely due to the lack of health insurance (Anderson et al., 1981; Estrada, Treviño, and Ray, 1990; Roberts and Lee, 1980; Solis et al., 1990; Treviño et al., 1991; Treviño et al., 1996). In addition, people who lack resources and health insurance are particularly vulnerable when they become elderly, a stage in their lives when health care needs increase significantly. Again, research has shown that Mexican elderly are less likely to

---

* Paper presented at the Second Conference on Aging in the Americas: Key Issues in Hispanic Health and Health care Policy Research. Austin, TX. September 22, 2005.

181

have any form of health insurance coverage compared to their peers from other racial and ethnic groups (Angel and Angel, 1996; Angel, Angel, and Markides, 2002; Harris, 1999). Moreover, Mexican foreign-born elderly—especially those who immigrated to the U.S. later in life—tend to have especially low levels of health insurance coverage. Furthermore, the lack of access to health care is likely to have cumulative effects on not only the well-being of the elderly, but also on their survivability. As such, it is important to assess the links between the absence of health insurance and mortality.

The health care needs of Mexican—and, more generally, Latino—elderly will become an increasingly serious concern as this population ages. While the Latino population is quite youthful, population projections suggest that the elderly will represent a rising share of the population over the coming decades. For example, while seniors accounted for 6% of the Latino population in 2000, they are projected to constitute 11% by 2030 (U.S. Census Bureau, 2004). The projected aging of the Latino population and more generally the U.S. population has serious implications related to the financing of health care for the elderly, variations in morbidity and mortality associated with type of health insurance coverage, and the criteria by which health insurance coverage is distributed to the elderly.

This paper has two research objectives. The first focuses on the identification of factors that contribute to the lack of health insurance coverage among Latino elderly. The second focuses on the relationship between lack of health insurance (along with other selected factors) and mortality among Latino elderly. These research objectives will be addressed with the use of data from the Hispanic Established Populations for Epidemiological Study of the Elderly (H-EPESE).

13.2. Comments from the Literature

Throughout the life course, Mexicans are less likely to report health insurance coverage compared to other racial and ethnic groups, including other subgroups of Latinos (e.g., Cubans and Puerto Ricans) (Amey, Seccombe, and Duncan, 1995; Angel and Angel, 1996; Angel, Angel, and Markides, 2002). The lack of coverage is related to low socioeconomic status, low educational attainment, and to employment in occupations that do not offer health insurance benefits (Mutchler and Burr, 1991; Anderson, Lewis, Giachello, Aday, and Chiu, 1981). Life-long patterns of inequalities cumulate in old age and chronic ailments (e.g., cardiovascular disease, diabetes, arthritis, etc.) become more pronounced. Research has demonstrated that despite a convergence in the mortality rates of Mexican and White elderly, Mexican elderly have significantly higher levels of disability (Angel and Angel, 1997; Angel and Angel, 1998; Hazuda and Espino, 1997; Jette et al., 1996; Lawrence and Jette, 1996; Lopez and Aguilera, 1991; Markides, et al. 1997; Markides and Wallace, 1996; Peek et al., 2003; Rudkin et al., 1997; Zsembik et al., 2000). It has been observed that the high rates of disability among Mexican elderly are aggravated by high rates of diabetes and obesity (Hazuda and Espino, 1997; Mutchler and Angel, 2000; Ostir et al., 1998), as well as low levels of medical insurance coverage (Mutchler and Angel, 2000).

The United States, unlike other developed countries, lacks universal health coverage and long-term care for the aged or disabled. Retirement age, typically 65, becomes a time when the state provides health insurance and medical care through the Medicare program (Angel and Angel, 1996). Since 1964 Medicare has provided persons 65 and older access to physicians and hospital services, regardless of the individual's ability to pay. This, however, does not mean that older Americans receive services for free. Rather older Americans may be required to pay a substantial portion of physician and hospital care costs (Angel and Angel, 1997). In addition, Medicare does not cover any catastrophic illnesses that require hospitalization or care in a nursing home. The lack of supplemental insurance to Medicare can be a barrier to health care for the elderly (Rowland and Lyons, 1987).

To complicate matters, individuals of Mexican-origin—especially the foreign-born—have high rates of non-participation in Medicare (Angel and Angel, 1996). Some of the reasons cited for this low level of participation in Medicare among Mexican elderly include lack of proficiency in the English language, low participation in Social Security, and not possessing U.S. citizenship (Angel and Angel, 1997). These reasons, however, do not fully capture the low participation among native-born Mexican-origin individuals.

A growing body of empirical studies shows that elderly Mexican Americans and other minority elderly individuals have great levels of unmet health care needs (Angel and Angel, 1996). An often cited reason for this unmet needs involves the disproportionate number of Latinos who had no health insurance coverage while employed in marginal jobs (National Council of La Raza, 1992). Jobs that do not provide insurance to working-age individuals will not provide health insurance coverage to workers upon retirement. Hence, employment and types of fringe benefits (e.g., health insurance) available to employees through retirement age is vital to the health care access of the elderly.

In sum, research has shown that certain segments of the Latino elderly are more likely to not have health insurance. This includes Mexicans, those with lower levels of education, and those who are foreign-born and who have been in the U.S. for a shorter period of time (Angel, Angel, and Markides, 2002). In this study, we will identify factors that are associated with the lack of any form of health insurance coverage among Latino elderly at two periods of time (1993–1994 and 1995–1996).

13.2.1. Health Insurance and Health Status

Access and utilization of health care services is critical in the treatment of the elderly with disability and declined functions (Hebert, Brayne, and Spiegelhalter, 1999). However, there is no conclusive evidence of the association between access to health care and health outcomes for elderly Mexican-origin individuals. For example, a study by Treviño et al. (1996) found that poor Mexican Americans in San Antonio with health insurance had higher health care access rates than their counterparts without insurance. They further found that for some health care access improved their health outcomes. In contrast, Harris (2001) in a study of diabetic adults using national samples of Whites, Blacks, and Mexicans, found that health status did not appear to be influenced by access to health care.

Although there may be conflicting results about the role of access to health care and health outcomes, it is clear that the lack of adequate health care insurance and inconsistent medical care have serious consequences. Mexicans and other people of color who are uninsured or underinsured often enter the health care system at a point when their condition has advanced (Rubio and Williams, 2001). This often translates into longer and more expensive hospital stays. Furthermore, delayed detection often means poorer survival rates (Muñoz, 1988, Morris et al., 1989).

In addition to delayed detection and treatment, individuals with limited or no health insurance coverage may be misdiagnosed or be prescribed medication that is inappropriate for their ailments. For example, Raji et al. (2003) in a cross-sectional study of Mexican elderly over the age of 65 found that four medications—chlorpropamide (used to treat type-2 [non-insulin-dependent] diabetes), propoxyphene (used to relieve mild to moderate pain), amitriptyline (used as an anti-anxiety medication in the treatment of mental illness and often used for the treatment of chronic pain), and dipyridamole (used to prevent dangerous blood clots from forming in individuals who have had a stroke)—accounted for 54% of all inappropriate prescriptions. However, they found that the overall percentage of inappropriate prescriptions is lower for Mexicans than is the case for Whites and Blacks. Without a doubt both access to and quality of care is paramount, but they are of particular importance for the elderly.

In this study, we assess the relationship between lack of health insurance coverage (and other selected factors) at two time periods and mortality as of December 2004 among Latino elderly.

13.3. Methods

Data from the Hispanic Established Epidemiological Study of the Elderly (H-EPESE) are used to conduct the analysis. The H-EPESE is a longitudinal study based on Latinos 65 years of age and older in 1993–1994 living in the Southwest (Arizona, California, Colorado, New Mexico, and Texas). We use three waves of data to conduct the analysis reported here: Wave 1 (1993–1994; N = 3,050); Wave 2 (1995–1996; N = 2,438); and Wave 4 (deaths as of December 2004 occurring to persons in Waves 1 and 2). These three waves of data allow us to carry out our two research objectives: 1) the identification of factors that contribute to the lack of insurance coverage at Waves 1 and 2; and 2) the relationship between the lack of health insurance (as well as other selected variables) in Waves 1 and 2 and mortality as of Wave 4. These two research objectives represent the two segments of the analysis presented below.

13.3.1. Analytical Plan: Part 1

The first part of the analysis involves the prediction of the lack of health insurance in Waves 1 and 2. Logistic regression is used to carry out the analysis. The dependent variable for this segment of the analysis is the lack of health insurance coverage

TABLE 13.1. Operationalization of dependent and independent variables used in the logistic regression analyses.

Dependent variables
Insurance coverage:
No = 1; Yes = 0
Death as of December 2004:
Yes = 1; No = 0
Independent variables
Ethnicity (1 dummy variable):
1) Mexican = 1; Other Latino = 0 (reference group)
Age (4 dummy variables):
1) 65–69 = 1; Else = 0; 2) 70–74 = 1; Else = 0; 3) 75–79 = 1; Else = 0; and 4) 80–84 = 1; Else = 0.
Reference group: 85 and older
Sex (1 dummy variable):
1) Female = 1; Male = 0 (reference group)
Nativity/years in U.S. (6 dummy variables):
1) Foreign-born, less than 10 years in U.S. = 1; Else = 0; 2) Foreign-born, 10–19 years in U.S.;
3) Foreign-born, 20–29 years in U.S.; 4) Foreign-born, 30–39 years in U.S.; 5) Foreign-born, 40–49
years in U.S.; and 6) Foreign-born, 50 or more years in U.S.
Reference group: U.S.-born
Marital status (2 dummy variables):
1) Married = 1; Else = 0; and 2) Widowed = 1; Else = 0.
Reference group: Other
Educational level (3 dummy variables):
1) 9–11 years of education = 1; Else = 0; 2) High school graduate/GED = 1; and 3) Post-high school
education = 1;
Else = 0
Reference group: 0–8 years of education
Employment status (1 dummy variable):
1) Employed = 1; Else = 0 (reference group)
Perceived level of economic difficulty (3 dummy variables):
1) A great deal = 1; Else = 0; 2) Some = 1; Else = 0; and 3) A little = 1; Else = 0
Reference group: None
Self-reported health status (3 dummy variables):
1) Excellent = 1; Else = 0; 2) Good = 1; Else = 0; and 3) Fair = 1; Else = 0
Reference group: Poor
State of residence (2 dummy variables):
1) California = 1; Else = 0; and 2) Texas = 1; Else = 0.
Reference group: Other Southwest (Arizona, Colorado, and New Mexico.

(1 = lack of any form of health insurance; 0 = presence of any form of health insurance). Ten variables are used as predictors of the lack of health insurance: 1) national origin; 2) age group; 3) gender; 4) nativity/years in the United States among the foreign-born; 5) marital status; 6) employment status; 7) economic difficulty; 8) self-assessed health; 9) education; and 10) state of residence. The operationalization of all variables used in the two sections of the analysis is shown in Table 13.1.

13.3.2. Analytical Plan: Part 2

The second part of the analysis involves the prediction of mortality as of December 2004 among the original respondents in Waves 1 and 2. Logistic regression is used to conduct this part of the analysis. The dependent variable is mortality (1 = mortality; 0 = alive). The predictors of mortality include the lack of health insurance coverage in Waves 1 and 2 along with the ten predictors used in the first part of the analysis.

13.4. Results

Before examining the results of the multivariate analysis, we begin with a descriptive overview of the prevalence of lack of health insurance coverage among Latino elderly in Waves 1 and 2 broken down by the predictor variables used in the analysis (see Table 13.2). In Wave 1 (1993–1994), approximately 10% of Latino elderly lacked any form of health insurance coverage. By the second wave (1995–1996) about 6% did not have any insurance. Given that health insurance coverage is fairly uniform among the elderly, it is clear that a significant portion of Latino elderly lack insurance. Note that almost three-fourths of Latino elderly have either Medicare or Medicaid while approximately one-fifth have private insurance and Medicare coverage.

The descriptive analysis also suggests that persons who lack any form of health insurance coverage are not randomly distributed across the Latino elderly population. Those without health insurance tend to be Mexican, younger, foreign-born persons who have resided in the United States for a relatively short period of time, persons with 9 to 11 years of education, employed individuals, those who indicate greater economic difficulty, persons with more favorable health, and people living outside of Texas.

While these patterns are instructive in obtaining a profile of Latino elderly who lack health insurance coverage, we need to move the analysis to a multivariate framework. Table 13.3 shows the results of the logistic regression analysis predicting the lack of any form of health coverage. Latino elderly lacking insurance have distinct characteristics. Two variables are signicantly related to the lack of health insurance in both waves (1 and 2). Foreign-born individuals who have been in the United States fewer than ten years are between eight times (Wave 2) and 11 times (Wave 1) more likely than the native-born to lack health insurance. Foreign-born elderly who immigrated late in their lives are at great risk for not having health insurance coverage. In addition, Latino elderly who are employed are about four times more likely to not have health insurance compared to their counterparts without employment in each wave. It is likely that individuals without health insurance lack resources more generally, which requires them to continue to work despite their advanced age.

Furthermore, there are clearer patterns involving significant relationships between the independent variables and lack of health insurance in Wave 1 than in

TABLE 13.2. Percentage of Latino elderly with different types of insurance by selected characteristics, Waves 1 and 2.

	Wave 1			Wave 2		
Selected characteristics	No insurance	Private insurance and Medicare	Medicare/ Medicaid	No insurance	Private insurance and Medicare	Medicare/ Medicaid
Ethnicity:						
Mexican	10.7	16.0	73.3	6.0	20.5	73.5
Other Latino	6.0	31.0	62.9	4.1	31.7	64.1
Age:						
65–69	14.6	17.1	68.3	6.7	21.0	72.3
70–74	8.7	17.1	74.3	6.0	22.3	71.7
75–79	7.5	15.2	77.3	4.7	24.6	70.7
80–84	6.1	22.2	71.7	5.5	21.4	73.1
85 and older	7.4	13.4	79.3	5.9	9.9	84.2
Sex:						
Male	9.7	20.0	70.3	6.8	24.9	68.3
Female	10.9	14.9	74.2	5.2	18.6	76.3
Nativity/years in U.S.:						
Foreign-born, less than 10 yrs	39.6	1.5	58.9	24.0	5.8	70.3
Foreign-born, 10–19 yrs	21.6	4.2	74.2	6.3	8.0	85.7
Foreign-born, 20–29 yrs	8.3	2.3	89.4	4.4	5.8	89.8
Foreign-born, 30–39 yrs	9.3	9.5	81.1	7.9	11.4	80.7
Foreign-born, 40–49 yrs	8.4	11.6	80.0	4.7	16.7	78.7
Foreign-born, 50 or more yrs	9.9	14.9	75.2	4.0	21.2	74.8
U.S.-born	6.8	24.2	69.1	4.7	27.3	68.0
Marital status:						
Married	10.5	21.1	68.4	6.0	27.4	66.6
Widowed	12.0	12.4	75.7	5.9	15.5	78.6
Other	6.2	11.3	82.5	5.5	10.2	84.3
Educational level:						
0–8 yrs	10.1	13.0	76.9	5.7	15.5	78.8
9–11 yrs	13.4	24.3	62.2	8.4	40.3	51.3
High school graduate/GED	10.8	40.0	49.2	5.1	57.6	37.3
Post-high school education	9.8	49.9	40.3	6.8	62.2	31.1
Employment status:						
Employed	31.6	25.5	42.9	20.8	32.1	47.2
Not employed	9.5	16.8	73.8	5.5	21.0	73.5
Level of economic difficulty:						
A great deal	15.0	7.9	77.1	5.9	10.2	83.9
Some	11.6	9.6	78.8	3.2	15.8	81.1
A little	5.0	14.9	80.1	5.3	27.8	66.9
None	8.2	40.0	51.8	9.7	48.3	42.1
Self-reported health status:						
Excellent	19.7	15.5	64.8	6.8	28.5	64.8
Good	9.4	23.6	67.0	7.5	24.3	68.2
Fair	9.7	17.0	73.4	5.2	20.4	74.4
Poor	6.1	8.8	85.1	4.2	13.8	82.0
State:						
California	13.9	11.8	74.3	8.8	22.5	68.7
Texas	6.4	15.5	78.1	4.3	18.4	77.4
Other Southwest	11.1	40.8	48.1	8.5	36.6	54.9
N	317	521	2,212	143	517	1,778
Percent	10.4	17.1	72.5	5.9	21.2	72.9

TABLE 13.3. Odds ratios representing the relationship between selected variables and lack of insurance among Latino elderly.

Selected characteristics	Wave 1: No insurance		Wave 2: No insurance	
Mexican	0.875		1.705	
Age 65–69	4.342	**	1.046	
Age 70–74	2.726	*	1.070	
Age 75–79	1.932		0.898	
Age 80–84	1.583		0.861	
Female	1.017		0.804	
Foreign-born, less than 10 yrs	11.153	**	7.82	**
Foreign-born, 10–19 yrs	4.252	**	1.411	
Foreign-born, 20–29 yrs	1.106		1.050	
Foreign-born, 30–39 yrs	1.139		1.650	
Foreign-born, 40–49 yrs	0.938		0.862	
Foreign-born, 50 or more yrs	2.173	**	1.147	
Married	1.978	*	1.050	
Widowed	2.549	**	1.113	
Employed	3.894	**	4.244	**
Great Economic Difficulty	2.087	**	0.468	**
Some Economic Difficulty	1.464		0.255	**
Little Economic Difficulty	0.607		0.449	**
Self-assessed health				
Excellent	2.833	**	1.467	
Good	1.321		1.983	*
Fair	1.228		1.364	
9–11 yrs education	1.465		1.648	
High school graduate	1.517		0.609	
Post-high school graduate education	1.325		0.845	
California	0.888		0.761	
Texas	0.413	**	0.630	
Chi-square likelihood ratio	317.145	**	102.978	**
df	26		26	
N	2,704		2,165	

*Statistically significant at the 0.05 level.
**Statistically significant at the 0.01 level.

Wave 2. Seven independent variables are significantly associated with the lack of health insurance coverage in Wave 1 (but not in Wave 2). In Wave 1, persons who lack health insurance coverage are more likely to be those who are 65–74 years of age, those who are foreign-born and have been in this country either 10–19 years or 50 or more years, those who are married or widowed, those who indicate that they have great economic difficulty, those who report their health as excellent, and persons living outside of Texas (Arizona, California, Colorado, and New Mexico).

Four independent variables are significantly related to the absence of health insurance in Wave 2 (but not in Wave 1). Latino elderly lacking health insurance

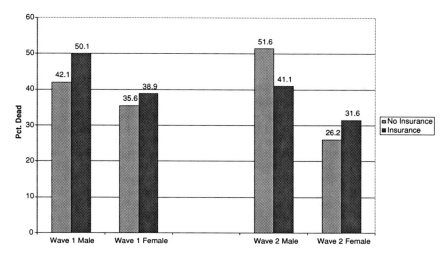

FIGURE 13.1. Percentage of Latino Elderly in Waves 1 and 2 Who Died by December 2004 by Sex and Insurance Coverage Status.

coverage in Wave 2 are more likely to be persons who self-report their health as good and those who do not have economic difficulties.

We now turn our attention to the second part of the analysis involving the mortality by 2004 among Latino elderly alive in Waves 1 and 2. We examine the prevalence of mortality among Latino elderly by presence/absence of health insurance coverage and sex in Waves 1 and 2. In only one case (males in Wave 2), do individuals lacking health insurance have higher mortality rates (51.6%) compared to their counterparts who do have insurance (41.1%) (Figure 13.1). In the other three cases, persons who lack health insurance tend to have lower mortality rates compared to those who have insurance. This paradoxical pattern may be associated with individuals who are in worse health being more likely to need and, thus, having health insurance compared to their peers who are in better health and who can afford to forego health insurance at least for a certain period of time.

Selected other independent variables are associated with mortality. For example, across both waves those who died by December 2004 were more likely to have been those who were 85 or older, males, and who reported their health as poor. Furthermore, certain variables are significantly related to mortality in Wave 1 (but not in Wave 2). In particular, in Wave 1, Latino elderly who died by December 2004 were more likely to be those who were foreign-born living in the U.S. between 10 and 19 years, those who were married or widowed, and those who were employed.

Table 13.4 presents the results of the logistic regression analysis predicting mortality among persons in Waves 1 and 2. We are especially interested in the relationship between the lack of health insurance in Waves 1 and 2 and mortality by December 2004. In both waves, lack of health insurance is positively but not significantly associated with mortality. This pattern may reflect the selective nature of those with and without health insurance relative to their health status as well as potential differences in health and mortality by specific type of insurance coverage.

Selected characteristics	In Wave 1: Death		In Wave 2: Death	
No insurance	1.199		1.294	
Mexican	1.033		1.123	
Age 65–69	0.066	**	0.085	**
Age 70–74	0.093	**	0.153	**
Age 75–79	0.151	**	0.217	**
Age 80–84	0.214	**	0.421	**
Female	0.542	**	0.471	**
Foreign-born, less than 10 yrs	1.340		1.096	
Foreign-born, 10–19 yrs	0.519	**	0.865	
Foreign-born, 20–29 yrs	0.794		1.209	
Foreign-born, 30–39 yrs	0.890			
Foreign-born, 40–49 yrs	0.882		1.133	
Foreign-born, 50 or more yrs	0.997		0.856	
Married	0.600	**	0.739	
Widowed	0.660	**	0.958	
Employed	0.466	*	0.761	
Great economic difficulty	1.224		1.068	
Some economic difficulty	1.002		0.903	
Little economic difficulty	1.194		0.863	
Self-assessed health:				
Excellent	0.385	**	0.341	**
Good	0.393	**	0.391	**
Fair	0.451	**	0.458	**
9–11 education	0.978		1.273	
High school graduate	0.955		1.010	
Post-high school graduate education	1.432		1.455	
California	1.166		0.823	
Texas	0.962		0.753	
Chi-Square Likelihood Ratio	348.450	**	295.611	
df	27		27	
N	2,425		2,049	

*Statistically significant at the 0.05 level.
*Statistically significant at the 0.01 level.

13.5. Conclusions

Latino elderly, particularly those of Mexican-origin, have the lowest level of health insurance coverage compared to their counterparts from other racial and ethnic groups. Research has shown that this group is vulnerable due to its lack of health insurance coverage and social and economic resources.

TABLE 13.4. Appendix

Selected characteristics	Percent dead
Personal income:	
Below $5,000	47.2
$5,000–$9,999	39.8
$10,000–$14,999	42.1
$15,000 and above	43.2
Self-reported health status:	
Excellent	
Good	30.2
Fair	34.5
Poor	37.5
	56.2
State:	
California	44.0
Texas	40.7
Other Southwest	49.4
N	2,563

Our findings provide a profile of the Latino elderly who lack health insurance coverage. Latino elderly who lack health insurance coverage tend to be younger, recent immigrants, those with greater economic difficulties, those who are employed, and those who report more favorable health conditions. The findings suggest that a segment of the Latino elderly tend to postpone acquiring health insurance coverage until they need it. While this may be a survival strategy associated with using limited economic resources to meet many needs varying by degree of immediate priority, lack of health insurance coverage is negatively associated with sound health maintenance. Without regular health checkups among the elderly, it becomes more difficult to detect health problems that can be treated before they become worse.

Our findings indicate that there is not a significant association between the lack of health insurance coverage and mortality among Latino elderly. However, this finding may be consistent with the notion that Latino elderly may be postponing health insurance coverage until they really need it. Hence, as we have seen, persons who report worse health conditions are the most likely to have health insurance coverage. These are also the individuals with the highest risk of mortality.

The findings from our study have important implications related to the health and health needs of Latino elderly. For example, Latinos continue to have significant unmet health needs including the need for adequate health care insurance and access to health care. While there are certainly many barriers associated with language and culture, the most immediate needs are economic in nature (e.g., lack of access to affordable and adequate health care) (Estrada, Treviño, and Ray, 1990; Solis et al. 1990). Furthermore, Latino elderly who do not have any form of health insurance, as well as those that lack adequate health insurance coverage, are likely to have worked in jobs that did not provide such coverage in earlier stages of life or to have immigrated to the United States at later stages in the life course. The poor and the working poor lack economic resources and health insurance coverage to obtain at the least preventive health care and even more so, catastrophic health care. Moreover, Latinos are often viewed as a highly familistic group that care for elderly relatives. Under such perceptions, pressure is placed on Latino families and extended families to meet the health care needs of aged relatives. However, generally Latino elderly who lack resources to access health care are part of families and extended families that have very limited resources. Even routine health care needs of the elderly place tremendous burdens on their families and extended families. People at the margins of society cannot afford to get ill.

Latino elderly have inadequate access to health care. These needs will become even more magnified as the Latino population ages. The relative size of the elderly in the Latino population is projected to nearly double from 6% in 2000 to 11% in 2030 (U.S. Census Bureau, 2004). Without increasing access to health care among Latinos including the elderly, the health care needs of the Latino population— the nation's largest minority group—will increase dramatically. The construction of public policy for decreasing racial and ethnic health disparities requires long-term thinking that weighs the investment in providing adequate and affordable health care to people throughout their lives and the greater costs that must be paid when health problems are detected and treated at late stages. Failure to invest in maintaining the health of the most vulnerable segments of our society commonly results in major health care costs in the future.

References

Amey, C., Seccombe, K. and Duncan, R.P. (1995). Health insurance coverage of Mexican American families in the U.S.: The effect of employment context and family structure in rural and urban settings. *Journal of Family Issues, 16*, 488–510.

Anderson, R., Lewis, S.Z., Giachello, A.L., Aday, L.A., and Chiu, G. (1981). Access to medical care among the Hispanic population of the Southwestern United States. *Journal of Health and Social Behavior, 22*, 78–89.

Angel, J.L., and Angel, R.J. (1998). Aging trends—Mexican Americans in the southwestern USA. *Journal of Cross-Cultural Gerontology, 13*, 181–290.

Angel, R.J., and Angel, J.L. (1997). *Who will care for us? Aging and long-term care in multicultural America.* New York: New York University Press.

Angel, R.J., and Angel, J.L. (1996). The extent of private and public health insurance coverage among adult Hispanics. *The Gerontologist, 36*, 332–340.

Angel, R.J., Angel, J.L., and Markides, K.S. (2002). Stability and change in health insurance among older Mexican Americans: Longitudinal evidence from the Hispanic Established Populations for Epidemiological Study of the Elderly. *American Journal of Public Health, 92*, 1264–1271.

Estrada, A.L., Treviño, F.M., and Ray, L.A. (1990). Health care utilization barriers among Mexican Americans: Evidence from HHANES 1982–84. *American Journal of Public Health 80 (Supplement)*, 27–31.

Harris, M.I. (1999). Racial and ethnic differences in health insurance coverage for adults with diabetes. *Diabetes Care, 22*, 1679–1682.

Harris, M.I. (2001). Racial and ethnic differences in health care access and health outcomes for adults with type 2 diabetes. *Diabetes Care, 24,* 454–459.

Hazuda, H.P. and Espino, D. (1997). Aging, chronic disease, and physical disability in Hispanic elderly. In K.S. Markides and M. Miranda (Eds.), *Minorities, aging, and health.* Newbury Park, CA: Sage Publications.

Hebert, R., Brayne, C. and Spiegelhalter, D. (1999). Factors associated with functional decline and improvement in a very elderly community-dwelling population. *American Journal of Epidemiology, 150,* 501–510.

Jette, A.M., Crawford, S.L. and Tennstedt, S.L. (1996). Toward understanding ethnic differences in late-life disability. *Research on Aging, 18,* 292–309.

Lawrence, R.H., and Jette, A.M. (1996). Disentangling the disablement process. *Journal of Gerontology: Social Sciences, 51B,* S173–S182.

Lopez, C. and Aguilera, E. (1991). *On the sidelines: Hispanic elderly and the continuum of care.* Washington, DC: National Council of La Raza.

Markides, K.S., Rudkin, L., Angel, R.L., and Espino, D.V. (1997). Health status of Hispanic elderly in the United States. In L. Martin and B. Soldo (Eds.), *Racial and ethnic differences in the health of older Americans* (pp. 217–235). Washington, DC: National Academy Press.

Markides, K.S., and Wallace, S.P. (1996). Health and long-term care needs of ethnic minority elderly. In R. Coe, J. Romeis, and J. Morley (Eds.), *Planning for long-term services for the elderly in the future.* New York: Springer Publishing Company.

Morris, D.L., Lucero, G.T., Joyce, E.V., Hannigan, E.V., and Tucker, E.R. (1989). Cervical cancer, a major killer of Hispanic women: Implications for health education. *Health Education, 20,* 23–28.

Muñoz, E. (1988). Care for the Hispanic poor: A growing segment of American society. *Journal of the American Medical Association, 260,* 2711–2712.

Mutchler, J.E., and Burr, J.A. (1991). Racial differences in health and health care service utilization in later life: The effect of socioeconomic status. *Journal of Health and Social Behavior, 32,* 342–356.

Mutchler, J.E., and Angel, J.L. (2000). Policy development and the older Latino population in the 21st century. *Journal of Aging and Social Policy, 11,* 177–188.

National Council of La Raza. (1992). *Hispanics and health insurance.* Washington, DC: Labor Council for Latin American Advancement.

Ostir, G., Markides, K.S., Black, S.A., and J.S. Goodwin. (1998). Lower body functioning as a predictor of subsequent disability among older Mexican Americans. *Journal of Gerontology: Medical Sciences, 54A,* M491–M495.

Peek, M.K., Ottenbacher, K.J., Markides, K.S., and Ostiri, G.V. (2003). Examining the disablement process among older Mexican American adults. *Social Science and Medicine, 57,* 413–425.

Raji, M.A., Ostir, G.V., Markides, K.S., Espino, D.V., and Goodwin, J.S. (2003). Potentially inappropriate medication use by elderly Mexican Americans. *Annuals of Pharmacotherapy, 37,* 1197–1202.

Roberts, R.E., and Lee, E.S. (1980). Medical care use by Mexican-Americans: Evidence from the Human Population Laboratory Studies. *Medical Care, 18,* 266–281.

Rowland, D. and Lyons, B. (1987). *Medicare's poor: Filling the gap in medical coverage for low-income elderly Americans.* Baltimore, MD: The Commonwealth Fund.

Rubio, M. and Williams, D. (2001). Race as a social phenomenon. In B.M. Beech, V. Setlow, and M. Roohani (Eds.), *Race and research in focus: Perspectives on minority participation in health studies* (pp. 1–26). Washington, DC: Publication Board of the American Public Heath Association.

Rudkin, L., Markides, K.S., and Espino, D.V. (1997). Functional disability in older Mexican Americans. *Topics in Geriatric Rehabilitation, 12,* 38–46.

Saenz, R. (2004). *Latinos and the changing face of America.* New York: Russell Sage Foundation.

Solis, J.M., Marks, G., Garcia, M., and Shelton, D. (1990). Acculturation, access to care, and use of preventive services by Hispanics: Findings from HHANES 1982–84. *American Journal of Public Health 80 (Supplement),* 11–19.

Treviño, F.M., Moyer, M.E., Valdez, R.B., and Stroup-Benham, C.A. (1991). Health insurance coverage and utilization of health services by Mexican Americans, mainland Puerto Ricans, and Cuban Americas. *Journal of the American Medical Association, 265(2),* 233–237.

Treviño, R.P., Treviño, F.M., Medina, R., Ramirez, G., and Ramirez, R.R. (1996). Health care access among Mexican Americans with different health insurance coverage. *Journal of Health Care for the Poor and Underserved, 7,* 112–21.

U.S. Census Bureau. (2004). *U.S. interim projections by age, sex, race, and Hispanic origin.* Washington, DC: U.S. Census Bureau. Internet release date: March 18, 2004. Webpage address: http://www.census.gov/ipc/www/usinterimproj/.

Zsembik, B.A., Peek, M.K., and Peek, C.W. (2000). Race and ethnic variation in the disablement process. *Journal of Aging and Health, 12,* 229–249.

14
Access Issues in the Care of Mexican-Origin Elders: A Clinical Perspective

David V. Espino and Liliana Oakes

14.1. Introduction

The burgeoning of the aging population has been well documented (Rice and Feldman, 1983; Manton, 1991). In 1990, 12.7% of Americans were over the age of 65. By the year 2030, the total number of elderly over 65 is expected to be 64.6 million. The over 85-group is the fastest growing segment in the population with the number of centenarians (age 100 and older) also rapidly increasing. By the year 2050, this older cohort, 85 and older, is expected to comprise of 5.2% of the total population (U.S. Census, 1983).

The nation's Hispanic population reached 41.3 million as of July 1, 2004, according to national estimates. Hispanics accounted for about one-half of the national population growth of 2.9 million between July 1, 2003, and July 1, 2004. The Hispanic growth rate of 3.6% over the 12-month period was more than three times that of the total population. Over 2 million Hispanics are over 65 years of age, with the oldest-old (over 85) constituting the fastest growing subgroup.

The growing numbers of Mexican origin elders will directly affect the nation's current overextended service system. Equitable distribution of resources to provide adequate access to services is the challenge legislators and society as a whole will face in the near future.

There are three key areas that need further examination: 1) Financial barriers to health care, 2) communication issues, and 3) quality issues. The following case illustrates various aspects of these areas as they relate to access and care.

14.2. Case Presentation: Part 1

Sra. GM is an 83-year-old monolingual Mexican-origin female who originally arrived with her parents in the United States during the Mexican revolution. Her husband of 50 years recently died and her daughters make an appointment with a primary care physician for initial evaluation for weight loss. She is 45 minutes late in arrival due to transportation problems. She had been seen by a variety of physicians prior to her visit with us and was most concerned about the costs of the visit. To date, she has not seen a Spanish-speaking physician. She has Medicare Part A and B as her primary insurance and has no prescription drug benefit. The office staff makes her a new appointment for one month later as she arrived late for her scheduled appointment.

14.2.1. Financial Barriers

The major financial access issues faced by Mexican-origin elders when they attempt to utilize health care are (i) affordability and (ii) availability and accessibility. In this age group, affordability relates to issues beyond health insurance costs. In Sra. GM's situation, although she seems to have excellent health care insurance with Medicare Part A and B, lack of prescription drug support has placed her in a position of financial instability due to costs. While Medicare A is universal, a significant proportion of the Mexican-origin population lacks this basic coverage. They have not been able to apply, either due to communication barriers or ineligibility for not having ever become American citizens. Likewise, those elders who lack Part B either do not qualify or have not applied to have premiums deducted from their monthly social security entitlement. Medigap and secondary insurance use is unusual in the Mexican-origin population.

While the Medicare Prescription Drug, Improvement, and Modernization Act, which took effect in the fall of 2005. Offers some relief in this area, the rules are complex. The renaming of Medicare+Choice programs to Medicare Advantage programs is no doubt confusing, as is medication coverage under both Part C prescription plans (MA-PDs) and Part D (PCPs) (American Pharmacists Association, 2004). Complexities with the new system combined with lack of clear guidelines no doubt has caused confusion for both elders and their care practitioners and no doubt many have and will continue to choose the least advantageous plan for them.

Ethnic-appropriate, quality health care availability remains a major barrier. Practitioners that are truly bicultural, have an interest in caring for frail elders, and are able to shoulder the financial strain for caring for a Medicare population are rarities. Sra. GM has seen many physicians, but has yet to see one that speaks her language. Medicare reimbursements to physicians are unreasonably low for the time and effort required to care for a frail elder patient. (Elon, 2003) Those bicultural practitioners willing to care for frail elders are able to pick and choose which patients they wish to care for, leaving many families on their own caring for elders with multiple complex, interacting illnesses with limited options in finding quality care.

Geographic accessibility can be a deterrent for accessing health care by Mexican origin elders. The financial burdens of transportation impact patient care even when it is available. Frail elders are unable to use public transportation. Public transportation for the disabled is extremely limited and elders must compete with other disabled persons for these scarce resources. Sra. GM would obviously not qualify for this type of program. In Sra. GM's case, she was unable to get a timely ride, and subsequently was not able to see her physician of choice due to constraints in the number of patients that the physician had to that or any other day.

Particularly, those elders living in rural areas may find it more difficult to access a rural health care system that is already limited. An elder who is Spanish-language monolingual and lives in a rural area may be at a disadvantage for physically accessing appropriate elder health care services, especially if he or she has a limited informal family support system.

14.3. Case Presentation: Part 2

A week later, Sra. GM begins to have early morning headaches and discusses this with her daughter who makes an appointment with a neurologist. Sra. GM arrives at the neurologist's office and signs all documents that are presented to her, including a living will. Sra. GM is put in a room with her daughter, who is minimally functional in Spanish. An English monolingual physician enters the room. The practitioner introduces himself in Spanish then proceeds interviewing the daughter in English. After a brief physical exam without removal of any cloth-ing, a diagnosis is made and a variety of tests are ordered. The purposes for the tests are discussed with the daughter who agrees with the plan. The neurologist diagnoses various illnesses: increased cholesterol, high blood pressure and tension headaches. After discussing the medication recommendations with the daughter he places her on three medications. After they leave the office, Sra. GM com-ments to the daughter that she thought that he was not a good physician. She subsequently does not take the prescribed medication despite the efforts of the daughter.

14.3.1. Communication Dynamics

The practitioner-patient interaction includes complexities of nuances in language, kinesthetics, and cultural communication style. A failure to recognize the dynam-ics of these factors during the health care encounter reduces effectiveness and worsens outcomes. Health care practitioners are inadequately or poorly prepared for multicultural work, resulting in poor intercultural skills with subsequent inade-quate clinical evaluations and care (Clark, 1983; Galanti, 1991). While Sra. GM's physician, who had cultural sensitivity training, used Spanish to gain rapport with her, his subsequent discussion with the daughter undermined his intentions to be culturally "in-tune" with the patient. Mini language courses that are designed to provide the health practitioner with basic Spanish language skills might actually do more harm than good as they lull both the health practitioner and the health care system that they work in into a false sense of security.

Kinesthetics and communication styles incorporate four elements seen in practitioner-patient encounters: a) entry/trust building, b) assessment, c) care plan development, and d) support mobilization (Zerwekh, 1991).

Entry and trust building are two areas that are most susceptible to threats caused by variations in cultural communication styles (Espino, 2004). Mexican-origin elders strongly prefer a longer period of socializing before reaching the purpose of the health care encounter (Kuipers, 1991; Maduro, 1983). Language barriers and demands on practitioner time inhibit not only access to health information but hinder the socialization process. Kinesthetics are critical, as patients must perceive the practitioner as unhurried and attentive. Body language indicating otherwise can sever whatever trust has been built. Failure to observe conventions in establishing social relationships can easily stress the practitioner-patient relationship, something that was seen in Sra. GM's response after the encounter.

Effective communication styles are critical during the clinical assessment. Most health care practitioners, regardless of their ethnic background, training, and the particular role they play, have been socialized to search for orthodox health problems. Medical problems that have cultural overtones have significant effect on patient and family functioning, but may be dismissed, ignored or misinterpreted by practitioners. Additionally, the common belief in the interpretation of natural or supernatural (Maduro, 1983; Trotter and Chavira, 1981) phenomena is easily dismissed or ignored by health care practitioners not trained to incorporate spirituality into their evaluations. Sra. GM's doctor did not address cultural areas.

The major consequence of inadequate health care plan development is lack of patient compliance, a significant health care barrier. Inadequate attention by the health care team to individual language, kinesthetics, or communication styles is common. When health care practitioners fail to carefully assess cultural perspectives of illness and signs of these perspectives are absent from the care plan, there is no logical reason for the elder or the family to comply. In Sra. GM's situation, her response after the encounter indicates that compliance might be a significant issue in this patient.

Support mobilization, a crucial aspect of elder care, is strongly influenced by culture. Practitioners' knowledge of who the socially and culturally appropriate caregivers are is essential to increasing compliance and improving health care access. More important, the ability to decrease conflict among multiple siblings, the spouse, and development of a functional decision-making process are crucial to decreasing perceived poor outcomes, which are a critical factor in malpractice actions.

14.4. Case Presentation: Part 3

Sra. GM begins to take echinacea for her headaches. Her daughter finally convinces her to take the cholesterol medication. Her headaches improve slightly but she begins to complain of muscle aches. Her daughter takes her to the local acupuncturist who treats her with some resolution of the symptoms. She subsequently is found unresponsive by her daughter and is taken to the emergency room. The daughter expresses that she wishes everything done for her mother. The staff is less than supportive of the decision and asks for an ethics consult. The patient becomes unresponsive, develops kidney failure and is transferred to the intensive care unit where she expires. The cause of death is determined to be due to the medications.

14.4.1. Quality Issues

Care quality is an issue that has been poorly explored in ethnic elders. Quality is directly related to health care outcomes, with the desired outcome in elders being decreased disease burden, increased quality of life, and, to a much lesser extent, decreased mortality. Even with adequate financial access, such as that available through managed care plans, health disparities in quality still persist (Brown, 2003). Issues of quality can revolve around practitioner care, system care, ethical decision-making, and racism.

Elder Mexican-origin patients perceive older physicians as providing more quality and tend to prefer long standing relationships with older practitioners. However, older physicians who practice longer may provide lower-quality care compared to younger physicians (Choudry, 2005). Therefore, an inherent risk may be the continued desire to see their older, less qualified physician in comparison to a younger, more knowledgeable physician. Families may consciously and unconsciously aid this process.

Specialists are not trained in providing primary care or elder care, or to recognize diseases outside their areas of expertise (Starfield, 2005a). Medicare patients and/or their caregivers have the tendency to "doctor shop" for frail elders, often trying to match symptoms to specialist (Starfield, 2005b). This behavior provides for further care fragmentation and promotes iatrogenesis and may increase mortality. The decision to put an 83-year-old woman on a cholesterol medication, while not malpractice, may not have been indicated in this patient's case. Furthermore, in Sra. GM's situation, she chose nontraditional medicine to augment traditional medicine, a choice that was a contributing factor to her death.

Ethical difficulties seem inherent when there are differences in the cultural definition of a situation, or if there are differences between the perceived values and right of health care practitioners and the needs of patients and their significant others. (Bedolla, 1995). Questions have even been raised as to the cultural appropriateness of some of the cornerstones of ethical decision-making used regularly in health care, such as the principle of individual autonomy, commonly used in decisions regarding truth telling and consent. In the case of Hispanics, the paradigm of individual autonomy is not clearly applicable (Espino, 2004). Family and shared decision-making take precedence over individual autonomy. This could create issues of perceived (by the family) decrease in quality of life as decisions are forced on individual families.

Ethnic differences may result in racist reactions, ranging from overt hostility or antipathy to grudging acceptance on the part of either practitioner or patient (Yeatts, Crow, and Folts, 1992). Racism may be further inflamed by language issues, which further lead to misunderstanding and poor access (Kuo, 2005). The consequences of racism reinforce poor attitudes toward Mexican-origin elders leading to poor care. The approach to Sra. GM's family was counterproductive and not consistent with her cultural values. It would not be surprising if a malpractice action arose from the situation.

14.5. Conclusions

Health care access issues for Mexican-origin elders are multiple and complex. While financial barriers are substantive, they are by no means the predominant barriers to obtaining quality health care. Issues of communications and quality predominantly affect outcomes and promote health disparities. Further evaluation of simple and effective methods to reduce these barriers will likely yield significantly decreased health disparities.

References

American Pharmacists Association. (2004). Understanding Medicare reform: What pharmacists need to know: #3, Navigating the Medicare drug benefit. Monograph, p. 2. American Pharmacists Association: Washington, DC.

Bedolla, M.A. (1995). Ethics in Hispanic elders. In D.V. Espino (Ed.), *Ethnogeriatrics.* Philadelphia, PA: W.B. Saunders.

Chouhdry, N.K., Fletcher, R.H., and Soumerai, S.B. (2005). Systematic review: The relationship between clinical experience and quality of health care. *Annals of Internal Medicine, 142,* 260–273.

Clark, M. (1983). Cultural context of medical practice. *The Western Journal of Medicine, 139(6),* 806–811. In M. Clark (Ed.) Special Issue: Cross-Cultural Medicine.

Elon, R.D. (2003). Reforming the care of our elders: Reflections on the role of reimbursement. *Journal of the American Medical Directors Association, 4(2),* 117–120.

Espino, D.V., Oakes, L.O., Amaya-Grever, I., Olivares, O., Alford, C.A., and Mouton, C.P. (2004). *Health care for Hispanic elder patients.* Kansas City, MO: AAFP Press.

Galanti, G.A. (1991). Caring for patients from different cultures. Philadelphia: University of Philadelphia Press.

Kuipers, J. (1991). Mexican Americans. In J.N. Giger and R.E. Davidhizar (Eds.), *Transcultural nursing: Assessment and intervention* (pp. 185–215). St. Louis, MO: Mosby Year Book.

Ku, L., and Flores, G. (2005). Pay now or pay later: Providing interpreter services in health care. *Health Affairs, 24(2),* 435–444.

Maduro, R. (1983). Curanderismo and Latino views of disease and curing. *The Western Journal of Medicine, 139(6),* 868–875. In M. Clark (Ed.), Special Issue: Cross-Cultural Medicine.

Manton, K.G. (1991). The dynamics of population aging: Demography and policy analysis. *Milbank Quarterly, 69,* 309–340.

Rice, D.P., and Feldman, J.J. (1983). Living longer in the United States: Demographic changes and health needs of the elderly. *Milbank Memorial Fund Quarterly/Health and Society, 61(3),* 362–396.

Starfield, B., Lemke, K.W., Bernhardt, T., Foldes, S.S., Forrest, C.B., Weiner, J.P. (2005). Co-morbidity: implications for the importance of primary care in "case" management. *Annals of Family Medicine, 1,* 8–14.

Starfield, B., Lemke, K.W., Herbert, R., Pavlovich, W.D., and Anderson, G. (2005). Comorbidity and the use of primary care and specialist care in the elderly. *Annals of Family Medicine, 3,* 215–222.

TRIAD Study Group (Brown, A.F., Brusuelas R., Gerzoff, R.B., Karter, A.J., Gregg, E., Mangione, C.M., Safford, M., Waitzfelder, B., et al.). (2003). Health behaviors and quality of care among Latinos with diabetes in managed care *American Journal of Public Health, 93(10),* 1694–1698.

Trotter, II, R.T., and Chavira, J.A. (1981). Cuanderismo: Mexican American folk healing. Athens: University of Georgia Press.

U.S. Census Bureau. (1992). Hispanic population passes 40 million. Census Bureau Report. http://www.census.gov/Press-Release/www/releases/archives/population/005164.

Yeatts, D.E., Crow, T., and Folts, E. (1992). Service use among low-income minority elderly: Strategies for overcoming barriers. *The Gerontologist, 32(1),* 24–32.

Zerwekh, J.V. (1991). A family caregiving model for public health nursing. *Nursing Outlook, 39(5),* 213–217. http://www.aafp.org/afp/20050415/newsletter.html

15
Cross Border Health Insurance and Aging Mexicans and Mexican Americans

David C. Warner

As the number of persons of Mexican descent in the United States has grown to more than 25 million, of whom more than 10 million were born in Mexico (U.S. Census. Bureau, 2005), the availability of medical care to this population and their relatives in the U.S. and in Mexico has increased in importance. Both in the U.S. and in Mexico there is increased attention to the issue of providing health insurance coverage for persons across borders and for families who may be separated by international borders or who commute across borders to work or who choose to live part of their lives in both countries. The availability of coverage may determine not only use of health care provider but also may have a significant impact on where individuals choose to live and to retire. Accordingly, it is worth reviewing some of the literature on cross border utilization of care and some of the cross border coverage initiatives that both the private sector and the Mexican government have initiated. The development of more seamless and transparent entitlements and responsibilities may be one necessary initiative to enhance the quality of life of aging populations in the U.S. and in Mexico.

Typically, the vast majority of users of health services across the U.S.–Mexico border have paid for their services out of pocket. There have been a number of studies of persons who live on the U.S. side of the border who, for cost, convenience, or by preference, use Mexican pharmacies, dentists, or physicians (Arrendondo-Vega, 1999; Homedes, et al., 1994; San Diego Dialogue, 1994). And, although this coverage has been somewhat attenuated by the growth of managed care and required gatekeepers, it has been common for insurers who write coverage for groups on the U.S. side to permit out-of-network utilization of certain hospitals or physicians on the Mexican side. Elderly Medicare beneficiaries on the border, both of Mexican-origin and otherwise, have long used pharmacists and dentists on the Mexican side because those services were not covered by Medicare and price was of great importance. Many studies have documented that low-income persons without insurance on the U.S. side frequently use services on the Mexican side as well. Similarly, there are a number of studies that document use of U.S. providers by Mexican border residents (Guendelman and Jasis, 1990; Zinnecker, 1990; Potter et al., 2003) as well as substantial use of world-class tertiary facilities such as MD Anderson or Methodist hospital in Houston or the Mayo Clinic by

Mexican patients. A continuing issue with much of this research is the limited level of cooperation between providers in providing cross border care and the difficulty of managing emergencies that occur to individuals when they are across the border (Moss, 1992).

Over the last decade or more a number of cross border health insurance arrangements have developed to serve one or another population. It is worth detailing these and then discussing what additional initiatives might be possible to make care more accessible and decisions on where to live more rational, especially for aging persons in both Mexico and the U.S. These plans are summarized in Table 15.1.

Cross border plans were developed on an ad hoc basis in California where a number of U.S. and Mexican companies sold health plans that covered care on the Mexican side of the border to employers primarily in the San Diego area. By 1999 California required such plans to obtain a license from the California Department of Corporations, which at that time regulated managed care entities (Maguire, 2004). One of these plans, SIMNSA, obtained such approval and continues to sell policies through employers and unions, which are limited to care in Mexico with the exception that care not obtainable in Mexico and emergencies taking place in the U.S. are covered in the United States. SIMNSA owns two large multi specialty clinics in Tijuana and has contracts with several hospitals there as well (Maguire, 2004). Virtually all the people who have initially purchased SIMNSA coverage either lived in Tijuana or had family who did. In fact, it was a condition of SIMNSA's certification that it sell policies only to Mexican citizens (Lenert and Cohen, 2004).

Health Insurers who were already licensed in California had the right to develop cross border plans as long as they were sold to groups in conjunction with a dual choice plan that offered enrollees a U.S.-based option. In 1979 Blue Shield of California developed Access Baja, which offered such a dual choice plan to employers and was available to workers who lived within 50 miles of the California–Mexico border in San Diego or Imperial counties or in Baja California Norte. Blue Shield has contracted with three hospitals in Tijuana and also with the Generales Seguros network of providers. Blue Shield has also developed a product for parts of the state further away from the border whereby the employee obtains the standard Blue Shield coverage in California, but his or her dependents living in Mexico or within 50 miles of the border can choose to exclusively receive care in Mexico (Savio, 2004).

Health Net, a large health insurer based in California but operating in a number of states, has also developed several products (both PPO and HMO) where enrollees can use their affiliated clinics and hospitals in Los Angeles or the SIMNSA network in Tijuana. This is attractive for families in which some members live in Los Angeles and the remainder in Tijuana or close to the border (Maguire, 2004).

A final and longer established cross border arrangement was developed by Western Growers Association, which is a membership association of agricultural businesses in California and Arizona offering a package of benefits from which its members can choose. In 1972 Western Growers added a cross border option to

TABLE 15.1. Cross border health insurance for aging Mexicans and Mexican Americans.

Type of coverage	Plan	Characteristics	Issues and applicability
Employer coverage for employed Mexicans and Mexican Americans in the U.S. and possibly family in Mexico	1. Normal employment-based plan	Covers worker and, if affordable, the worker's spouse and children. Generally covers care in the U.S. within network and only on an emergency basis in Mexico. Growing number of Latino initiatives may develop networks for family in Mexico	Elderly relatives not covered unless spouse or the worker him/herself. Spouse and child(ren) living in Mexico are generally covered in U.S. only if there is dependent coverage and increasingly through a U.S.-based gatekeeper
	2. Mexico-only plans	Covers U.S. worker and generally family in Mexico only, except for emergencies and services not available in Mexico. Usually part of a dual choice—SIMNSA, Access Baja, and some versions of Salud con Health Net	Elderly relatives not covered unless spouse. Sold in San Diego and Imperial counties through employer groups as well as Los Angeles
	3. Choice: Mexico–U.S. plans dual	Covers U.S. worker and generally family in Mexico only, except for emergencies and services not available in Mexico. Usually part of a dual choice—SIMNSA, Access Baja and some versions of Salud con Health Net (i) Workers can choose Mexico only, U.S. only, both U.S. and Mexico for self and dependents (Salud con Health Net) (ii) Worker lives away from the border and chooses the U.S.-based plan but family lives on the border and is enrolled in Mexico-only plan (Access Baja) iii. Farm worker covered with limited benefits in U.S. but worker and family covered with greater benefits in Mexico (Western Growers)	In all of these only elderly workers and their elderly spouses are likely to be the only elderly persons covered. But these plans do exist

(Continued)

TABLE 15.1. (*Continued*)

Type of coverage	Plan	Characteristics	Issues and applicability
	4. Health savings accounts	These are increasingly an option for employees and if a family is young and not likely to incur much in the way of expenses; not a bad option	For workers on the border or whose family is in Mexico this might be an attractive option since costs are lower and expenses in Mexico may be covered while they are usually not by most employer plans. It can also be a way to generate savings
Mexican programs for Mexicans in the U.S. with families in Mexico	5. Instituto Mexicano del Seguro Sociál (IMSS) family health insurance for Mexicans abroad	Since 1990 the IMSS permits Mexicans living in the U.S. to pay an annual premium for coverage in Mexico for members of one's nuclear family including parents. There is an increasing premium based on age and limits on coverage of pre-existing conditions.	There are limits and quite a lot of red tape in purchasing it through the consulates in L.A., Houston, or Chicago. Very few have signed up and since in recent years it has become possible to sign up in Mexico directly it is possible many take this option
	6. Sistema de Protección Social in Salud (Seguro Popular)	This is a relatively new program being phased in across the states of Mexico. It permits Mexicans to buy into a package of services for a sliding premium scale with the lowest incomes paying no premium.	The government hopes that immigrants to the U.S. will buy themselves and their families into this or have employers as part of Immigration Reform do so. Immigrants might find it preferable to send the funds home directly and let family members there make their own decisions
Programs for Mexicans who wish to visit or retire to the U.S.	7. Private health insurance in Mexico or the U.S.	Private health insurance in Mexico is very difficult to acquire for older persons unless they have held it for an extended period. It is possible to buy coverage that includes coverage in the U.S. but it can be subject to limits and conditions	The private health insurance industry is limited in size in Mexico with only about $2\frac{1}{2}\%$ of expenditures and more than half of all health expenditures out of pocket.

(*Continued*)

TABLE 15.1. (*Continued*)

Type of coverage	Plan	Characteristics	Issues and applicability
Options for Mexicans and Mexican Americans living in the U.S. who wish to retire to or visit Mexico	8. Medicare and Medicaid	Neither Medicare nor Medicaid covers care in Mexico except in very limited circumstances where a Mexican hospital is the closest appropriate provider for an emergency that occurs in the U.S.	Several of the Medigap policies do cover care for insured who are traveling on trips of less than 60 days duration. One initiative worth pursing is to do a research and demonstration waiver to investigate the effect of Medicare covering care of retirees to Mexico in Mexico
Other possibilities	9. Discount cards and bi-national PPOs	Discount plans are for sale in a number of cities in the U.S. and some are beginning to include providers in Mexico	See Sekure Health Care Program for an example of a product that is being developed.

the ERISA plan coverage available to its members. Western Growers developed individual contracts with physicians, pharmacies, hospitals, and other providers in six cities in Mexico bordering California or Arizona (Shanahan, 2004). Western Growers also offers the "Mexico Panel" rider to members who wish to add it. Generally workers' families are also covered. There is also the option to cover workers for six months when they are not in the fields, but that option does not seem to be taken too often. In terms of numbers of persons covered, the Western Growers plan is the largest that offers cross border options. It was estimated that in 2002 at least 95,000 persons were covered by the Mexican panel rider (Shanahan, 2004).

None of these plans cover elderly Mexicans and Mexican Americans unless they or their spouses happen to be employed by employers who provide such coverage. What is interesting is that there is an attempt to develop commercial initiatives to take care of the needs of these bi-national families and that Mexican providers are becoming familiar to some extent with a U.S. system of insurance and the related certification and documentation requirements. Similarly, U.S. insurers are becoming far more sophisticated about the Mexican health care delivery system.

Other developing initiatives that may have some impact on utilization of services in Mexico are the wider availability of health savings accounts and discount cards. Health savings accounts have become much more widely available with the passage of the Medicare Modernization Act of 2003, which, as a condition of passage, a number of congressmen required all employers be able to offer such plans and that some other restrictions on their financing be removed. Some employees on the border may find it possible to not have to spend the savings account portion by going to lower cost providers in Mexico. IG try do decids to use the savings account portion they may need to book it will remain before tax dollars. But, many workers will not have such coverage and will not be able to purchase it for their dependents. Another growing initiative is the sale of discount cards, which offer "discounts" for those who use certain providers. One entity that has developed a network of providers in California and Mexico is Sekure Health Care.

The Mexican government has developed two plans through which workers in the U.S. can buy themselves and their family members, including parents, coverage in Mexico. One is the Instituto Mexicano de Seguro Sociál (IMSS), which is the main health insurance coverage for workers in the private formal sector who contribute to the Social Security system. This plan is administered out of the consulates in Los Angeles, Houston, and Chicago. In order to cover nuclear family members in Mexico (including parents), a Mexican living in the U.S. may pay an annual premium for each member of the family, which increases with age. According to an application form from the Houston consulate examined online in 2004, the premium for a parent or spouse 60 years or older was $256. There are limits on certain pre-existing conditions. For instance, IMSS will not cover any degenerative chronic illness or condition known or diagnosed by a physician and/or evident at the time the insurance is contracted. Insured members that are diagnosed with AIDS, HIV, diabetes or cancer in their first year at IMSS will not be treated and will lose their annual fee (Martinez, 2004). Further, there are waiting periods of one year for most scheduled surgeries that do not put one's life at risk and two years for orthopedic surgery.

The premium must be paid in a lump sum annually and there is quite a lot of documentation and correspondence involved. And each individual must be enrolled at a particular IMSS location. In the early years there was an arrangement with the United Farm Workers that did not work too well because many farm workers were from rural areas not well served by the IMSS. In recent years the number of persons enrolled has declined still further. In June 2003 the number of enrollees was 1,366 subscribers, 2009 beneficiaries, and 317 additional family members (personal communication, Daniel Valdez, January 27, 2005). The IMSS has had financial difficulties in recent years, and it does not seem to have promoted this plan aggressively. Also, since individuals in Mexico can buy into the IMSS on the same terms, it seems likely that immigrants to the U.S. would think that sending funds home for people to buy in at the location where they plan to use services makes more sense.

Another initiative that the Mexican government is currently developing is the Sistema de Protección Sociál en Salud, or "Seguro Popular," which permits Mexicans to buy into a package of services, pharmaceuticals, and procedures on a sliding fee scale in which the two lowest income deciles in the population receive the services for free. This initiative is being rolled out on a state-by-state basis. I believe there is interest in encouraging immigrants to the U.S. to buy themselves and their dependents into this plan. Resources, however, are limited, and as of six months ago I understand there had been limited buy-in from citizens in Mexico and the U.S. who are required to pay the premiums to any extent. This may have been because of the limited services available through the public health system clinics in the past. In 2003 out of pocket direct expenditure for health care in Mexico accounted for 51% of total expenditure (OECD, 2003). This was primarily for drugs and services in the private sector since the public systems have small, if any, co-pays.

Elderly Mexicans and Mexican Americans who do have Medicare coverage and/or Medicaid coverage cannot use it in Mexico since services there are not covered to any significant extent. It is possible to purchase Medigap coverage, which will pay for emergency medical needs during trips of less than 60 days duration. One initiative that could be helpful to elderly Mexican-origin retirees who are eligible for Medicare or Medicaid would be availability of coverage for them in Mexico after retiring to Mexico, and if developed properly such initiatives could substantially improve the standard of living of many, while possibly reducing publicly funded medical and social welfare costs in the United States. We have studied this issue extensively (Warner, Jahnke, 2001) and over the next two years hope to develop the basis for a waiver application to Center for Medicarr and Medicaid services (CMS) to at least cover retirees under Medicare on some basis in one or two retiree locations such as Jalisco, where it might be possible to pick up more affluent non Latino retirees in the Lake Chapala area while also covering a higher percentage of Latino retirees in Guadalajara and towns in other parts of the state. The idea would be that after such a waiver were in place and evaluated for several years, the reform might be made permanent. Because of increased numbers of persons choosing to retire abroad from the countries in the Organization for Economic Cooperation & Development (OECD), lack of portability of medical entitlements may become a major block to such movement and keep needed resources from the medical care systems of the receiving countries.

Our prospective evaluation will need to examine whether such coverage will lead to improved quality of care and access to care for beneficiaries while not costing additional funds. Consequently, if such a waiver were to be granted there would be a good deal of work to be done to assure that these conditions were achievable. We have worked through some of the dimensions of such a process already (Policy Research Project, Medicare Payment, 1999). One point to be made is that such a benefit might not only provide less expensive medical care than if such retirees had to return to the United States for medical care, but also that a number of people probably do not retire to Mexico because medical care there is not covered by Medicare. Finally, if some home- and community-based payments were available in lieu of Medicaid for persons who would qualify for Medicaid coverage and nursing home placement in the U.S., then much more preferable and much less costly options might be chosen by many in Mexico. Elsewhere I have detailed how perhaps the federal government could provide such coverage at less than half the cost of normal coverage, thus saving money and taking the state out of the equation entirely or at least substantially (Warner, 1991). It is noteworthy that the Government of the Philippines has launched an initiative to lobby for Medicare to cover Filipinos who are beneficiaries back in the Philippines.

In conclusion, as the integration of North America continues rapidly on a number of fronts it has become increasingly important to integrate or at least coordinate service entitlements and standards for the provision of services. To date, adjustments in the area of health services have been particularly parochial, limited, and generally shortsighted. In this context, three overriding issues are as follows:

1. There is little integration between the two health delivery systems. NAFTA notwithstanding, health insurance and entitlements are generally not portable between the two countries, the right to practice medicine is not portable, and there are major barriers to becoming licensed in the other nation (Warner, 2003). Pharmaceuticals are regulated and priced very differently in the two countries, health professional training is quite different, and there is little communication between medical providers or health professional schools in the two countries.

2. There is a need to make medical services available to immigrants from Mexico and their families in the United States and in Mexico's sending communities . Another high priority is the development of a better functioning public health system on the U.S.–Mexico border and between the two countries.

3. Finally, Mexico must find a way to modernize its health system while extending health services to more citizens without short changing other vital needs. And the U.S. is under increasing pressure to develop a system of assurance to health care that provides broader protection without putting so many of its citizens at severe financial risk.

Initiatives to develop increased availability of cross border health insurance for elderly Mexican Americans and Mexicans on a number of fronts are extremely appropriate steps in beginning to address some of these issues.

References

Arrendondo-Vega, J.A. (1999). The use of Mexican private medical services by American nationals in the border city of Tijuana. Doctoral dissertation, University of London.

Guendelman, S., and Jasis, M. (Nov–Dec 1990). Measuring Tijuana residents' choice of Mexican or U.S. health Services. *Public Health Reports, Vol. 105(6)*, 575–579.

Homedes, N., Sosa, C., F., Nichols, A., Otalora-Solar, M., Le Brec, P., Vazquez, L.-A.(Nov–Dec 1994). Utilization of health services along the Arizona–Sonora border, *Salúd Pública de México, Vol. 36(6)*, 633–645.

Lenert, A.J., and Cohen, Y. (2004). California legislation and regulation. In D. Warner and P. Schneider (Proj. Dirs.), *Cross border health insurance: Options for Texas* (p. 20). U.S. Mexican Policy Series, Policy Report No. 12, LBJ School of Public Affairs, University of Texas at Austin.

Maguire, R.C. (2004). SIMNSA and Salúd con Health Net. In D. Warner and P. Schneider (Proj. Dirs.), *Cross border health insurance: Options for Texas* (49–66). U.S. Mexican Policy Series, Policy Report No. 12, LBJ School of Public Affairs, University of Texas at Austin.

Martinez, C. (2004). Appendix J. In D. Warner and P. Schneider (Proj. Dirs.), *Cross border health insurance: Options for Texas* (p. 310). U.S. Mexican Policy Series, Policy Report No. 12, LBJ School of Public Affairs, University of Texas at Austin.

Moss, K.D. (1992). *Improving patient transfers across the U.S. –Mexico border.* MA thesis, LBJ School of Public Affairs, University of Texas at Austin.

OECD. (2005). *Reviews of health systems—Mexico.* Figure 1.1. The health system in Mexico, main flows (p. 22), Paris, Organization for Economic Co-operation & Development.

Policy Research Project on Medicare Payment for Medical Services in Mexico. (1999). Three technical papers on a research and demonstration waiver for Medicare coverage in Mexico. U.S.–Mexican Policy Studies Program, Occasional Paper No. 7, LBJ School of Public Affairs, University of Texas at Austin.

Potter, J., Moore, A., and Byrd, T. (2003) Cross-border procurement of contraception: Estimates from a postpartum survey in El Paso, Texas. *Contraception, 68(4),* 281–287.

San Diego Dialogue. (1994). Crossing the border for medical or dental services. San Diego Dialogue, Border Fact Sheet, No. 30.

Savio, P. (2004). Blue Shield Access Baja. In D. Warner and P. Schneider (Proj. Dirs.), *Cross border health insurance: Options for Texas* (p. 35–48). U.S. Mexican Policy Series, Policy Report No. 12, LBJ School of Public Affairs, University of Texas at Austin.

Shanahan, K. (2004). Western Growers Association. In D. Warner and P. Schneider (Proj. Dirs.), *Cross border health insurance: Options for Texas* (p. 66–77). U.S. Mexican Policy Series, Policy Report No. 12, LBJ School of Public Affairs, University of Texas at Austin.

U.S. Census Bureau. (April 20, 2005). *Facts for features: Cinco de Mayo 2005.* Retrieved September 12, 2005, from http://www.census.gov/Press-Release/www/releases/archives/facts_for_features_special_editions/004707.html.

Warner, D.C. (Proj. Dir.). (1999). *Getting what you paid for: Extending Medicare to eligible beneficiaries in Mexico.* Austin, Texas: U.S.–Mexican Policy Series, Policy Report No. 9, LBJ School of Public Affairs, University of Texas at Austin.

Warner, D.C. (1991). Mexican provision of health and human services to American citizens: Barriers and opportunities. In S. Diaz-Briquets and S. Weintraub (Eds.), *Regional and sectoral development in Mexico as alternatives to migration* (pp. 133–154). Series of Development and International Migration in Mexico, Central America, and the Caribbean Basin, Volume II, 1st edition, Boulder: Westview Press.

Warner, D.C. (Proj. Dir.). (1997). *NAFTA and trade in medical services between the U.S. and Mexico.* Austin, Texas: U.S.–Mexican Policy Series, Policy Report No. 7, LBJ School of Public Affairs, University of Texas at Austin.

Warner, D.C. and Jahnke, L. (Jan–Feb, 2001). Toward better access to health insurance coverage for U.S. retirees in Mexico. *Salúd Pública de México, Vol. 43(1),* 59–66.

Zinnecker, A.K. (1990). *Health along the Texas–Mexico Border: Insights on the utilization of health services by Mexican nationals.* Professional Report, LBJ School of Public Affairs, University of Texas at Austin.

16
Cultural Myths and Other Fables About Promoting Health in Mexican Americans: Lessons Learned from Starr County Border Health Intervention Research

Sharon A. Brown

16.1. Introduction

In 2000, 35 million Hispanics resided in the U.S., an increase of 50% since 1990; 60% were of Mexican origin. By 2035, the Hispanic population is expected to comprise 20% of the U.S. population and 60% of the Texas population, making it the largest minority group (U.S. Census Bureau, 2001). Twenty-one percent of the U.S. population lives in border states. More than a third of these individuals, mostly minorities, live in medically underserved border communities characterized by poverty, pollution, and deprivation (U.S. Department of Health and Human Services, 2003). Starr County, the site of our previous diabetes intervention studies, is a Texas–Mexico border community that is the poorest county in Texas and one of the poorest in the U.S. (Texas Department of Health (TDH), 2002b). Starr County is representative of many border areas: The unemployment rate is 22%, almost six times that of the rest of the state. The ratio of population per general/family practice MD is 7,657:1, compared to 3,789:1 for the rest of the state; the registered nurse ratio is 851:1, compared to 159:1 for the state (TDH, 2002b).

Type 2 diabetes affects more than 16 million people nationwide; costs $132 billion annually; and affects 14% of Hispanic adults, ranking as the 4th leading cause of death in females and the 8th in males (American Diabetes Association (ADA), 2003; Diabetes Research Working Group, 1999). Increasing obesity among Mexican Americans and other groups will make diabetes the most prevalent health problem of the future (Burke, Williams, Gaskill, Hazuda, Haffner, and Stern, 1999). In Texas Hispanics, diabetes is currently responsible for 7% of total deaths, two to three times the rate for non-Hispanic Whites. By the age of 60, 50% of Mexican

The Second Aging in the Americas Conference, "Key Issues in Hispanic Health and Health Care Policy Research," September 21–22, 2005, LBJ School of Public Affairs, The University of Texas at Austin.

Americans have diagnosed diabetes, undiagnosed diabetes, or impaired fasting glucose (American Diabetes Association (ADA), 2001). Hispanics not only tend to be diagnosed at younger ages but they also exhibit higher fasting blood glucose (FBG) levels, lower insulin sensitivity and higher insulin response, and higher rates of diabetes complications, e.g., retinopathy and peripheral vascular disease, than non-Hispanic Whites (ADA, 2001; Black, Ray, and Markides, 1999; Harris, Klein, Cowie, Rowland, and Byrd-Holt, 1998). For example, in our previous Starr County studies, mean baseline glycosylated hemoglobin (HbA_{1c}) levels have consistently been at 12% (Brown, Blozis, Kouzekanani, Garcia, Winchell, and Hanis, 2005; Brown, Garcia, Kouzekanani, and Hanis, 2002); and it is estimated that HbA_{1c} levels greater than 10% are associated with three-year health care costs that are 11% higher than HbA_{1c} levels below 6% (Gilmer, O'Connor, Rush, Grain, Whitebird, Hanson, and Solberg, 2005).

Traditional approaches to managing diabetes with this population have often been culturally insensitive and ineffective. A recent review of the research literature on behavioral interventions employed with diabetes indicated that typical programs are community-based. Although studies of minority populations are increasing, there continue to be few research studies conducted with Mexican Americans in particular (Brown, Garcia, and Winchell, 2002) (see Table 16.1). Since 1988 we have developed and tested effective culturally competent, community-based diabetes self-management interventions in Starr County offered in Spanish by bilingual Mexican American nurses, dietitians, and community workers from the border area. During the early focus groups that we held in Starr County in preparation for these studies and also during the conduct of these studies, we identified a number of cultural barriers that interfere with health promotion programs in Mexican American communities (Benavides-Vaello, Garcia, Brown, and Winchell,

TABLE 16.1. Summary of recent research findings on effects of behavioral strategies with minority groups*

Types of behavioral strategies**	Research findings
Knowledge + behavioral strategy	Number of ethnic minorities participating in studies remains low
Empowerment	Number of studies involving minority groups has increased
Support groups	Mexican Americans/Hispanics tend to be the least studied group
Counseling	Focus groups are a major strategy
Motivation	Few minority health professionals have been prepared
Problem solving	Most behavioral programs for minority groups are
Contracting	community-based
	Research suggests that frequent contact and follow-up is necessary to achieve improved health outcomes
	Family involvement and social support are key factors in improving health
	Most studies focus on physiologic outcomes
	Community demands may dictate characteristics of the behavioral intervention and of the strategies employed

Source: Brown, S.A., Garcia, A.A., and Winchell, M. (2002). Reaching underserved populations and cultural competence in diabetes education. *Current Diabetes Reports, 2,* 166–176.
*All interventions achieved statistically significant effects.

2004). Our experience has demonstrated that many of these suspected barriers are based upon cultural myths held by many individuals, including both Mexican American and non-Mexican American health care providers.

The purpose of this paper is to provide evidence from our Starr County studies, as well as from other research, that many of these commonly held "barriers" represent cultural stereotypes that become "self-fulfilling prophesies." These perceived "barriers" limit, if not completely undermine, the well-intentioned efforts of everyone involved in improving the health of Mexican Americans with diabetes. The cultural myths discussed below represent barriers that are either non-existent or that can be overcome with appropriate, culturally competent interventions.

16.2. Cultural Myths

16.2.1. Recruitment and Retention of Mexican Americans in Research Studies

A commonly held notion is that Mexican Americans, as well as other minority populations, are difficult, if not impossible, to recruit and retain as participants in research studies. However, during recruitment into our Starr County intervention studies, less than 5% of the potential subjects who were contacted declined to participate. And we were able to achieve an overall average data collection retention rate of 80% to 90% across all of our studies (Brown et al., 2005; Brown, Garcia, et al., 2002).

To achieve this level of recruitment and retention, we used a comprehensive, culturally competent approach. Initial contacts with subjects were made by telephone by bilingual staff employed in the Research Field Office in the targeted community. Communication was in the preferred language, most often Spanish. The staff who recruited study participants and scheduled appointments for data collection used a non-judgmental approach towards individuals' ability to make and keep appointments and to follow recommended health behaviors. Flexible scheduling allowed individuals to come by the office in the morning before work to participate in the portion of the data collection requiring fasting blood specimens and return later in the day to complete the remainder of the session involving questionnaires. In addition, the staff accepted the possibility of missed appointments and, when necessary, was prepared to reschedule at the participant's convenience. In some instances, home visits were arranged when study participants did not have transportation to the Research Field Office. Several telephone reminders were made the week before the scheduled appointment. Transportation to all data collection sessions was provided when necessary. The staff was trained to conduct efficient data collection processes so each subject spent only one and one half hours at each session. Laboratory results and graphs of downloaded home glucose monitoring data were provided as immediate feedback during the exit interviews at the completion of the data collection session. Healthy snacks of Mexican American foods were available for individuals to eat after their fasting blood samples were drawn. Thus, carefully designed recruitment of potential subjects and scheduling of data collection sessions enabled excellent recruitment and retention of study participants, thus demonstrating that Mexican Americans will participate in research if attention is given to their specific needs and cultural issues.

16.2.2. Genetic Predisposition for Type 2 Diabetes

Native American genetic admixture has been linked to type 2 diabetes in Mexican Americans (ADA, 2001; Cox, Frigge, Nicolae, Concannon, Hanis, Bell, and Kong, 1999; McDermott, 1998; Hanis et al., 1996). The current estimates are that approximately 30% of the Mexican American gene pool derives from Native American sources (Hanis, Hewett-Emmett, Bertin, and Schull, 1991). The three-fold increased risk of diabetes in siblings of the Mexican Americans in Starr County who have diabetes is consistent with a genetic effect on diabetes prevalence (Horikawa et al., 2000; Hanis et al., 1996). The current diabetes epidemic, however, is not due to genetic influences alone because genetic modification is an extremely slow process (Bamshad, 2005). The current diabetes crisis is more likely due to environmental factors, many of which are modifiable, such as sedentary lifestyles, low socioeconomic status, barriers to accessing health care, poor diet, and lack of health education (Jankowski, Ben-Ezra, Kendrick, Moriss, and Nichols, 1999).

We have demonstrated that a non-pharmacological, culturally competent diabetes self-management intervention was able to overcome the genetic effects and improved glycemic control of impoverished Mexican American border residents with type 2 diabetes (Brown, Garcia, et al., 2002; Brown et al., 2005). In the original study reported in 2002, comparing a one-year intervention with a wait-listed control group, we showed a 1.4%-age point lower HbA_{1c} at six months in the experimental group compared to the control group. In the next study, we found a decrease in two interventions, "compressed" (22 contact hours) and "extended" (52 contact hours), with 0.7%-age point and 1.0%-age point reductions in HbA_{1c}, respectively. There was a significant "dosage effect" of attendance in the most intensive "extended" intervention, with the largest reductions in HbA_{1c}—a 1.0%-age point reduction—achieved by those who attended at least 50% of the sessions. The top 10% of the attendees achieved an average 6%-age point HbA_{1c} reduction. At twelve months, HbA_{1c} reductions achieved by the less intensive "compressed" group had diminished (Brown et al., 2005).

National trials have demonstrated that for every 1.0%-age point reduction in HbA_{1c}, significant reductions are achieved in microvascular and cardiovascular complications. Clearly, with health interventions designed specifically for the culture, health improvements can be attained and genetic influences can be overcome.

16.2.3. "Fatalism"

"Fatalism" is a belief that diabetes is a punishment for past shortcomings and can prevent individuals from assuming personal responsibility for their health. We have conducted numerous focus group interviews with diabetic residents of Starr County and with key informants—nurses, dietitians, physicians, and county health workers. We learned that by a number of health beliefs held by the residents of this community, one overriding belief was that one could not "control" diabetes. The high diabetes prevalence had contributed to a perception that everyone would eventually "get" the disease. During data collection sessions research staff asked participants if they had been diagnosed with diabetes; common answers were "yes" or "not yet" (Brown, Becker, Garcia, Barton, and Hanis, 2002).

Little research has been conducted on health beliefs such as "fatalism" and their relationships to health outcomes. A meta-analytic synthesis of the literature involving causal modeling of health beliefs as determinants of diabetes outcomes found that health beliefs have "... direct and indirect effects on diabetes metabolic control ..." (Brown and Hedges, 1994; Brown, Becker, et al., 2002); and beliefs about the ability to "control" the disease predicted levels of fasting blood glucose and HbA$_{1c}$ in subjects (Surgenor, Horn, Hudson, Lunt, and Tennent, 2000; Watkins, Connell, Fitzgerald, Klem, Hickey, and Ingersoll-Dayton, 2000). The few studies that have been conducted on health beliefs in minority populations demonstrated that control and social support were particularly important (Fitzgerald, Gruppen, Anderson, Funnell, Jacober, Grunberger, and Aman, 2000). In Starr County, analyses of health belief data showed that a belief of "control" over diabetes consistently predicted HbA$_{1c}$ levels (Brown, Becker, et al., 2002). Consequently, strategies aimed at instilling a sense of "control" over diabetes through knowledge of the disease and effective self-management strategies significantly improve health, in particular, metabolic control.

16.2.4. Gender Roles

Traditional gender roles can influence health beliefs and the effects of lifestyle interventions directed at improving diabetes. In traditional societies, women in the household are expected to prepare meals for their family and care for ill family members, sometimes disregarding their own dietary and health needs. In our studies of Mexican Americans, we have found that males, compared to females, expressed stronger perceptions of *control* over their diabetes ($F = 4.1$, $p = 0.05$) and *social support for diet* ($F = 6.1$, $p = 0.01$) (Brown, Becker, et al., 2002). Men, on average, had lower HbA$_{1c}$ levels ($t = -3.11$, $p = 0.002$), and those men who had greater attendance at lifestyle educational programs achieved greater improvements in HbA$_{1c}$ levels (Brown et al., 2005). However, in all instances both males and females achieved significantly improved health outcomes as a result of lifestyle interventions. Specific attention to differential gender needs should be addressed in future research and clinical programs to enable all individuals to achieve the maximum health benefits.

16.2.5. Acculturation and Dietary Practices of Mexican Americans

Traditional interventions, particularly with Hispanics, are sometimes provided in the wrong language and may be based on incorrect and/or insensitive assumptions about the cultural characteristics of the population. Minority populations have frequently been labeled "noncompliant" and received different, inadequate, or inappropriate diabetes treatment (Brown, Garcia, et al., 2002; Haffner, Fong, Stern, Pugh, Hazuda, Patterson, van Heuven, and Klein, 1988). In some instances, health care providers may have misadvised Mexican Americans by recommending that they discard cultural lifestyles, food preferences for example, for a healthier Anglo lifestyle (Alcozer, 2000). This recommendation is not only culturally insensitive but also factually wrong. Past studies of Alaskan Natives, for example, have documented that while this population has a high rate of diabetes, returning them to their traditional cultural diet of fish and marine mammals significantly improved glucose intolerance and insulin resistance (Ebbesson, Kennish, Ebbesson, Go, and Yeh, 1999). Since diet can be considered the cornerstone of diabetes treatment, dietary recommendations must be accurate and based upon recognition of cultural and social factors.

16.2.6. Weight Loss and Type 2 Diabetes

Obesity is a common pathway for type 2 diabetes in all populations but is exacerbated in U.S. minority populations, particularly along the Texas—Mexico border, where higher rates of overweight/obesity and lower rates of physical activity have been observed (TDH, 2002a; Ravussin and Bogardus, 2000). In the Diabetes Prevention Program (Diabetes Prevention Program Research Group, 2003), moderate levels of physical activity, that is, walking 30 minutes per day, and losing 5% to 7% of body weight through reduced caloric intake, resulted in the largest reduction in diabetes risk (58%). Our experience in Starr County with individuals already diagnosed with type 2 diabetes, as well as the experience of other researchers, has documented that, while weight loss is the most difficult health goal to achieve, a moderate level of weight loss (a 5% to 10% reduction) and increasing physical activity improved insulin sensitivity and glycemic control (Bunt, Hanson, Salbe, Tataranni, and Harper, 2003; Brown et al., 2002; Wing and Jeffery, 1995). A meta-analysis of studies on types of interventions designed for promoting weight loss in diabetes and subsequent reductions in HbA$_{1c}$ indicated that diet was the most effective strategy (Brown, Upchurch, Anding, Winter, and Ramirez, 1996), resulting in an average weight loss of 20 pounds and an HbA$_{1c}$ reduction of 2.4%-age points. The recommended diet is a low-calorie diet that results in a deficit of 500 to 1000 kilocalories per day (National Institutes of Health, 1998). A complicating factor when providing lifestyle interventions for Mexican Americans is the need to adapt recommendations for healthy eating, as well as for increasing physical activity, to cultural preferences and norms in order to be effective with this population (Brown et al., 2005; Brown, Garcia, et al., 2002).

16.2.7. Community Health Workers ("Promotoras")

A common trend in the health care of Mexican Americans who live in rural and/or border communities is to engage community health workers ("promotoras") to fill the health care gap in these medically underserved communities. With such a dearth of health care providers in Starr County, we originally intended to employ community workers to conduct support group sessions; however, study participants preferred to have a nurse or dietitian—a perceived authority figure—directing each intervention session. So, we adjusted our approach and engaged community workers to provide organizational services for the project, rather than directing the sessions. We employed eight community workers who met the following criteria: (1) high school graduate (minimum); (2) bilingual; (3) resident of Starr County, Texas; (4) licensed to drive; and (5) diagnosed with type 2 diabetes (preferable). Community workers were responsible for arranging intervention sites, contacting patients and their families weekly, organizing equipment and supplies, providing transportation when necessary, assisting dietitians with food preparation, distributing home glucose monitors and monitoring supplies, and managing attendance records and home glucose monitoring activities. As county residents, each had knowledge of and relationships with persons with diabetes and their families and provided important linkages with the local Mexican American community. The major lesson learned is that researchers must assess communities to determine if community workers are appropriate as health care extenders rather than assume that these workers will be automatically accepted. The key factor is the specific role of the community workers; that is, perhaps they will be accepted as linkages to the community but not as substitutes for health care providers.

16.3. Discussion

Diabetes constitutes a severe health and economic threat for Mexican Americans, who are now younger than other ethnic groups. As the Mexican American population grows and ages, diabetes will become an even more serious public health problem (Burke et al., 1999). We have demonstrated in Starr County research that a culturally competent behavioral intervention could improve glycemic control sufficient enough to significantly reduce morbidity and mortality. The strategies we have employed were designed to address the cultural characteristics of the population and thus have been effective in overcoming cultural myths that had been promulgated as barriers to improving health in Mexican Americans with diabetes. The significance of these research findings cannot be overstated because strategies commonly employed with persons diagnosed with diabetes have been designed for White populations and have been relatively ineffective with minority groups.

Discussions of diabetes self-management interventions have always led to concerns about costs and the belief that the low rates of compliance with lifestyle programs are not worth the associated costs. Typical diabetes lifestyle programs range from 4 to 15 hours of self-management education and behavioral intervention over a two- to three-month time period and cost between $95 and $125 per one-hour session; group instruction costs slightly less (Braiotta, 2004). The most intensive Starr County intervention ("extended") was considerably more intensive than typical programs—52 hours over a 12-month time period—and cost significantly less, $7.39 per person per hour. In all of our intervention programs, the cost was estimated to be less than that associated with a year's prescription of a single medication; it is not excessive when the costs of diabetes morbidity and mortality are considered (Brown et al., 2005).

In spite of the fact that there is a widespread recognition of the increased risk for diabetes of Mexican Americans, that is, diabetes constitutes a major health disparity between Whites and minority populations such as Mexican Americans, there remains a paucity of diabetes research conducted with Mexican Americans. The Starr County studies, and a few other similar studies conducted in other Mexican American communities, are significant steps in the overall plan to develop effective culturally competent diabetes self-management strategies that can be translated into other border and non-border Mexican American communities. Meta-analyses have supported the positive effects of diabetes self-management interventions (Brown, 1988, 1990, 1992). Current self-management interventions are designed according to National Standards and involve two major components: 1) education regarding the day-to-day self-management of diabetes involving topics of home glucose monitoring, medications, dietary principles, physical activity, symptom management; and 2) behavioral strategies to foster incorporation of the recommended lifestyle changes into individuals' daily activities. In the Diabetes Control and Complications Trial (DCCT, 1993), UK Prospective Diabetes Study (UKPDS Group, 1998), and Diabetes Prevention Program (Diabetes Prevention Program Research Group (DPP), 2003), diabetes self-management education resulting in lifestyle changes and intensive medical treatment resulting in tight glucose control delayed diabetes onset or reduced morbidity. "... [T]he incidence of microvascular and macrovascular endpoints of diabetes at increasing HbA_{1c} ... showed a log-linear relationship, indicating any reduction in glycemia ... would be advantageous ... " (UKPDS Group, 1998; DCCT, 1993). As stated above, for every 1%-age point reduction in HbA_{1c}, rates of diabetes complications decreased by 30 to 75 percent. Starr County research goals are in accordance with priority objectives of Healthy People 2010 and address *the* major public health problem of minority populations, especially along the U.S.-Mexico border (U.S. Department of Health and Human Services (DHHS), 2000). Achieving positive long-term results with this population that has few personal resources is essential for addressing the growing diabetes prevalence in Mexican Americans wherever they reside.

In conclusion, interventions designed to address cultural myths and barriers are effective and not excessively costly when compared to other therapeutic strategies. Interventions that are not culturally competent are ineffective and thus contribute to myths of "fatalism," lack of control over health, and unwillingness of minority populations to participate in lifestyle research and clinical programs. Researchers and clinicians must recognize that there may be as much heterogeneity within a culture as there is between cultures and resist labeling individuals based on stereotyped views (Bamshad, 2005; Brown et al., 2002). Future research must address remaining barriers and address variables that explain within-culture differences in responsiveness to behavioral interventions. Such variables include gender, cultural roles, and differences in socioeconomic status, to name a few.

References

Alcozer, F. (2000). Secondary analysis of perceptions and meanings of type 2 diabetes among Mexican American women. *Diabetes Educator, 26*, 785–795.

American Diabetes Association. (2003). Economic costs of diabetes in the U.S. in 2002. *Diabetes Care, 26*, 917–932.

American Diabetes Association. (2001). Implications of the United Kingdom Prospective Diabetes Study (UKPDS). *Diabetes Care, 24 (Suppl.)*, S28–S32.

Bamshad, M. (2005). Genetic influences on health: Does race matter? *JAMA, 294*, 937–946.

Benavides-Vaello, S., Garcia, A.A., Brown, S.A., and Winchell, M. (2004). Using focus groups to plan and evaluate diabetes self-management interventions for Mexican Americans. *Diabetes Educator, 30*, 238, 242–244, 247–250, 252, 254, 256.

Black, S.A., Ray, L.A., and Markides, K.S. (1999). The prevalence and health burden of self-reported diabetes in older Mexican Americans: Findings from the Hispanic established populations for epidemiologic studies of the elderly. *American Journal of Public Health, 89*, 546–552.

Braiotta, R. (2004). Diabetes education programs are essential. *National Federation of the Blind, Voice of the Diabetic Quarterly Newspaper, August 2004*. Retrieved November 13, 2004 from http://www.nfb.org/vod/vfal9912.htm

Brown, S.A. (1992). Meta-analysis of diabetes patient education research: Variations in intervention effects across studies. *Research in Nursing and Health, 15*, 409–419.

Brown, S.A. (1990). Studies of educational interventions and outcomes in diabetic adults: A meta-analysis revisited. *Patient Education and Counseling, 16*, 198–215.

Brown, S.A. (1988). Effects of educational interventions in diabetes care: A meta-analysis of findings. *Nursing Research, 37*, 223–230.

Brown, S.A., Blozis, S.A., Kouzekanani, K., Garcia, A.A., Winchell, M., and Hanis, C.L. (2005). Dosage effects of diabetes self-management education for Mexican Americans: The Starr County Border Health Initiative. *Diabetes Care, 28*, 527–532.

Brown, S.A., Becker, H.A., Garcia, A.A., Barton, S.A., and Hanis, C.L. (2002). Measuring health beliefs in Spanish-speaking Mexican Americans with type 2 diabetes: Adapting an existing instrument. *Research in Nursing and Health, 25*, 145–158.

Brown, S.A., Garcia, A.A., Kouzekanani, K., and Hanis, C.L. (2002). Culturally competent diabetes self-management education for Mexican Americans: The Starr County Border Health Initiative. *Diabetes Care, 25*, 259–268.

Brown, S.A., Garcia, A.A., and Winchell, M. (2002). Reaching underserved populations and cultural competence in diabetes education. *Current Diabetes Reports, 2*, 166–176.

Brown, S.A., Upchurch, S., Anding, R., Winter, M., and Ramirez, G. (1996). Promoting weight loss in type II diabetes. *Diabetes Care, 19*, 613–624.

Brown, S.A., and Hedges, L.V. (1994). Predicting metabolic control in diabetes: A pilot study using meta-analysis to estimate a linear model. *Nursing Research, 43*, 362–368.

Bunt, J.C., Hanson, R.L., Salbe, A.D., Tataranni, P.S., and Harper, I.T. (2003). Weight, adiposity, and physical activity as determinants of an insulin sensitivity index in Pima Indian children. *Diabetes Care, 26*, 2524–2530.

Burke, J.P., Williams, K., Gaskill, S.P., Hazuda, H.P., Haffner, S.M., and Stern, M.P. (1999). Rapid rise in the incidence of type 2 diabetes from 1987 to 1996: Results from the San Antonio Heart Study. *Archives of Internal Medicine, 159*, 1450–1456.

Cox, N.J., Frigge, M., Nicolae, D.L., Concannon, P., Hanis, C.L., Bell, G.I., and Kong, A. (1999). Loci on chromosomes 2 (NIDDM1) and 15 interact to increase susceptibility to diabetes in Mexican Americans. *Nature Genetics, 21*, 213–215.

DCCT Research Group. (1993). The effect of intensive treatment of diabetes on the development and progression of long-term complications in insulin-dependent diabetes mellitus. *New England Journal of Medicine, 329,* 977–986.

Diabetes Prevention Program Research Group (DPP). (2003). Within-trial cost-effectiveness of lifestyle intervention or metformin for the primary prevention of type 2 diabetes. *Diabetes Care, 26,* 2518–2523.

Diabetes Research Working Group. (1999). *Conquering diabetes: A strategic plan for the 21st century.* Bethesda, MD: National Institutes of Health.

Ebbesson, S.O., Kennish, J., Ebbesson, L., Go, O., and Yeh, J. (1999). Diabetes is related to fatty acid imbalance in Eskimos. *International Journal of Circumpolar Health, 58,* 108–119.

Fitzgerald, J.T., Gruppen, L.D., Anderson, R.M., Funnell, M.M., Jacober, S.J., Grunberger, G., and Aman, L.C. (2000). The influence of treatment modality and ethnicity on attitudes in type 2 diabetes. *Diabetes Care, 23,* 313–318.

Gilmer, T.P., O'Connor, P.J., Rush, W.A., Grain, A.L., Whitebird, R.R., Hanson, A.M., and Solberg, L.I. (2005). Predictors of health care costs in adults with diabetes. *Diabetes Care, 28,* 59–64.

Haffner, S.M., Fong, D., Stern, M.P., Pugh, J.A., Hazuda, H.P., Patterson, J.K., van Heuven, W.A.I., and Klein, R. (1988). Diabetic retinopathy in Mexican Americans and non-Hispanic Whites. *Diabetes, 37,* 878–884.

Hanis, C.L., Boerwinkle, E., Chakraborty, R., Ellsworth, D.L., Concannon, P., Stirling, B., Morrison, V.A., Wapelhorst, B., Spielman, R.S., Gogolin-Ewens, K.J., Shepard, J.M., Williams, S.R., Risch, N., Hinds, D., Iwasaki, N., Ogata, M., Omori, Y., Petzold, C., Rietzch, H., Schroder, H.E., Schulze, J., Cox, N.J., Menzel, S., Voriraj, V.V., Chen, X., et al. (1996). A genome-wide search for human non-insulin-dependent (type 2) diabetes genes reveals a major susceptibility locus on chromosome 2. *Nature Genetics, 13,* 161–166.

Hanis, C.L., Hewett-Emmett, D., Bertin, T.K., and Schull, W.J. (1991). Origins of U.S. Hispanics. Implications for diabetes. *Diabetes Care, 14,* 618–627.

Harris, M.I., Klein, R., Cowie, C.C., Rowland, M., and Byrd-Holt, D.D. (1998). Is the risk of diabetic retinopathy greater in non-Hispanic Blacks and Mexican Americans than in non-Hispanic Whites with type 2 diabetes? A U.S. population study. *Diabetes Care, 21,* 1230–1235.

Horikawa, Y., Oda, N., Cox, N.J., Li, X., Orho-Melander, M., Hara, M., Hinokio, Y., Lindner, T.H., Mashima, H., Schwarz, P.E., del Bosque-Plata, L., Horikawa, Y., Oda, Y., Yoshiuchi, I., Colilla, S., Polonsky, K.S., Wei, S., Concannon, P., Iwasaki, N., Schulze, J., Baier, L.J., Bogardus, C., Groop, L., Boerwinkle, E., Hanis, C.L., and Bell, G.I. (2000). Genetic variation in the gene encoding *calpain*-10 is associated with type 2 diabetes mellitus. *Nature Genetics, 26,* 163–175.

Jankowski, C., Ben-Ezra, V., Kendrick, K., Moriss, R., and Nichols, D. (1999). Effect of exercise on postprandial insulin responses in Mexican American and non-Hispanic women. *Metabolism: Clinical and Experimental, 48,* 971–977.

McDermott, R. (1998). Ethics, epidemiology and the thrifty gene: Biological determinism as a health hazard. *Social Science in Medicine, 47,* 1189–1195.

National Institutes of Health. (1998). *Clinical guidelines on the identification, evaluation, and treatment of overweight and obesity in adults: The evidence report.* NIH publication No. 98–4083.

Ravussin, E., and Bogardus, C. (2000). Energy balance and weight regulation: Genetics versus environment. *British Journal of Nutrition, 83,* S17–20.

Surgenor, L.J., Horn, J., Hudson, S.M., Lunt, H., and Tennent, J. (2000). Metabolic control and psychological sense of control in women with diabetes mellitus. Alternative considerations of the relationship. *Journal of Psychosomatic Research, 49,* 267–273.

Texas Department of Health. (2002a). The health of Texans: Texas state strategic health plan. Austin, TX: Texas Dept. of Health.

Texas Department of Health. (2002b). *Selected facts for Starr County—2000.* Austin, TX: Texas Dept. of Health.

UKPDS Group. (1998). Intensive blood glucose control with sulphonylureas or insulin compared with conventional treatment and risk of complications in patients with type 2 diabetes (UKPDS 33). *Lancet, 352,* 837–853.

U.S. Department of Commerce Census Bureau. (2001). *The Hispanic population. Census 2000 brief.* Retrieved June 28, 2005 from http://www.census.gov/prod/2001pubs/c2kbr01-3.pdf

U.S. Department of Health and Human Services. (2000). Healthy People 2010. Available at: http://www.healthypeople.gov. Accessed June 28, 2005.

U.S. Department of Health and Human Services (2003). *HRSA Facts about US/Mexico Border Health.* Retrieved December 28, 2003 from http://bphc.hrsa.gov/bphc/borderhealth/

Watkins, K.W., Connell, C.M., Fitzgerald, J.T., Klem, L., Hickey, T., and Ingersoll-Dayton, B. (2000). Effect of adults' self-regulation of diabetes on quality-of-life outcomes. *Diabetes Care, 23,* 1511–1515.

Wing, R.R., and Jeffery, R.W. (1995). Effect of modest weight loss on changes in cardiovascular risk factors: Are there differences between men and women or between weight loss and maintenance? *International Journal of Obesity, 19,* 67–73.

17
Health Insurance Coverage and Health Care Utilization along the U.S.−Mexico Border: Evidence from the Border Epidemiologic Study on Aging*

Elena Bastida, H. Shelton Brown, and José A. Pagán

17.1. Introduction

One-fifth of the U.S. adult population does not have health insurance coverage and it is projected that the ranks of the uninsured will continue to grow due to increasing health care costs and rising health insurance premiums (DeNavas-Walt, Proctor and Lee, 2005; Gilmer and Kronick, 2001; Rowland, 2004). The U.S. uninsured population is not only relatively large (almost 46 million people) but it is not homogenously distributed across states and communities. Incidentally, the four Southwestern border states, California, Arizona, New Mexico, and Texas, are also the only states where the percentage of the total state population without health insurance coverage exceeds 18%. These states account for 30% of the total U.S. uninusured population, with approximately 12 million uninsured persons (U.S. Census Bureau, 2005a). Texas has the highest percentage of uninsured persons is the U.S. (26.0%) and uninsurance rates are disproportionately high in the Texas counties bordering Mexico (Fronstin, 2002; Fisher, 2005). More specifically, the south Texas border counties of Hidalgo, Cameron, and Starr have some of the highest rate of uninsurance in the country (32.8%, 30.8%, and 37.9% in 2000, respectively; see U.S. Census, 2005a). It is no coincidence that the poverty levels in these counties are also among the highest in the country (33.0%, 30.6%, and 39.0% in 2002, respectively; see U.S. Census, 2005b).

The existence of relatively high uninsured populations at the community level communities could potentially have significant economic effects in these areas, particularly affecting local hospitals and clinics, physician practices, public health, and the overall effectiveness of the local health care system (Hadley and Cunningham, 2004). A recent Institute of Medicine (IOM) report concludes that "uninsured residents have worse access to health care in communities with high uninsured rates than they do in communities with relatively low rates" (IOM, 2003:

---

* This study was funded by a National Institutes of Health, National Institute of General Medical Sciences (NIH 2S06 GM 08038–32) grant to Elena Bastida.

p. 7). Health care providers may face serious financial difficulties when they have to provide mostly uncompensated care to a relatively large uninsured population. Consequently, many outpatient providers and hospitals may be faced with lower revenue and they may be forced to reduce the quantity and quality of health services provided in the local community. Local and state governments are also likely to be faced with more responsibilities of providing health care services for the uninsured (Thorpe, 2004). Indeed, this is very much the case along the Texas border counties where local governments continuously struggle to provide an ever increasing number of services for their uninsured residents (e.g., Laredo, Texas; see Landreck and Garza, 2002). At a 2003 meeting of Hidalgo and Cameron counties commissioners on the medically underserved population of these counties, commissioners expressed frustration at their limited resources and how even these were dwindling in the face of continuously increasing demand of county financed health care and sharp increases in cost (Border Issues on Health Care, October 2003).

The unique situation of the U.S.-Mexico border region, however, provides an interesting twist to the past and current policy discourse on health care and the unisured at the national level. For it is in this region— along the 2,000-mile international border between a developed and less developed country—that throughout the years an alternative scenario developed in response to the health care needs of the uninsured population in this region. Hence, empirical studies that examine predictors of health insurance coverage and its association with the utilization of the Mexican health care system in border states and along border counties will add valuable and provocative information likely to expand and enhance our current national discourse on the uninsured.

Studies of uninsured border residents and their routine border crossings when they need health care must be framed within the unique socioeconomic and demographic context of the border and the Mexican, national health system in which private health care is an important, and frequently encouraged, alternative sector (Ward, 1985; Lasset et al., 1996; Pan American Health Organization, 1998). Crossing the border to obtain health care services in Mexico is a decades-old alternative, well documented in the literature (Macias, 2001; Warner, 1991, 2005). Much of the existing literature, however, is based on limited samples (Macias, 2001; Calvillo and Lal, 2003; Casner and Guerra, 2003), the targeting of a specific condition (Potter et al., 2003), customs declarations (McKeithan and Shepherd, 1996), or observational studies (Thompson, 1993; Calvillo and Lal, 2001). Even less is known about the factors underlying the purchase of health care services across the border from the perspective of a health care demand framework.

In this study we use survey data from Wave 3 of the Border Epidemiologic Study on Aging (BESA) (n = 1,048) to identify significant predictors of health insurance coverage for U.S. border residents. These predictors are then used to analyze how factors related to U.S.-based health insurance coverage are associated with the demand for medical care south of the border. Finally, we explore the associaiton between U.S. or Mexican nativity, and years of residence in the U.S. for the Mexican-born group and the demand for medical care in Mexico.

The paper is organized as follows. Section 17.2 reviews current policy research on the multiple effects generated by the absence of health insurance coverage and discusses multiple issues related to cross border health care purchases. Section 17.3 describes the model and the data; Section 17.4 presents results; and Section 17.5 discusses policy implications and provides some concluding remarks.

17.2. Policy Research on the Consequences of Inadequate Health Insurance Coverage

As indicated above, the tens of millions of U.S. adults without health insurance coverage are not homogenously distributed across the country. The U.S. Census Bureau has developed the Small Area Health Insurance Estimates (SAHIE) program to provide model-based estimates of health insurance coverage for counties and states. The 2000 uninsured and insured population estimates were released in July 2005 and are based on a mixed effects linear regression model wherein the log proportion of the insured population is a function of several predictors (Fisher, 2005). These predictors mostly include variables constructed from administrative records, and the dependent variable is the three-year average of county-level observations from the Annual Social and Economic Supplement (ASEC) of the Current Population Survey (CPS). According to SAHIE program estimates, there were 39.2 million uninsured adults in the U.S. (about 14% of the U.S. adult population). The proportion of the total population without health insurance at the county level varies substantially across the U.S., from a high of 37.9% in Starr County, Texas, to percentage lows in the high teens in many counties in the Eastern and Midwestern states (U.S. Census, 2005a).

The absence of public or private health insurance could have health, economic, and social effects on individuals, families and communities by influencing different aspects of the health care system (IOM, 2002; 2003; 2004; Thorpe, 2004). Pagan and Pauly (2006) discuss several mechanisms by which relatively high rates of community uninsurance can have detrimental effects on the health care access and availability for everyone in the community. One mechanism is related to provider financing. The lower use of health services by the uninsured population leads to lower revenue for health care providers, while the use of health services by the uninsured population increases the uncompensated care burden faced by local health care providers. There is also the possibility that high local uninsurance could increase the price of health care to the insured. If their insurance coverage is incomplete (high deductibles or copays) and if rising medical care prices lead to less generous coverage, the insured may face higher out-of-pocket payments. Another possibility is that a higher proportion of uninsured means fewer customers with insurance that will cover high prices, and if providers are subject to economies of scale or scope, providers may be forced to cut the intensity and availability of care for everyone in the community.

The level of uninsurance can have adverse effects on primary care, emergency medical services, specialty services and hospital care (IOM, 2003). Local public health services (e.g., community health centers) could be adversely affected by high uninsurance rates as well (Lewin and Altman, 2000). Physicians practicing in communities with increasing rates of uninsurance are more likely to decrease unprofitable services and hours, or they may move to another area. There is also empirical evidence that high community uninsurance rates are associated with overcrowed hospital emergency departments (Derlet, 2002; GAO, 1993) and the closing of hospital trauma services (Dailey et al., 1992; Selzer et al., 2001). Hospitals are particularly likely to be substantially affected by a high or increasing local uninsured population. Higher community-level uninsurance rates are associated with a lower availability of hospital-based health services (e.g., trauma and some psychiatric services) as well as hospital beds per capita (Gaskin and Needleman, 2002).

The consequences of uninsurance in the U.S. in general and in the U.S.–Mexico border region in particular provide an interesting turn in the current debate on uninsurance because of the particular characteristics of the region. More specifically, U.S. border residents can access the Mexican health care system because this system includes a private care sector catering to both U.S. residents and wealthier Mexicans. Border residents also have access to inexpensive pharmaceuticals in Mexico, which they can bring across the border for personal use (Macias, 2001). One can also literally see inexpensive dental and physician offices from the U.S. side of the border. Even when border crossing may not occur, Mexican health care providers compete with American providers, perhaps limiting prices on the U.S. side of the border. This is evidenced, for example, in specialty practices, such as plastic surgery and dermatology, where coverage for many procedures is highly unlikely by U.S. insurance companies and public programs.

The intersection of high unemployment, low-paying jobs with few benefits, and the proximity of an affordable private health care option within a short proximity contribute to the viability of the Mexican health care system for many border residents. Certainly, health insurance demand will affect provider and country choice. Nativity in Mexico and number of years of residence in the U.S. may also play an important role in the demand for health care services south of the border.

17.3. Data and Model

The data for this study come from the third wave of a panel study of adults 37 years of age and older residing in the south Texas border region known as the Lower Rio Grande Valley (Hidalgo, Cameron, and Starr counties). The BESA is an ongoing population-based panel study of middle-aged and older Mexican Americans. The 1996–1997 baseline wave of this panel study used area probability sampling that resulted in a final sample of 1,089 households with at least one member age 45 and older who agreed to complete an in-home, face-to-face interview in either Spanish or English, with a response rate of 89%. Using proportionate representation based on U.S. Census tracts, randomly drawn numbers were used to identify streets and houses for canvassing purposes. Survey participants were administered a two-hour face-to-face health survey instrument every two years. Further details on the sampling design can be obtained elsewhere (Bastida and Pagán, 2002; Brown, Pagán, and Bastida, 2005). For Wave 3, however, an additional and younger cohort 37 to 45 years old was drawn in 2001 for inclusion in the 2002 data collection effort. Data presented here are from Wave 3 only and, therefore, the reported number of primary respondents, despite an approximately 8% attrition between waves, is approximately the same in total numbers as the initial sample.

BESA includes extensive socioeconomic, demographic, and health information. For the purpose of this study, the key advantage of BESA is that it also includes information about the use of health care services and whether these services were obtained in the U.S. or Mexico. The nearest Mexican facilities are in neighboring municipalities of Reynosa, Río Bravo, and Miguel Alemán in Tamaulipas, Mexico. Data from BESA respondents indicate that 75% of the uninsured and 24.4% of insured participants had obtained medical attention in these three municipalities in Mexico.

Our empirical specification of the determinants of cross border health care use is based on the idea that utilization depends on the perceived need for services (self-reported health status and activities of daily living), individual predisposing characteristics (age, years of education, marital status, and gender) and enabling factors (health insurance and household income). These factors were grouped and included in the model based on previous research suggesting that health care utilization and access vary substantially across these dimensions (Andersen and Davidson, 2001; Andersen et al., 2002).

The independent variable of greatest interest was whether the respondent had private or public health insurance coverage at the time of the interview. As such, we estimate a logistic regression model where visiting a doctor in Mexico is a function of perceived need, predisposing characteristics, and enabling factors. We use "visiting a doctor in Mexico" here as a better discriminant indicator of health care utilization in Mexico for the uninsured population than "purchase of medications" because many among the U.S. insured population indicate purchasing medications in Mexico. We also estimate a logistic regression of the factors that are related to health insurance coverage.

Household income variables potentially affect health status as well as health care demand (see, e.g., Grossman, 1972). Although health status increases with household income, it is not clear that health care generally is a luxury good, or even a normal good. It is also not clear that wealthier and/or healthier people, if they do demand more health care, will demand Mexican care. It may be that those with higher income on the U.S. side view Mexican care as inferior. Therefore, household income variables may or may not be correlated with the use of Mexican health care.

"Number of years of education" is also included in our model. While education and income are highly correlated, Grossman theorizes that the educated are better producers of health. Under the Grossman model, doctor visits per se are not desired demand per se, but are "inputs" used to produce the desired level of health. Exercise, diet, and public health practices may substitute for physician care. In a border context, it is not clear whether education will increase or decrease the use of Mexican services. It is clear that relative to non-border areas in the U.S., these health care inputs are cheaper. Perhaps the relatively low-priced Mexican fees for healthcare may be substituted for health inputs such as diet or exercise.

Health status is measured by self-rated health and self-reported activities of daily living(ADLs). Generally, poorer health status should lead to greater health care demand, which, in a border context, would include Mexican health care. We also included indicators on whether a hospitalization occurred in the last 12 months or the last 5 years.

Lastly, in a border context some demographic variables partly indicate knowledge and familiarity with the U.S. in general and the U.S. health care system in particular by Mexican immigrants. For instance, for analytical purposes we assume that for those born in Mexico, the number of years in the U.S. is likely to be a good indicator of their familiarity with the U.S. health care system. We also assume that although most Mexican Americans born on the U.S. side of the border share many cultural beliefs and practices with the Mexican born, being born in the U.S. indicates on average greater familiarity with the the U.S. health care system, since, for example, their birth entitles them to certain benefits not always available to the Mexican-born.

17.4. Results

Table 17.1 presents the distribution of socio-demographic characteristics for Wave 3 of the BESA sample (2002) by health insurance status. In this study, 30% of respondents in Wave 3 were not covered by a health insurance plan at the time of the interview. However, 57% of the population under 50 years of age was uninsured compared to 46% of adults between the ages of 50 to 64 years of age and only 2% of adults ages 65 and above. Uninsurance rates were also relatively high for men (33%) when compared to women (23%), married adults (36%), those with a household income of $15,001–30,000 per year (41%), those with 12 years or less of schooling (29–40%), and adults born in Mexico (41%, compared to 19% for those born in the U.S.). For those adults born in Mexico, there is also a strong negative association between the number of years in the U.S. and the uninsurance rate. For example, 84% of Mexican-born adults that have resided in the U.S. less than 10 years were uninsured compared to 17% of those who have resided in the U.S. for 40 years or more, further validating our earlier assumption that the longer the Mexican immigrant lives in the U.S., the greater the expected familiarity with the health care system. Not suprisingly the socioeconomic profile of those who report health insurance coverage in the third wave of the BESA point to the most advantaged in terms of income and education. In terms of demographics, the elderly as a group are the most advantaged, since 98% of BESA participants over the age of 65 were covered by public insurance (Medicare) and adults under the age of 50 were the most disadvantaged, with the highest percentage of uninsurance in the study.

TABLE 17.1. Socio-demographic characteristic: Uninsured and insured BESA
participants (N = 1,037), Wave 3.

Characteristics	Uninsured, row (%)	Insured, row (%) +	p-value
Age			
Under 50 years	41.0 (57)	13.4 (43)	
50–64 years	56.1 (46)	28.0 (54)	
65+	2.9 (2)	58.6 (98)	***
Sex			
Female	75.8(23)	66.2 (77)	
Male	24.2(33)	33.8 (67)	***
Marital status			
Married	74.4 (36)	56.6 (64)	
Single	15.2 (27)	17.9 (73)	
Household income			
Under $7,000	24.5 (27)	28.2 (73)	
$7,000–15,000	41.5 (32)	37.7 (67)	
$15,001–30,000	27.5 (41)	17.1 (59)	***
>$30,000	6.5 (14)	17.1 (86)	
Education			
Below 7 years	53.9 (29)	57.5 (71)	
7–8 years	11.6 (39)	7.9 (61)	
9–12 years	27.4 (40)	17.7 (60)	***
13+ years	7.1 (15)	16.9 (85)	
Country of birth			
Mexico	67.4 (41)	41.1 (59)	***
U.S.	32.6 (19)	58.9 (81)	
Length of residence in U.S.			
(for those born in Mexico)			
<10 years	6.9 (84)	0.6 (16)	
10–19 years	15.9 (69)	2.9 (31)	
20–29 years	22.1 (54)	8.0 (46)	
30–39 years	12.2 (39)	8.0 (61)	***
40+ years	10.6 (17)	21.7 (83)	
Native	32.7 (19)	58.8 (81)	

* $p < .10$, ** $p < .05$, *** $p < .01$

Table 17.2 presents the distribution of health and health care characteristics for
BESA's Wave 3 by health insurance status. Uninsured adults had better health
than insured adults, as indicated by their self-rated health and the activities of
daily living. For example, 33.9% of uninsured adults were in fair or poor health
compared to 54.6% of insured adults. Insured adults were much more likely to
report the use of hospital services than uninsured adults, but uninsured adults were
more likely to report that they had utilized medical care in Mexico. For example,
57% of adults receiving medical attention in Mexico are uninsured compared to
only 12% of adults who do not receive medical attention in Mexico.

Table 17.3 reports results from logistic regression models of the factors associated with health insurance coverage and visiting a doctor in Mexico within the

TABLE 17.2. Health and healthcare utilization: Uninsured and insured BESA participants (N = 1,037), Wave 3.

Characteristics	Uninsured, row (%)	Insured, row (%) +	p-value
Self-rated health			
Excellent	29.3 (42)	17.1 (58)	
Very good	36.8 (36)	28.2 (64)	
Fair	28.0 (21)	44.1 (79)	
Poor	5.9 (19)	10.5 (81)	***
Activity of daily living (ADL)			
Dependency			
Yes	10.4 (15)	25.4 (85)	
No	89.6 (34)	74.6 (66)	***
Instrumental ADL (IADL)			
Dependency			
Yes	25.6 (19)	47.5 (81)	
No	74.4 (38)	52.5 (62)	***
Hospitalized in the last 12 months			
Yes	5.2 (10)	19.3 (90)	
No	94.8 (33)	80.7 (67)	***
Hospitalized in the last 5 years			
Yes	11.1 (14)	30.1 (86)	
No	88.9 (35)	69.9 (65)	***
Have you been a patient in a nursing home (last 12 months)			
Yes	0.0 (00)	1.6 (100)	
No	100.0 (31)	98.4 (69)	*
Received medical attention in Mexico			
Yes	75.6 (57)	24.4 (43)	
No	24.4 (12)	75.6 (88)	***

$* p < .10, ** p < .05, *** p < .01$

last year. Adults aged 65 and over were more likely to be insured (e.g., Medicare) than their younger counterparts. Women were also more likely to be insured than men. There is also a clear negative relation between household income and health insurance coverage. For example, adults with a household income under $7,000 per year were 86% less likely to have health insurance coverage than those with annual household incomes of over $30,000. A similar result is evident for years of education, with those with more schooling being more likely to report health insurance coverage.

Mexican birth was negatively related to being insured, but the regression coefficient for this variable was statistically insignificant at conventional levels. For Mexican-born adults, there was no statistically significant connection between length of residence in the U.S. and the likelihood of being covered by health insurance.

Table 17.3 also reports the results from the logistic regression model that specifically used visiting a doctor in Mexico as the dependent variable. Adults with

TABLE 17.3. Logistics regressions for health insurance coverage and doctor visits in Mexico.

Independent variables	Insured		Visited Doctor in Mexico	
	B	Odds Ratio	B	Odds Ratio
Age				
Under 50 years	−4.2***	0.015	−0.39	1.47
50–64 Years	−4.0***	0.018	−0.03	1.03
65+ Years	RC		RC	
Sex				
Male	0.53**	1.7	−0.06	0.94
Female	RC		RC	
Marital status				
Married	−0.37	0.69	0.02	1.02
Single	0.53	1.71	0.31	1.14
Widowed	RC		RC	
Household income				
Under $7,000	−2.0***	0.14	−0.73*	0.49*
$7,000–15,000	−1.6***	0.21	0.24	1.28
$15,001–30,000	−0.92***	0.40	0.41	1.51
>$30,000	RC		RC	
Education				
Below 7 years	−1.23***	0.29	−0.05	0.95
7–8 years	−1.23***	0.29	−0.28	0.77
9–12 years	1.04***	.035	0.11	1.12
13+ years	RC		RC	
Birth country				
Mexico	−1.00	0.39	0.82	2.28
U.S.	RC		RC	
Self-rated health				
Excellent	−0.70	0.50	−0.68*	0.50
Very good	−0.93**	0.39	−0.26	0.78
Fair	−0.45	0.64	−0.26	0.78
Poor	RC		RC	
ADL Instrumental				
No	−0.53***	0.59	0.21	1.23
Yes	RC		RC	
ADL				
No	−0.39	0.68	0.29	1.34
Yes	RC		RC	
Length of residence in U.S.				
Under 10 years	RC		RC	
10–19 years	−0.80	0.45	1.29	3.64
20–29 years	0.10	1.11	−0.09	0.92
30–39 years	0.73	2.07	0.07	1.07
40+	0.92	2.52	−0.54	0.58
Insured				
No	—	—	1.96***	7.09
Yes	RC		RC	

** $p < .05$
*** $p < .01$

household incomes under \$7,000 per year are 52% less likely to have visited a doctor in Mexico than all other adults. For these very low income adults even the relatively inexpensive Mexican private care sector lies beyond their economic means; many reported regular Medicaid coverage; however, those who did not qualify for regular Medicaid coverage indicated that in a medical emergency they rely on the hospital emergency room and/or free or reduced cost medical assistance available in the U.S., as most U.S. adults do in similar circumstances elsewhere. Adults with excellent health were also less likely to visit a Mexican doctor than those with very good, fair, or poor health. The most significant finding, however, was that uninsured adults were seven times more likely to have used physician services in Mexico than their insured counterparts.

17.5. Conclusion

Health insurance coverage seems to be the most important determinant of the demand for Mexican physician services by U.S. border residents. After controlling for health insurance coverage, neither household income nor length of residence in the U.S. is a significant factor in the use of Mexican medical care. Thus, it appears that U.S. border residents utilize the U.S. health care system if they are insured and the Mexican health care system if they are uninsured and able to afford its required cash payment at the time of service.

Summarizing the results above, an interesting profile emerges of those U.S. residents who are most likely to utilize Mexican medical care. Those at the two opposite ends of the income categories are less likely to utilize Mexican medical care for entirely different reasons. Participants reporting household incomes below \$7,000 report either Medicaid coverage or reliance on free or reduced medical care in the U.S. Conversely, those reporting incomes over \$30,000 represented the income group most likely to have private health insurance. However, participants in the two middle income categories, that is those reporting incomes between \$7,000 and \$15,000, and \$15,001 and \$30,000 were more likely to utilize Mexican medical care. Moreover, participants who subjectively rated their health as excellent are two times less likely to utilize Mexican medical care than those who indicate poor health. Household income, number of years of education, and health status are important determinants of health insurance coverage. Mexican birth and length of residence in the U.S. are not significantly related to having health insurance coverage. Absence of any form of insurance, private or public, was the best and most significant predictor of receiving Mexican medical care, and this finding persisted after controlling for other predictors.

It is clear that U.S. health insurance coverage drives the demand for health care in Mexican border communities, particularly medical care, more than Mexican birth or familiarity with the Mexican health care system drives it. However, there is a need to develop better structural models to assess the factors that are related to the choice of health insurance coverage. Data collection for Wave 4 of BESA currently in progress will include, when completed, a large number of questions on health insurance coverage and cross border health care utilization. Preliminary results from the data suggest that when the uninsured were asked to provide a reason that best explains their lack of insurace coverage, two thirds report that they could obtain low cost health care in Mexico. Of the remaining respondents, 16% indicated that they can obtain free or low cost medical assistance in the United States. Hence, future research should continue to explore how best to serve the health care needs of this disadvantaged population while taking into account the major role that the Mexican health care system plays as an alternative to the U.S. health care system that seems unable to fully meet the medical needs of a large and growing economically disadvantaged population north of the Rio Grande.

Finally, we note that for the "poorest of the poor," as Maril (1989) referred to the poorest residents of the south Texas border region, the availability of inexpensive Mexican private health care does not provide a plausible option as indeed it may for those with slightly higher incomes. These persons remain medically underserved, unless covered by Medicaid, usually waiting for months to be seen by a doctor when chronic care is regularly needed, or rushing to the emergency room when presented with an unaticipated health crisis. For the most economically disadvantaged border residents, the Mexican private health care system—which along the border regulates itself by maintaining a competitive cost advantage over the U.S. system but, nonetheless, requires cash payments at the time of service—is priced well beyond their actual means.

References

Andersen, R., Davidson, P. (2001). Improving access to care in America: Individual and contextual indicators. In R. Anderson, T. Rice, and G. Kominski (Eds.), *Changing the U.S. health care system: Key issues in health services, policy and management* (pp. 3–30). San Francisco: Jossey-Bass.

Andersen, R., Yu, H., Wyn, R., Davidson, P., Brown, E., and Teleki, S. (2002). Access to medical care for low-income persons: How do communities make a difference? *Medical Care Research and Review, 59(4)*, 384–411.

Bastida, E., and Pagán, J. (2002). The impact of diabetes on adult employment and earnings of Mexican Americans: Findings from a community based study. *Health Economics, 11*, 403–413.

Brown, S., Pagán, J., and Bastida, E. (2005). The impact of diabetes on employment: Genetic IVs in a bivariate probit. *Health Economics, 14*, 537–544.

Calvillo, J., and Lal, L. (2003). Pilot study of a survey of U.S. residents purchasing medications in Mexico: Demographics, reasons, and types of medications purchased. *Clinical Therapeutics, 25(2)*, 561.

Casner P, and Guerra L. (1992). Purchasing prescription medication in Mexico without a prescription. The experience at the border. *Western Journal of Medicine, 156(5)*, 512–516.

Dailey, J., Teter, H., and Cowley, R. (1992). Trauma center closures: A national assessment. *Journal of Trauma, 33(4)*, 539–546.

DeNavas-Walt, C., Proctor, B., and Lee, C. (2005). *U.S. Census Bureau, current population reports, P60-229, income, poverty, and health insurance coverage in the United States: 2004.* Washington, DC: U.S. Government Printing Office.

Derlet, R. (2002). Overcrowding in emergency departments: Increased demand and decreased capacity. *Annals of Emergency Medicine, 39(4)*, 430–432.

Fisher, R. (2005). Health insurance coverage estimation. County and state working paper (July 2005). Washington, DC: U.S. Census Bureau. Available at: http://www. census.gov/hhes/www/sahie/publications.html Accessed September 5, 2005.

Fronstin, P. (2002). *Sources of health insurance and characteristics of the uninsured: Analysis of the March 2002, Current Population Survey. Issue Brief 252.* Washington, DC: Employee Benefit Research Institute.

GAO (General Accounting Office). (1993). *Emergency departments: Unevenly affected by growth and change in patient use, Publication No. B-251319.* Washington, DC: U.S. Government Printing Office.

Gaskin, D., and Needleman J. (2002). *The impact of uninsured populations on the availability of hospital services and financial status of hospitals.* Baltimore, MD: The Johns Hopkins Bloomberg School of Public Health.

Gilmer, T., and Kronick, R. (2001). Calm before the storm: Expected increase in the number of uninsured Americans. *Health Affairs, 20(6),* 207–210.

Grossman, M. (1972). On the concept of health capital and the demand for health. *Journal of Political Economy, 80,* 223–255.

Hadley, J., and Cunningham, P. (2004). Availability of safety net providers and access to care of uninsured persons. *Health Services Research, 39(5),* 1527–1546.

Hidalgo County Border Health Symposium. (2003). International trade and technology building. The University of Texas Pan American, October 2003.

IOM (Institute of Medicine). (2004). *Insuring America's health: Principles and recommendations.* Washington, DC: National Academies Press.

IOM (Institute of Medicine). (2003). *A shared Destiny: Community effects of uninsurance.* Washington, DC: National Academies Press.

IOM (Institute of Medicine). (2002). *Care without coverage: Too little, too late.* Washington, DC: National Academies Press.

Landeck, M., and Garza, C. (2002). Utilization of physician health care services in Mexico by U.S. Hispanic border residents. *Health Mark Quaterly, 20(1),* 3–16.

Lewin, M., and Altman, S. (Eds.). (2000). *America's health care safety net. Intact but endangered.* Washington, DC: National Academies Press.

Maril, R. (1989). *Poorest of Americans: The Mexican American of the Lower Rio Grande Valley of Texas.* Notre Dame, IN: University of Notre Dame Press.

McKeithan, E., and Shepherd, M. (1996). Pharmaceutical products declared by US residents on returning to the United States from Mexico. *Clinical Therapeutics, 18(6),* 1242–1251.

Pan American Health Organization (PAHO). (1998). Health situation analysis and trends summary: Country chapter summary from health in the Americas. Country Health Profile. http:// www.paho.org/english/sha/prflmex.htm

Pagán, J., and Pauly, M. (2006). Community-level uninsurance and the unmet medical needs of insured and uninsured adults. *Health Services Research, 41*(3), 788–803.

Potter, J., Moore, A., and Byrd, T. (2003). Cross-border procurement of contraception: Estimates from a postpartum survey in El Paso, Texas. *Contraception, 68(4),* 281.

Rowland, D. (2004). *Uninsured in America* [Statement for U.S. House of Representatives, Committee on Ways and Means, Subcommittee on Health, March 9]. Menlo Park, CA: Kaiser Family Foundation.

Selzer, D., Gomez, L., Jacobson, T., Wischmeyer, R., Sood, R., and Broadie, T. (2001). Public hospital-based level I trauma centers: Financial survival in the new millennium. *Journal of Trauma-Injury, Infection and Critical Care, 51(2),* 301–307.

Thompson, W. (1993). Utilizacion y practicas de servicios de salúd por los Mexico-Americanos de la frontera del valley bajo del Rio Bravo. *Revista de Salúd Fronteriza/Border Health, 9*(2), 1–10.

Thorpe, K. (2004). Protecting the uninsured. *New England Journal of Medicine, 351(15),* 1479–1481.

U.S. Census. (2005a). *Small area health insurance estimates.* [http://www.census.gov/hhes/www/sahie/data.html]. Accessed October 29, 2005.

U.S. Census. (2005b) *Small area health insurance estimates.* [http://www.census.gov/hhes/www/saipe/saipe.html]. Accessed October 29, 2005.

Ward, P. (1985). *Welfare politics in Mexico.* London: Allen & Unwin.

Warner, D. (1991). Health issues in the U.S.-Mexican border. *JAMA, 265,* 242–247.

Warner, D. (2005). Viability of cross border health insurance in helping to meet the needs of immigrant families. Annual Meeting of the American Public Health Association, December. Philadelphia, PA.

Section 4
Options for Health Care Reform

18
Impoverishing and Catastrophic Household Health Spending Among Families with Older Adults in Mexico: A Health Reform Priority*

Felicia Marie Knaul, Héctor Arreola-Ornelas, Oscar Méndez-Carniado, and Ana Cristina Torres

18.1. Introduction

One of the most important challenges facing health systems is population aging. International estimates suggest that the aging process will increase the cost of health by 41% between 2000 and 2050, so that health spending could reach 11% of work GDP (United Nations, 2002).

As is the case in many countries, this has implications for the need for investment in health in the future as the demand for health care becomes ever more focused on the more costly interventions associated with chronic illness and aging. In Canada, the relative cost by population group shows health care spending is 4.6 times higher

* Various parts of this text are summaries of materials previously developed and published and appropriate citations are given throughout the text. The paper is part of the research presented at the Global Development Network Conference in St. Petersburg, FL, that was awarded a medal in the category "Institutional Development and Changes in the Health Sector" (Knaul, Arreola, Méndez and Miranda, 2006). We are grateful for comments received through this selection process, at the Second Conference on Aging in the Americas held at the University of Texas at Austin, and in seminars at the Instituto Nacional de Salud Pública, the Harvard University/Ellison Institute meeting in Cuernavaca, and iHEA and from Cristian Baeza, Stefano Bertozzi, Octavio Gómez, Pablo Escandón, Julio Frenk, Emmanuela Gakidou, Paul Gertler, Martha Miranda, Rachel Nugent, Gustavo Olaiz, Miguel Angel Lezana, and Fernando Montenegro. We also thank Fundación Mexicana para la Salud and its Consejo Promotor Competitividad y Salud, CONACYT (Ref. 38391-D and SALUD-2004-C01-191) and the World Health Organization (through National Institute on Aging, National Institutes of Health, Office of Global Health Affairs, Dept. of Health and Human Services, Grant Y1-AG-9421 OPHS-9-062 Cross-country Differences in Health Systems: Economic Outcomes and Medical Care Utilization) for institutional and financial support; the World Bank Regional Study on *Risk Pooling*, Ahorro y Prevención; INEGI for data and Efrén Motta, Javier Dorantes, Monica Hurtado, Fabrizio Almazan, Vanesa Leyva, Rachel Maguire, Sonia Peña, Bronwyn Underhill and Maja Pleic for support with text and research. The research in this paper is the sole responsibility of the authors and does not necessarily reflect the views of their respective institutions.

for 65- to 74-year olds relative to children aged 0 to 14. The figure is 8.7 for the 75- to 84-year old group, and 18.8 for the group over age 84 (Romanov, 2002).

Faced with the complex health care demands of an aging population, one of the most important challenges facing health systems is the development of fair, equitable, efficient, and sustainable financing mechanisms. Despite this, financial protection for health continues to be segmented and fragmented, particularly in developing countries. As a result, access to insurance coverage through social security is regressive, there is an over-reliance on out-of-pocket spending to finance the health system, and impoverishing health spending is common, particularly among the poor and the uninsured. A policy priority is to develop forms of organizing health financing that will permit developing countries to face the challenges of aging before their populations age. This will help to institute incentives that promote equity, healthy lifestyles, and efficient organizations that will be able to promote better investment in health and discourage spiraling health care costs.

In the case of Mexico the transition experience has been rapid, complex, and profound. Projections show that in half a century the country will have reached levels of aging that took 200 years to achieve in European countries. Life expectancy rose from 40 years in 1943 to 75 in 2000. By 2030 more than 15 million Mexicans will be aged 65 or older and by 2050, the figure will be 1 in 4—a fourfold increase over 2000 (CONAPO, 1999; Lozano, 1997; Secretaría de Salud, 2001; Partida, 2002; Partida and García, 2002).

At the same time, the country is experiencing a rapid, polarized epidemiological transiton (Frenk, Bobadilla, Sepúlveda, et al., 1989). Between 1950 and 2000, the proportion of deaths attributable to non-comunicable diseases grew from 44% to 73%. The concentration of health care utilization among the aged and the chronically ill is beginning (Secretaría de Salud, 2001; Knaul, Arreola, Borja, et al., 2004). As a result of the interaction between these processes, the health system must simultaneouly meet the health needs of poverty and underdevelopment that are concentrated in specific population groups and in the poorest states, and those of chronic disease and population aging that are distributed among both rich and poor populations and require ever increasing budgetary allocations (Frenk, Lozano, González-Block, et al., 1994; Secretaría de Salud, 2001).

The fragmented health care system created in Mexico in the early 1940s offered financial protection only to the salaried work force and was thus ill-equipped to meet the challenges of aging and epidemiologic transition (Frenk, González-Pier, Gómez-Dantés, and Knaul, 2006; Knaul and Frenk, 2005; Frenk, Knaul, Gómez-Dantés, et al., 2004; Secretaría de Salud, 2005). As of 2000 more than 50% of care was financed directly by families through out-of-pocket payments. As a result catastrophic and impoverishing health expenditures are common and concentrated among the poor and the uninsured.

To address these challenges, the 2003 structural reform of the Mexican health system was designed to increase financial protection by offering subsidized, publicly provided health insurance to the 50 million Mexicans who are not covered by social security and are concentrated among the poor. One of the most important expected results of the introduction of this insurance scheme is a reduction in out-of-pocket spending among previously uninsured families and hence in the incidence of catastrophic and impoverishing spending.

This study examines the determinants of catastrophic and impoverishing health spending among Mexican households, before and after the reform. In order to document the evolution of financial protection the paper analyzes indicators of "absolute" (referred herein as "impoverishing") and "relative" (referred as "catastrophic") impoverishment from health spending and equity of health system financing between 1992 and 2004. The econometric work focuses on the factors associated with differences in health spending across families with a particular emphasis on family composition and the presence of older adults. The research uses seven rounds of the Encuesta National da Ingresos & Gastos de los Hegares (ENIGH) undertaken every two years between 1992 and 2004. The results generate a series of recommendations for research and for policy in terms of extending financial protection in the context of population aging in Mexico.

The methodology builds on the framework for measuring health system performance, and specifically fairness of finance, put forward by the World Health Organization (World Health Organization, 2000; Murray and Frenk, 2000). It also builds on studies of absolute impoverishment from health spending such as Wagstaff and Van Doorslaer (2003) for Vietnam. This work also draws on previous studies by the authors compiled in Knaul, Arreola, Méndez, and Miranda (2006). Torres and Knaul (2003) study catastrophic spending between 1992 and 2000 using the ENIGH with a focus on family composition, and these results are incorporated into this study (INEGI, 1992, 1994, 1996, 1998, 2000 and 2004). The econometric research benefited from previous studies such as Parker and Wong (1997).

This document has six sections. Section two provides a framework for analyzing financial protection as an outcome of the organization of the health system's financing. Section three describes the Mexican health system prior to the reform and provides an overview of the health sector reform of 2003. The fourth part develops the indicators of catastrophic and impoverishing health expenditure and equity of health system financing, analyzes the econometric methodology, and describes the data. The fifth part presents the results. It analyzes trends in health spending between 1992 and 2004 and the econometric results with a focus on differences in health spending by family composition. The final section summarizes the main conclusions including suggestions for future research and policy recommendations.

18.2. Financial Protection in Health[1]

The model for measuring health system performance published by the World Health Organization in 2000 identifies three intrinsic goals: the health of the population, quality (responsiveness of the system), and financial protection. Based on

---

[1] This section summarizes parts of Murray, Knaul, Musgrove, et al. (2000).

this model a health system that offers financial protection is one where no family faces impoverishment from health spending and each member of the society contributes financially according to their financial capacity and independent of their health status or health care requirements.

The degree of financial protection a health system affords is reflected in the health spending of families. Health systems are financed through a mix of three main mechanisms: monies gathered by the state via specific and general taxes; contributions to social security via deductions or taxes; and private payments which can be made either out-of-pocket or for private insurance (Wagstaff and Van Doorslaer, 1998; 1999). Out-of-pocket financing of health is considered the most inefficient and inequitable means of financing a health system. It is also the most likely to characterize unfair distributions of health financing and to generate risk of impoverishment (Frenk, Lozano, González-Block, et al., 1994; Phelps, 2003; World Health Organization, 2000; Xu, Evans, Kawabata, et al., 2003; Knaul and Frenk, 2005).

Out-of-pocket payments are typically made at the point-of-service and the individual consumer chooses, as a function of income, how much they are willing and able to purchase. Some of the standard requirements for efficiency and competition—that the consumer can choose among providers to achieve a fair price and that the consumer has the same knowledge as the service provider—are violated, as asymmetries of information, illness itself, and the urgency of treatment limit the capacity of the patient to search among providers and to minimize price. Catastrophic, and potentially impoverishing, expenditures arise because the ceiling on cost is the individual's maximum capacity to pay at the time of purchase. The financing of out-of-pocket payments is limited by the individual or household access to credit and borrowing, which is often severely constrained by poverty. Necessary care is forgone if the cost of care exceeds the ability to pay at the time of service. Further, out-of-pocket payments are the most fragmented across individual consumers since there is no possibility of pooling risk. These factors begin to explain why health systems financed by out-of-pocket spending tend to be associated with poverty and lower levels of economic development (Knaul, Arreola, Méndez, and Miranda, 2006).

A number of Latin American countries support highly segmented health systems in which the poor have less access to medical attention and are at higher risk of suffering catastrophic and impoverishing health expenditures. In these systems, health care for families with workers in the formal sector is financed using payroll-based, social security models (Londoño and Frenk, 1997). Typically, the informal sector, independent workers, and those who are out of the labor force receive limited health benefits through a variety of underfunded public sector schemes without explicit rights to a health care package. Out-of-pocket payments are common among these families because they lack access to sufficient, quality health services. They are often forced to choose between satisfying other basic needs such as education, food and housing, or foregoing necessary health care. This typified the Mexican health system prior to the reform of 2003 that created the System for Social Protection in Health.

18.3. Health Financing and the Organization of the Health Sector in Mexico[2]

The need to improve the equity of health financing is one of the key challenges facing the Mexican health system. More than 50% of care is financed by out-of-pocket payments by families (Frenk, González-Pier, Gómez-Dantes and Knaul, 2006; Secretaría de Salud, 2005). Further, in 2000, between 2 and 4 million Mexican households suffered catastrophic or impoverishing health expenditures. These incidents were concentrated among the poor and more than four times more common among the uninsured (Knaul, Arreola, and Méndez, 2005). As has been highlighted in several existing studies on the Mexican health sector and most recently in Frenk, González-Pier, Gómez-Dantés, and Knaul (2006), this lack of financial protection for households was reflected in the rank of Mexico in the World Health Organization (2000) analysis of health system performance. Overall, Mexico was ranked in position 51 out of 191 countries. However, when ranked by financial protection, its position was heavily affected, occupying position 144.

The root causes of these problems are systemic and have also been identified in several studies (Frenk, González-Pier, Gómez-Dantés, and Knaul, 2006; Frenk, Sepúlveda, Gomez-Dantés, et al., 2003; Frenk, Knaul and Gómez-Dantés, 2004). From its inception in 1943 and through to the reform of 2003, this system was based on a segmented and vertically integrated model in which each institution was responsible for providing all functions (stewardship, financing, and service provision) to a particular population group. The segmentation of the original model is based on the separation of health rights between the families with access to to social security, and the rest of the population (the self-employed, the unemployed, non-salaried and informal sector workers, and those who do not work) who were excluded from formal social insurance schemes.

Recent analysis of the Mexican health system identified several financial imbalances including low overall spending on health, heavy reliance on out-of-pocket spending as a source of finance, and inequity in allocation between the insured and the uninsured and within and across states (Secretaría de Salud, 2001; Frenk et al., 2004; Frenk, González-Pier, Gómez-Dantés, Knaul, 2006). Regressive insurance coverage worsened the situation. While more than 60% of the richest quintile of the population was insured, the figure was approximately 10% for the poorest quintile.[3] Further, overall public spending and access to social security was lowest in the poorest states where the epidemiologic backlog (higher prevalence of communicable diseases than non-comunicable) and health needs are the greatest. For example, in *Chiapas* and *Oaxaca*, only one-fifth of households were insured in 2000 (Figure 18.1). These imbalances promoted a high degree of inequality and financial fragmentation of the system and left Mexico ill-equipped to meet the

---

[2] This section draws on a section of Frenk, et al. (2004) and on Knaul and Frenk (2005) .
[3] Authors estimations based on data from ENIGH 2000 and SSA 2003.

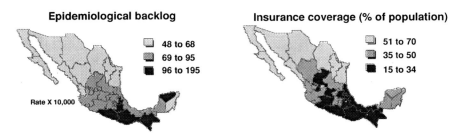

FIGURE 18.1. Insurance coverage in Mexico is inequeitable, 2000. Source: Authors estimations based on data from the Census 2000; ENIGH, 2000; and *Salud: México 2002*,
Secretaria de Salud (2003). Knaul, Arreola, Mendez and Borja (2003).

challenges of a middle-income country going through epidemiological transition
and facing an emergent aging process.

To address these challenges the 2003 structural reform of the Mexican health
system was designed to increase financial protection by offering publicly provided
health insurance to the 50 million Mexicans who are not covered by social security and are concentrated among the poor.[4] The new System for Social Protection
in Health (SSPH) operates through the Seguro Popular de Salud (Popular Health
Insurance). The reform was passed into law in April 2003, and SSPH went into operation on January 1st, 2004, with the goal of achieving universal health insurance
coverage by 2010.

The SSPH and the SPS are financed through a subsidized, progressive, and
triparite system to which affiliated families, state governements, and the federal
government each contribute per affiliated household. The reconfiguration of the
sources and allocation of funds via the reform seeks to increase the efficiency of
financing, as well as equity and financial protection for households. One of the
most important expected results is a reduction in out-of-pocket spending among
previously uninsured families and hence in the incidence of catastrophic and impoverishing spending, particularly among the poor (Knaul and Frenk, 2005; Frenk,
et al., 2004; Frenk, González-Pier, Gómez-Dantés and Knaul, 2006).

Between 2001 and 2003, prior to the reform, the Seguro Popular operated as
a pilot program and 614,000 families were affiliated (Secretaría de Salud, 2004).
The expansion of coverage has proceeded and by September of 2006, more than
4,000,000 families were affiliated (Secretaría de Salud, 2005). As stipulated by
law, the affiliation process has focused on the poor, and the nature of the progressive
subsidy is such that these families do not contribute financially.

The logic of the reform separates funding for personal health services between
an essential package of primary and secondary-level interventions in ambulatory

---

[4] We provide only very basic information on the reform that we consider essential to understanding this paper. For a more complete overview please see Knaul and Frenk (2005)
and Frenk, González-Pier, Gómez-Dantés, and Knaul (2006).

settings and general hospitals, and a package of high-cost tertiary-level services. The number of personal health interventions and services included in the Seguro Popular package and covered by the Fund is being gradually expanded (Secretaría de Salud, 2005). The list is updated annually using an explicit priority-setting mechanism based on the burden of disease, cost-effectiveness, and resource availability. In 2004 the Seguro Popular and the Fund covered 105 interventions focused on: cancers, cardiovascular problems, cerebra-vascular diseases, severe injury, long-term rehabilitation, HIV-AIDS, neo-natal intensive-care, organ transplants, and dialysis. In 2006, 249 interventions are included in the package of essential health care interventions and 17 more are covered in the Fund (Frenk, González-Pier, Gómez-Dantés, and Knaul, 2006).

18.4. Indicators, Methodologies and Data[5]

18.4.1. Indicators of Fairness in Financing and Catastrophic and Impoverishing Health Spending

Following Knaul, Arreola and Méndez (2005), this study makes use of four indicators to document and measure the effects on households of increasing financial protection in health through the reform, each of which is explained briefly below:

1. The Index of Fairness in Financial Contributions (IFFC) (World Health Organization, 2000; Murray et al., 2000)
2. The proportion of households with catastrophic health expenditures, measured as spending 30% or more of disposible income (total income less spending on basic needs approximated by food expenditure) (Murray et al., 2000)
3. The proportion of households with impoverishing health expenditures, defined as falling below the absolute poverty line due to health spending or significantly deepening their level of poverty for those that are below the poverty line (Knaul, Arreola and Méndez, 2005; Wagstaff and Van Doorslaer, 2003)
4. The proportion of families with either catastrophic or impoverishing health spending which is termed "excessive health expenditure."

According to the World Health Organization (2000) methodology for measuring the fairness of health financing (Murray et al., 2000), the Household Financial Contribution (HFC) of household h is:

$$HFC_h = \frac{HE_h}{DI_h} \tag{1}$$

where HE_h is the per capita expenditure on health of household h, and DI_h is per capita permanent income minus subsistence expenditure of household h. The

---

[5] This section is taken from Knaul, Arreola, and Méndez (2005).

numerator includes all financial contributions to the health system attributable to the household through taxes, social security contributions, private insurance, and direct, out-of-pocket payments. For taxes and social security contributions that are not earmarked for health, total household payments must be multiplied by the share of these revenues that ultimately goes to finance the health system using National Health Accounts. DI_h is approximated by total household per capita expenditure, net of household per capita food expenditure.

The distribution of HFC identifies how the burden of health system financing affects households and is the building block for the IFFC:

$$IFFC = 1 - 4 \left(\frac{\sum_{i=1}^{n} |HFC_h - \overline{HFC}|^3}{0.125n} \right) \qquad (2)$$

Values of IFFC closest to unity indicate systems that offer more financial protection in health. The cube of the absolute difference places strong emphasis on health expenditures that are a very high proportion of income. Further, by using disposable income the indicator places substantial weight on the poorest households, which are likely to have low nominal expenditure on health.

Catastrophic expenditures are defined as those for which a household spends more than 30% of their effective non-subsistence income on health following Murray et al. (2000). This is a *relative* measure of health care expenditure given as a proportion of disposable income, and thus emphasizes equity aspects and the question "What is *too much* spending for a household?"

Wagstaff and Van Doorslaer (2003) propose indicators of absolute impoverishment. In this paper their methodology is adapted and absolute impoverishment is measured as falling below, or falling further below, the poverty line of $1 dollar a day per person due to health expenditures. The definition of what constitutes "falling further below" the poverty line is somewhat difficult, since, for all families below a poverty line, any nominal expenditure on health, irrespective of how small it is, increases impoverishment in an empirical sense. In this paper, a solution is adopted based on available data and with the caveat that more research is required. In the ENAGS survey (described below and in INSP, FUNSALUD and Secretaría de Salud, 2001) families are asked if they had to modify their expenditure on food, education, or housing because of health spending. The level at which this occured was approximately 6-7% of a family's disposable income. Thus, this level is the cutoff for including an expenditure by a family below the poverty line as impoverishing.

This approach of combining the absolute and relative aspects[6] of health spending provides a composite indicator of impoverishment—excessive health spending. The total is less than the sum of the parts, as there are families with catastrophic spending (more than 30% of disposable income) that is not impoverishing (taking

---

[6] The work on absolute impoverishment and the combination of absolute and relative impoverishment was originally developed in Arreola, et al. (2004).

Absolute Poverty Status of Families (Poverty Line: < 1 USD per capita per day)	Health Expenditures as % of Disposable Income	Category of health spending	
		Relative	*Absolute*
Above the Poverty Line	< 30%	--	--
	>= 30%	Catastrophic	--
Cross poverty line due to health expenditures	< 30%	--	Impoverishing
	>= 30%	Catastrophic +	Impoverishing = Excessive
Already below the poverty line before spending on health	< 6%	--	--
	6-29%	--	Impoverishing
	>= 30%	Catastrophic +	Impoverishing = Excessive

FIGURE 18.2. Impoverishing and catastrophic health spending. Source: Knaul, Arreola, et al. (2004) & (2005), adapting methodologies in Wagstaff & Vandoorsaler (2003).

them below the poverty line), families with impoverishing expenditures that are not catastrophic, and families whose health spending is both catastrophic and impoverishing in the absolute sense. Considering both categories—catastrophic and impoverishing—guarantees the inclusion of families with very large nominal expenditures, although they do not become impoverished, as well as families with low nominal expenditures that do become impoverished. The subcategories of catastrophic and impoverishment health expenditures that make up total excessive events are presented in Figure 18.2 and discussed in greater detail in Knaul, Arreola and Méndez (2005).

The inclusion of both impoverishing and catastrophic health spending, as well as the IFFC, is particularly important in the context of aging and health reform in Mexico. First, aging and chronic illness are affecting both rich and poor families, meaning that is important to be able to identify health expenditures that are onerous for families regardless of their nominal size. Further, as the reform proceeds to cover segments of the population that are not poor, the associated demand for high-cost interventions associated with chronic care will increase. The health system must be able to identify the needs of these families in order to plan resources appropriately.

18.4.2. Econometric Methods

The regression analysis investigates the effect of household composition and the initial stages of health reform and extension of the Seguro Popular on catastrophic and impoverishing expenditure. The basic descriptive data for the variables are presented in Table 18.1. Overall, we observe that approximately only 43% of the househoulds in ENIGH 2004 report to have some health insurance. While the insured households are more likely to live in urban areas, rural households are most likely uninsured. Also, around 20% of households have people aged 65 or older and only about 12% of these are insured. By 2004 the Seguro Popular program had covered 10.7% of the uninsured population.

Five dependent variables are analyzed: 1) the probability that a household has a catastrophic expenditure; 2) the probability that a household suffers an impoverishing expenditure; 3) the probability that a household suffers excessive health spending (categories 1 and 2); 4) total health spending as a proportion of disposable income (HFC); and 5) family out-of-pocket spending per capita. Logit models are used for the first three dependent variables, which are categorical variables that take the value of unity if the household has a catastrophic, impoverishing or excessive health expenditure, respectively; and a value of cero if not. Tobit models (censored on zero) are used for the last two, since these refer to dependent variables which cannot take any value below zero.

The unit of analysis is the household. Each of the regressions is undertaken for: the whole population; families with access to social security; families with no access to social security and with or without Seguro Popular; and families without access to social security and with or without Seguro Popular for each one of the income quintiles. A selection of the regressions is presented in this document.

Each regression includes the following groups of control (independent) variables:

- Presence of a family member over 65 years or less than 5 years old (sign of these variables is expected to be positive).
- Gender and education of the household head, if the family includes a person with social security access or private medical insurance, and the level of urbanization of the location where the households resides. These variables measure the capacity of the household to cover health needs and access health services.
- In order to identify the needs of the state in terms of protecting the population from excessive health expenditure, three variables are included: the percentage of households under the poverty line in 2002, the percentage of households under the poverty line in 2004, and change between 2002 and 2004.
- The proportion of the population (0–100%) covered by the PHI in each state before the application of the 2004 ENIGH.

18.4.3. Data

The database used in this study is a time series of the Encuesta Nacional de Ingresos y Gastos de los Hogares-ENIGH (National Household Income and Expenditre Survey), undertaken by the Instituto Nocional de Estadística, Geografía e Informática-INEGI (National Institute of Statistics, Geography and Informatics) (INEGI) every two years since 1992 to 2004 (ENIGH, 1992, 1994, 1996, 1998, 2000, 2002 and 2004). For the descriptive analysis the full time series is

TABLE 18.1. Sample descriptive statistics.

	All		Non-insured		Insured Insured		Quintile 1 (per capita spending)	
	Mean	Std. dev.	Mean	Std. dev.	Mean	Std. dev.	Mean	Std. dev.
Household characteristics								
Older people (65+), no children, in the households (0 = no, 1 = yes)	16.3	0.4	21.5	0.4	9.4	0.3	16.4	0.4
Older people (65+) and children (less than 5) in the households (0 = no, 1 = yes)	2.9	0.2	2.9	0.2	3.0	0.2	6.4	0.2
Children (less than 5) and no older people (65+) in the household (0 = no, 1 = yes)	27.7	0.4	26.3	0.4	29.6	0.5	40.0	0.5
Without older people or children	53.0	0.5	49.2	0.5	58.0	0.5	37.2	0.5
Presence of older people (65 or plus) in the households (0 = no, 1 = yes)	19.3	0.4	24.5	0.4	12.4	0.3	22.9	0.4
Presence of children in the household (0 = no, 1 = yes)	30.7	0.5	29.2	0.5	32.5	0.5	46.4	0.5
Households with social security access	43.2	0.5	0.0		100.0		21.2	0.4
Number of people in the household	4.0	2.0	3.8	2.0	4.3	1.9	5.2	2.3
Residence in areas over 100,000 inhabs. (0 = no, 1 = yes)	49.5	0.5	38.2	0.5	64.3	0.5	20.7	0.4
Residence in areas from 15,000 to 99,999 inhabs. (0 = no, 1 = yes)	13.8	0.3	14.2	0.3	13.2	0.3	13.2	0.3
Residence in areas from 2,500 to 14,999 inhabs. (0 = no, 1 = yes)	13.8	0.3	14.9	0.4	12.4	0.3	17.6	0.4
Residence in areas with less than 2,500 inhabs. (0 = no, 1 = yes)	22.9	0.4	32.6	0.5	10.1	0.3	48.5	0.5
Gender of household head (0 = female, 1 = male)	23.3	0.4	26.4	0.4	19.1	0.4	18.1	0.4
Education of household head	7.3	5.0	5.8	4.6	9.2	4.9	4.0	3.5
Total household income per capita	2397.2	5532.7	1897.7	6366.6	3054.4	4098.7	614.2	653.6

(*Cont.*)

TABLE 18.1. (*Continued*)

	All		Non-insured		Insured		Quintile 1 (per capita spending)	
	Mean	Std. dev.	Mean	Std. dev.	Mean	Std. dev.	Mean	Std. dev.
State characteristics where the household is located								
% of households in the state under the poverty line in 2004	12.3	10.5	13.6	11.0	10.5	9.4	15.6	11.2
% change in households under the poverty line between 2002 and 2004	−4.2	13.1	−5.6	13.8	−2.5	12.0	−8.3	15.5
% of households in the state under the poverty line in 2002	16.5	16.6	19.2	17.3	13.0	14.9	23.9	18.5
% of the uninsured population covered by the Popular Health Insurance	10.7	13.0	10.7	12.8	10.7	13.3	12.2	13.9
No. of households in the sample	22,569		12,352		10,217		4,521	
No. of households with expansion factor	25,819,327		14,669,164		11,150,163		5,164,345	

Source: Estimations made by the author with data from the ENIGH 2004.

used. The regressions are based on the ENIGH 2004, which sample size is 22,595 households.[7]

The 1992 and 1994 ENIGH reflect the pre-crisis period (the 1994 ENIGH was undertaken before the crisis); the economic crisis period that goes from the end of 1994 to 1996 is measured with the 1996 and 1998 ENIGH; the post crisis, recovery period is reflected in the 1998 and 2000 ENIGH; and, the implantation of the Seguro Popular and the reform, covering the years 2001–2004, is included in the ENIGH 2002 and 2004 (Knaul, Arreola, and Méndez, 2005).

The ENIGH is a cross-sectional survey that includes a representative sample at the national and sub-national level. Household-level information is provided on all sources of income and expenditure. Individual information is included on labor force participation and social security rights, and basic sociodemographic characteristics. The survey is applied between August and November of each year and data on health are based on expenditures incurred during the three months prior to the survey. The base questionnaire has remained relatively intact since 1992.

In order to analyze the indicators over time, payments for social security and taxes were calculated based on the laws in place for each year. All the monetary variables were deflated using the National Consumer Prices Index to constant prices for 2000 (Banco de México, 2005). Further, a household is defined as "lacking access to social security" if there is no member with a right to insurance coverage of any type, public or private, other than Seguro Popular. This assumes that a single insurance or social security payment covers an entire family.

Another database that is used is the Encuesta Nacional de Aseguramiento y Gasto en Salud—ENAGS (National Insurance and Health Spending Survey) of 2001. The ENAGS is a national survey of almost 2000 households undertaken to determine the target population of the PHI. It includes specific questions on the willingness of persons to pay for subscription to a health insurance similar to the Seguro Popular. Coverage variables for Seguro Popular come from annual reports of the Mexican Ministry of Health (Secretaría de Salud, 2003, 2004, 2005) and from the 2004 coverage data reported by the National Comission for Social Protection in Health. All of the other variables come from the ENIGH 2004.

18.5. Results

18.5.1. Descriptive Trends in Catastrophic and Impoverishing Health Spending Between 1992 and 2004[7]

In 2000, based on data from the ENIGH, 3.4% of families suffered catastrophic health expenditures and 3.8% impoverishing health expenditures per trimester.

---

[7] The sample sizes of other ENIGH surveys used in the descriptive analysis are the following: 10,503 households in 1992; 12,815 in 1994; 14,042 in 1996; 10,952 in 1998; 10,108 in 2000; and 17,167 in 2002.

[7] This section draws on Knaul, Arreola, and Mendez (2005).

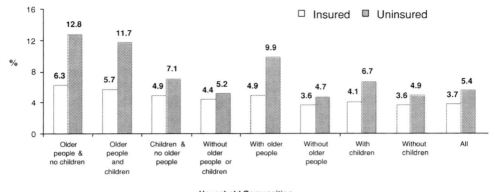

FIGURE 18.3. Out-of-pocket health spending as a proportion of disposable income by insurance coverage and family composition (2000). Source: Authors calculations based on ENIGH, 2000.

Overall, 6.3% of families had an incident of excessive health spending and the IFFC was 0.915.[8]

The rates are substantially higher among the poor and the uninsured. In 2000 more than 20% of households in the poorest quintile suffered a catastrophic or impoverishing health expenditure as compared to less than 5% in the wealthier quintiles. The rates are more than four times higher among the uninsured as compared to the insured at 9.6% and 2.2%, respectively.

Household composition is also associated with important differences in excessive health spending. Families that include persons over age 65 spend, on average, a higher proportion of their disposable income on health as compared to families with or without children (Torres and Knaul, 2003). This is true across income groups, yet the proportional difference is particularly large for the poorest quintile. It is also the case when the sample is divided between the insured and the uninsured. Families with people over age 65—be they insured or uninsured and with or without children—spend a higher proportion of disposable income on health (Figure 18.3). At the same time, the proportions are higher among the uninsured for all family groupings. Considering families with non-zero spending on health, uninsured families with adults over age 65 and without children spent 12.8% of disposable income on health in 2000, as compared to 6.3% for insured families. For families with older adults and children the figures are 11.7% and 5.7%, respectively. For families without older adults and with children the figures are 7.1% for the uninsured and 4.9% for the insured.

---

[8] These estimates are similar to the average over the period 1992–2004, and for this reason the ENIGH 2000 is used for describing the overall descriptive statistics.

TABLE 18.2. Evolution over time of catastrophic and impoverishing health expenditure (1992–2004). Source: Authors calculations using ENIGH 1992–2004. Originally published in Knaul, Arreola and Mendez, 2005.

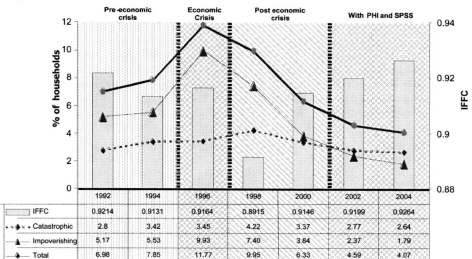

	1992	1994	1996	1998	2000	2002	2004
IFFC	0.9214	0.9131	0.9164	0.8915	0.9146	0.9199	0.9264
Catastrophic	2.8	3.42	3.45	4.22	3.37	2.77	2.64
Impoverishing	5.17	5.53	9.93	7.40	3.84	2.37	1.79
Total	6.98	7.85	11.77	9.95	6.33	4.59	4.07

The trends over time between 1992 and 2004 show certain patterns. All of the indicators of financial protection show a deterioration during the period of economic crisis and an improvement post-crisis that continues after 2000 (Table 18.2).[9] The deterioration is especially marked for impoverishing health spending during the crisis period, which affected both families with and without access to social security. By contrast, the improvements between 2000 and 2004 have been focused among households that lack access to social security and are either uninsured or have Seguro Popular. The improvements are also concentrated in the poorest quintile.

There has also been a change over time in the composition of households with excessive health spending. Although before and during the period of economic crisis the majority of households with excessive spending were below the poverty line and spent less than 30% of disposable income, in 2004 it is the group of households above the poverty line with catastrophic expenditures in health that predominate. This group has increased its participation from 26% to 56% of the total, while the group below the poverty line has fallen from 52% to 29% (Figure 18.4). Thus, catastrophic health spending is now more common than impoverishing health spending among households.

---

[9] Calculations that went into producing Table 18.2 (including confidence intervals) are available from the authors upon request. The poverty line is constant $1 U.S. In Knaul et al (2006) the trends are analyzed using different poverty lines. Although the peak in 1994–1996 is reduced somewhat, the overall pattern is robust to changing the poverty line.

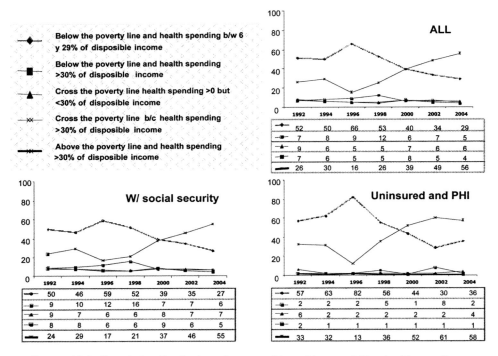

FIGURE 18.4. Cumulative distribution of catastrophic and impoverishing health spending by category. Source: Authors calculations using ENIGH 1992–2004.

During the time period under study there were significant changes in the patterns of catastrophic and impoverishing health spending across broad family groups. Families that include children under 5 years of age and/or adults over age 65 tend to have higher rates of catastrophic and impoverishing health spending (Figure 18.5).

The pattern over time for each family group is similar to the pattern for the the population as a whole. Still, the increase in the prevalence of excessive health spending during the period of economic crisis is particularly high for families with older adults and young children. Given that this is a small population group and hence a small sample, the results must be taken with caution, yet merit further analysis. Further, prior to 1998, excessive health spending events were more common among families with children, and since 2000 they are more common among families with older adults.

The distribution, by family composition, of catastrophic and impoverishing health expenditure are shown separately in Figure 18.6. The results suggest that catastrophic health spending is more common among families with older adults and impoverishing health spending among families with children. Further, the higher rates of catastrophic health spending among families with older adults, combined with the reduction in the incidence of impoverishing health spending, help to ex-

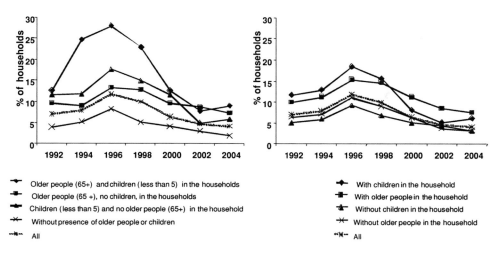

FIGURE 18.5. Excessive health spending by household composition (1992−2004). Source: Authors Calculations using ENIGH 1992–2004.

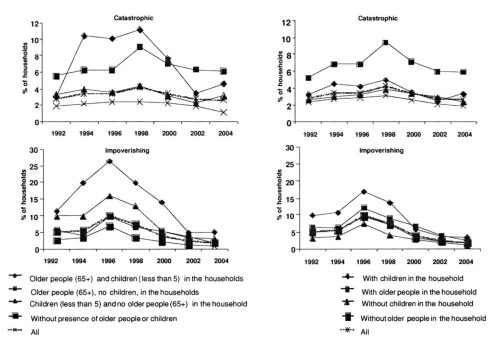

FIGURE 18.6. Catastrophic and impoverishing health spending by houshold composition (1992–2004). Source: Authors calculations using ENIGH 1992–4004.

plain the recent concentration in cases of health spending that are over 30% of disposable income.

18.5.2. Econometric Results of the Determinants of Catastrophic and Impoverishing Health Spending Among Households

The econometric analysis (Tables 18.3 and 18.4) shows that families with older adults and/or children are more inclined to suffer both catastrophic and/or impoverishing health expenditure, and to have higher levels out-of-pocket spending per capita and relative to their income level. The coefficients are significant in all of the regressions and are robust to changes in the specification and in the sample.[10]

Household size is positively correlated with excessive spending, impoverishment, and out-of-pocket spending, but negatively with catastrophic and out-of-pocket spending relative to disposable income. This variable is likely to be measuring a combination of wealth and the number of people using health services. All of the regressions (in Table 18.3) suggest that catastrophic and impoverishing health spending is less common among families living in the large urban areas than in smaller cities or in the rural areas. For the regressions of out-of-pocket spending, in which total household income is controlled, the variables for place of residence are insignificant. This suggests that residence is controlling for some aspects of income.

Catastrophic health spending is less common and out-of-pocket spending per capita and relative to income are lower among families with access to social security. Education of the household head is associated with a lower probability of catastrophic and impoverishing health spending (Table 19.3), although the opposite is true for out-of-pocket spending (Table 19.4). Some of the regressons show a negative relation between the extension of Seguro Popular coverage at the state level and out-of-pocket spending relative to disposable income and per capita out-of-pocket spending. This point is analyzed in greater detail and with community-level data in Knaul, et al (2006).

18.6. Conclusions

This study documents the evolution of catastrophic and impoverishing spending among Mexican households between 1992 and 2004. A large group of these households are below the poverty line or cross the poverty line due to health spending. At the same time, impoverishing health spending is much more common among the poorest deciles. Catastrophic health spending, which places more weight on

---

[10] The results varying the sample and the specification of the independent variables are available from the authors upon request.

TABLE 18.3. Logit estimations that a household suffers excessive, catastrophic, or impoverishing expenditure.

		Excessive spending (catastrophic and/or impoverishing)		Catastrophic health spending (30% of disposable family income)		Impoverishing health spending (cross poverty line or increase poverty if below poverty line)	
		All	Non-insured	All	Non-insured	All	Non-insured
Household characteristics							
Older people (65+), no children, in the households (0 = no, 1 = yes)	Coef	0.92	0.91	1.05	1.04	0.40	0.43
	z	(9.09)	(9.00)	(8.98)	(8.89)	(2.20)	(2.35)
Older people (65+) and children (less than 5) in the households (0 = no, 1 = yes)	Coef	1.07	1.07	1.17	1.16	0.89	0.82
	z	(6.45)	(6.44)	(4.93)	(4.92)	(4.00)	(3.71)
Children (less than 5) and no older people (65+) in the household (0 = no, 1 = yes)	Coef	0.74	0.74	0.77	0.78	0.77	0.63
	z	(8.02)	(8.07)	(6.51)	(6.56)	(5.65)	(4.64)
Households with social security access	Coef	−0.45	−0.45	−0.66	−0.66	0.03	0.37
	z	(4.90)	(4.89)	(5.82)	(5.79)	(0.23)	(2.40)
Number of people in the household	Coef	0.07	0.07	−0.09	−0.09	0.22	0.19
	z	(3.96)	(4.02)	(3.54)	(3.49)	(9.86)	(8.17)
Residence in areas from 15,000 to 99,999 inhabs. (0 = no, 1 = yes)	Coef	0.46	0.46	0.46	0.47	0.81	0.76
	z	(4.05)	(4.08)	(3.59)	(3.62)	(3.56)	(3.31)
Residence in areas from 2,500 to 14,999 inhabs. (0 = no, 1 = yes)	Coef	0.54	0.55	0.35	0.36	1.33	1.17
	z	(3.75)	(3.77)	(1.95)	(1.97)	(5.60)	(4.92)
Residence in areas with less than 2,500 inhabs. (0 = no, 1 = yes)	Coef	0.95	0.96	0.61	0.62	1.91	1.65
	z	(9.78)	(9.80)	(5.22)	(5.22)	(10.17)	(8.64)
Gender of household head (0 = female, 1 = male)	Coef	0.09	0.09	0.09	0.09	0.21	0.17
	z	(0.96)	(1.00)	(0.83)	(0.87)	(1.36)	(1.15)

(Cont.)

TABLE 18.3. (*Continued*)

		Excessive spending (catastrophic and/or impoverishing)		Catastrophic health spending (30%* of disposable family income)		Impoverishing health spending (cross poverty line or increase poverty if below poverty line)	
		All	Non-insured	All	Non-insured	All	Non-insured
Education of household head	Coef	**−0.05**	**−0.05**	**−0.03**	**−0.03**	**−0.11**	**−0.07**
	z	**−(5.00)**	**−(5.25)**	**−(2.38)**	**−(2.65)**	**−(6.39)**	**−(3.74)**
Total household income per capita (x1000)	Coef		**0.01**		**0.01**		**−0.77**
	z		**(2.49)**		**(2.23)**		**−(6.43)**
State-level Characteristics (x100)							
% of households in the state under the poverty line in 2004	Coef	**0.56**	**0.56**	0.03	0.04	**1.05**	0.60
	z	**(1.78)**	**(1.80)**	(0.09)	(0.09)	**(2.28)**	(1.29)
% change in households under the poverty line between 2002 and 2004	Coef	**−0.86**	**−0.86**	**0.63**	**0.63**	**−2.32**	**−1.85**
	z	**−(3.41)**	**−(3.43)**	**(1.83)**	**(1.82)**	**−(6.82)**	**−(5.36)**
% of households in the state under the poverty line in 2002	Coef						
	z						
% of the uninsured population covered by the Popular Health Insurance	Coef	**−0.42**	**−0.41**	−0.30	−0.30	−0.27	−0.27
	z	**−(1.72)**	**−(1.70)**	−(1.03)	−(1.00)	−(0.72)	−(0.73)
Constant	Coef	**−4.16**	**−4.17**	**−3.80**	**−3.81**	**−6.56**	**−5.59**
	z	**−(27.34)**	**−(27.40)**	**−(21.08)**	**−(21.11)**	**−(24.00)**	**−(18.44)**
Number of observations		22,569	22,569	22,569	22,569	22,569	22,569
LR chi2 (variables)		626.48	630.47	300.68	304.20	785.47	852.70
Prob > chi2		0.00	0.00	0.00	0.00	0.00	0.00
Pseudo R2		0.09	0.09	0.06	0.06	0.21	0.22
Log-likelihood		−3294.81	−3292.81	−2391.42	−2389.66	−1507.49	−1473.88

Note: The values in bold are significative at the 10% level.

TABLE 18.4. Tobit regressions on financial contribution of households and per capita out of pocket spending.

Household characteristics		Out of pocket health spending as a % of disposable income				Out-of-pocket per capita spending			
		All	Non-insured	All	Non insured	All	Non-insured	All	Non insured
Older people (65+), no children, in the households (0 = no, 1 = yes)	Coef	0.02	0.02	0.02	0.02	73.49	69.78	73.49	69.78
	t	(14.39)	(9.93)	(14.39)	(9.93)	(7.36)	(4.83)	(7.36)	(4.83)
Older people (65+) and children (less than 5) in the households (0 = no, 1 = yes)	Coef	0.01	0.01	0.01	0.01	65.75	65.70	65.75	65.70
	t	(5.44)	(3.89)	(5.44)	(3.89)	(3.13)	(1.95)	(3.13)	(1.95)
Children (less than 5) and no older people (65+) in the household (0 = no, 1 = yes)	Coef	0.01	0.01	0.01	0.01	52.21	46.64	52.21	46.64
	t	(7.53)	(5.05)	(7.53)	(5.05)	(6.42)	(3.48)	(6.42)	(3.48)
Households with social security access	Coef	-0.01		-0.01		-56.59		-56.59	
	t	-(11.84)		-(11.84)		-(7.66)		-(7.66)	
Number of people in the household (x100)	Coef	-0.37	-0.43	-0.37	-0.43	425.91	714.91	425.91	714.91
	t	-(18.00)	-(12.86)	-(18.00)	-(12.86)	(2.22)	(2.41)	(2.22)	(2.41)
Residence in areas from 15,000 to 99,999 inhabs. (0 = no, 1 = yes) (x100)	Coef	0.44	0.65	0.44	0.65	2208.04	2740.78	2208.04	2740.78
	t	(4.38)	(3.72)	(4.38)	(3.72)	(2.33)	(1.77)	(2.33)	(1.77)
Residence in areas from 2,500 to 14,999 inhabs. (0 = no, 1 = yes) (x100)	Coef	0.32	0.30	0.32	0.30	318.48	-103.74	318.48	-103.74
	t	(2.11)	(1.31)	(2.11)	(1.31)	(0.22)	-(0.05)	(0.22)	-(0.05)
Residence in areas with less than 2,500 inhabs. (0 = no, 1 = yes) (x 100)	Coef	0.35	0.32	0.35	0.32	29.05	-1168.20	29.05	-1168.20
	t	(3.59)	(2.04)	(3.59)	(2.04)	(0.03)	-(0.83)	(0.03)	-(0.83)
Gender of household head (0 = female, 1 = male) (x 100)	Coef	-0.45	-0.56	-0.45	-0.56	-1089.22	-671.88	-1089.22	-671.88
	t	-(5.18)	-(3.99)	-(5.18)	-(3.99)	-(1.33)	-(0.53)	-(1.33)	-(0.53)
Education of household head (x 1000)	Coef	0.36	0.25	0.36	0.25	8211.24	8609.50	8211.24	8609.50
	t	(4.51)	(1.79)	(4.51)	(1.79)	(10.74)	(6.83)	(10.74)	(6.83)
Total household income per capita (x 1000)	Coef					20.52	21.97	20.52	21.97
	t					(147.39)	(111.67)	(147.39)	(111.67)

(Cont.)

TABLE 18.4. (Continued)

		Out of pocket health spending as a % of disposable income				Out-of-pocket per capita spending			
		All	Non-insured	All	Non insured	All	Non-insured	All	Non insured
State-level variables									
% of households in the state under the poverty line in 2004 (x 10,000)	Coef	0.16	0.19	**0.90**	**1.37**	-3876.72	-4199.96	716.21	2525.43
	t	(0.48)	(0.36)	**(2.17)**	**(2.13)**	(-1.20)	(-0.87)	(0.18)	(0.44)
% change in households under the poverty line between 2002 and 2004 (x 10,000)	Coef	**0.73**	**1.18**			4592.93	**6725.40**		
	t	**(2.51)**	**(2.63)**			(1.65)	**(1.67)**		
% of households in the state under the poverty line in 2002 (x 10,000)	Coef			-0.73	-1.18			-4592.93	-6725.40
	t			(-2.51)	(-2.63)			(-1.65)	(-1.67)
% of the population covered by the Popular Health Insurance (x 100)	Coef	-0.08	-0.11	-0.08	-0.11	-75.39	-90.19	-75.39	-90.19
	t	(-3.55)	(-2.76)	(-3.55)	(-2.76)	(-3.54)	(-2.62)	(-3.54)	(-2.62)
Constant	Coef	**0.01**	**0.01**	**0.01**	**0.01**	-194.48	-231.89	-194.48	-231.89
	t	(10.64)	(6.53)	(10.64)	(6.53)	(-14.63)	(-11.51)	(-14.63)	(-11.51)
Residual standard error		0.05	0.06	0.05	0.06	461.32	541.60	461.32	541.60
Observations number		22525	12332	22525	12332	22569	12352	22569	12352
LR chi2 (variables)		990.86	410.47	990.86	410.47	2138.37	1287.38	2138.37	1287.38
Prob > chi2		0.00	0.00	0.00	0.00	0.00	0.00	0.00	0.00
Pseudo R2		-0.03	-0.03	-0.03	-0.03	0.01	0.01	0.01	0.01
Log-likelihood		17702.12	7979.39	17702.12	7979.39	-115107.73	-64596.00	-115107.73	-64596.00

Note: The values in bold are significant at the 10% level.

equity issues and a relative definition of impoverishment, occurs among both rich and poor households.

The reduction in the number of households with impoverishing health spending has been accompanied by an important change over time in the composition of the overall category that is termed "excessive health spending." This change is likely to be related to many factors including economic progress, poverty alleviation, recovery from financial crisis, and the epidemiological transition with the accompanying process of population aging. Before, during, and immediately after the period of economic crisis, impoverishing health spending was more common and constituted a larger proportion of households with excessive health spending. Economic recovery and the decrease in the number of households below the poverty line has generated a reduction in impoverishing health spending, while the prevalence of catastrophic health spending has experienced less change. For these reasons, at the beginning of the period impoverishing health spending was more common than catastrophic health spending. By 2000 the rates were similar and catastrophic spending now dominates the distribution. Thus, it is now catastrophic health expenditure that dominates the overall distribution, and these incidents are likely to be associated with higher levels of health spending.

The results on family composition show that before 1998 catastrophic and impoverishing expenditures were more common among families with young children, and since 2000 they have become more common among families with elderly members. Further, the regression analysis shows that the presence of a person over age 65 is a significant determinant of the probability that a family suffers an impoverishing or catastrophic health expenditure.

These results suggest several policy recommendations in the context of the transition to universal coverage in 2010. First, the change in the composition of households with excessive health expenditures indicates that in the future the Seguro Popular will have to focus on households with higher income and catastrophic rather than impoverishing health expenditure. It will be increasingly important in future research to isolate the nature and causes of these expenditures. Another priority is to affiliate particular population groups such as families that include older adults—almost 20% of all families—in order to achieve the goal of reducing catastrophic health spending.

Second, the results suggest the importance of expanding the package of insured services to cover chronic illness and the diseases associated with aging in subsequent phases of the reform. These tend to include the more expensive health problems that are becoming increasingly important in the burden of disease in Mexico as epidemiologic transition expands.

In order to meet the complex health care demands of an aging population, one of the most important challenges facing health systems is to develop fair, equitable, efficient, and sustainable financing mechanisms. In the case of developing countries, it is important to develop health financing institutions that will permit them to face the challenges of aging before their populations age.

The Mexican health reform that created the System for Social Protecion in Health in 2003 is a recent example of a reorganization of health financing that seeks to promote more and more efficient investment in health, while at the same time discouraging spiraling health care costs. The first years of the reform suggest substantial advances in implementing the reform and covering the poorest segments of the population. One of the key challenges for future stages of the reform will be addressing the needs of an aging population.

References

Arreola, H., Knaul, F., Méndez O., and Nigenda, G. (2004). Disposición a pagar por un seguro de Salud público en México. FUNSalud-INSP. Documento de trabajo. México, D.F.

Banco de México. (2005). Informe Anual 2004. BANXICO, México, D.F. Retrieved from http://www.banxico.org.mx/gPublicaciones/FSPublicaciones.html.

Consejo Nacional de Población (CONAPO). (1999). Envejecimiento demográfico de México: Retos y perspectivas por una sociedad para todas las edades. México, D.F.

Frenk, J., González-Pier E., Goméz-Dantés O., Lezana, M.A., and Knaul, F. (2006). Comprehensive reform to improve health system performance in Mexico. *The Lancet*. Vol. 38. Issue 9546, pp: 1524–1534.

Frenk, J., Knaul, F., and Gómez-Dantés, O. (2004). Closing the relevance-excellence gap in health research: The use of evidence in Mexican health reform. In S. Matlin (Ed.), *Global forum update on research for Health 2005: Health research to achieve the millennium development goals.* UK: Pro-Book Publishing.

Frenk, J., Knaul, F., Gómez-Dantés, O., *González-*Pier, E., Lezana, M.A., Hernández-Llamas, H., et al. (2004). Fair financing and universal social protection: The structural reform of the Mexican health system. Working Paper prepared for the International Conference, Innovations in Health Financing. Secretaria de Salud. México, D.F. April 20-21, 2004.

Frenk, J., Sepúlveda, J., Goméz-Dantés, O., and Knaul, F. (2003) Evidence-based health policy: Three generations of reform in México. *Lancet, 362(9396),* 1667–1671.

Frenk, J., Lozano, R., González-Block, M.A., Zurita, B., Knaul, F., Cruz, C., (1994) Economía y Salud: Propuesta para el avance del sistema de Salud en México. Informe Final. FUNSalud. México, D.F.

Frenk, J., Bobadilla, J.L., Sepúlveda, J., and López-Cervantes, M., (1989). Health transition in middle-income countries: New challenges for health care. *Health Policy and Planning, 4,* 29–39.

Instituto Nacional de Estadistíca, Geografía e Informática (INEGI). (1992). Encuesta Nacional de Ingresos y Gastos de los Hogares 1992. Aguascalientes, Ags. México.

Instituto Nacional de Estadistíca, Geografía e Informática (INEGI). (1994). Encuesta Nacional de Ingresos y Gastos de los Hogares 1994. Aguascalientes, Ags. México.

Instituto Nacional de Estadistíca, Geografía e Informática (INEGI). (1996). Encuesta Nacional de Ingresos y Gastos de los Hogares 1996. Aguascalientes, Ags. México.

Instituto Nacional de Estadistíca, Geografía e Informática (INEGI). (1998). Encuesta Nacional de Ingresos y Gastos de los Hogares 1998. Aguascalientes, Ags. México.

Instituto Nacional de Estadistíca, Geografía e Informática (INEGI). (2000). Encuesta Nacional de Ingresos y Gastos de los Hogares 2000. Aguascalientes, Ags. México.

Instituto Nacional de Estadistíca, Geografía e Informática (INEGI). (2002). Encuesta Nacional de Ingresos y Gastos de los Hogares 2002. Aguascalientes, Ags. México.

Instituto Nacional de Estadistíca, Geografía e Informática (INEGI). (2004). Encuesta Nacional de Ingresos y Gastos de los Hogares 2004. Aguascalientes, Ags. México.

Instituto Nacional de Salud Pública de México, Fundación Mexicana para la Salud and Secretariá de Salud. (2001). Encuesta Nacional de Aseguramiento y Gasto en Salud, 2001. INSP-Funsalud, Documento de trabajo. México, D.F., 2001.

Knaul, F., et al. (2006). Evidence is good for your health system: Policy reform to remedy catastrophic and impoverishing health spending in Mexico. *The Lancet.* Vol. 348. Issue 9549, pp: 1828–1841.

Knaul, F., Arreola, H., Méndez, O., and Miranda, M. (2006). Preventing impoverishment, promoting equity and protecting households from financial crisis: Universal health insurance through institutional reform in Mexico. Working Paper. Mexico City: Fundación Mexicana para la Salud. Available at: http://www.gdnet.org/pdf2/gdn_library/awards_medals/2005/medals_cat3_first.pdf. Paper presented at the Global Development Network Conference, St. Petersburg, Russia. January, 2006.

Knaul, F., and Frenk, J. (2005). Health insurance in Mexico: Achieving universal coverage through structural reform. *Health Affairs,* 24(6), 1467–1476.

Knaul, F., Arreola, H., and Méndez, O. (2005). Protección financiera en Salud: México 1992 a 2004. *Salud Pública de México, 47,* 430–439.

Knaul, F., Arreola, H., Méndez, O., and Martínez, A.,(2005). Fair health financing and catastrophic health expenditures: Potential impact of the coverage extension of the Popular Health Insurance in Mexico. *Salud Pública de México, 47,* S1:S54–S65.

Knaul, F., Arreola, H., and Méndez, O. (2004). Catastrophic and impoverishing health expenditure: Increasing risk pooling in the Mexican health system. Background Paper for: Baeza, C., and Packard, T. *Beyond survival, protecting households against the impoverishing effects of health shocks.* Regional World Bank Report, Latin America and the Caribbean Region, for the regional study *Risk pooling, ahorro y prevención: Políticas e instituciones para la protección de los más pobres de los efectos de los shocks de Salud.* Mimeo. México, D.F.

Knaul, F., Arreola, H., Borja, C., Méndez, O., Soberón, G., Escandón P., et al., (2004). Competitividad y Salud: La propuesta. *Economía y Salud, 19.* FUNSalud. México, D.F.

Knaul, F., Arreola, H., Borja C., Méndez, O., and Torres, A.C., (2003). El sistema de protección social en México: Efectos potenciales sobre la justicia financiera y los gastos catastróficos de los hogares. In F. Knaul and G. Nigenda (Eds.), *Caleidoscopio de la Salud* (pp. 275–291). FUNSalud. México, D.F.

Londoño, J.L., and Frenk, J. (1997). Structured pluralism: Towards an innovative model for health system reform in Latin America. *Health Policy, 41(1),* 1–36.

Lozano, R. (1997). El peso de la enfermedad en México: Avances y desafíos. In J. Frenk (Ed.), *El observatorio de la Salud: De la investigación a las políticas y de las políticas a la acción* (pp. 23–61). FUNSalud. México, D.F.

Murray, C.J.L., and Frenk, J. (2000). A World Health Organization framework for health system performance assessment. *Bulletin of the World Health Organization. 78(6),* 717–731.

Murray, C.J.L., Knaul, F., Musgrove, P., and Xu, K., (2000). Defining and measuring fairness of financial contribution. Global Programme on Evidence Working Paper (24). World Health Organization.

Parker, S., and Wong, R. (1997). Household income and health care expenditures in Mexico. *Health Policy, 40,* 237–255.

Partida, V. (2002). La situación demográfica actual. In Consejo Nacional de Población (CONAPO), *La situación demográfica en México* (pp. 11–15). México, D.F.

Partida, V., and García, J.E. (2002). El cambio epidemiológico reciente. In Consejo Nacional de Población (CONAPO), *La situación demográfica en México* (pp. 17–27). México, D.F

Phelps, C. (2003). *Health economics.* Boston: Addison-Wesley.

Romanov, R.J. (2002). Building values for the future of health care in Canada. Commission on the Future of Health Care in Canada. Canada, November.

Secretariá de Salud. (2005). Salud: México 2004. Información para la rendición de cuentas. Secretaria de Salud. México, D.F.

Secretariá de Salud. (2004). Salud: México 2003. Información para la rendición de cuentas. Secretaria de Salud. México, D.F.

Secretariá de Salud. (2003). Salud: México 2002. Información para la rendición de cuentas. Secretaria de Salud. México, D.F.

Secretariá de Salud. (2001). Programa Nacional de Salud 2001–2006. Secretaria de Salud. México, D.F.

Torres, A.C., and Knaul, F. (2003). Determinantes del gasto de bolsillo en Salud e implicacionaes para el aseguramiento universal en México. In F. Knaul and G. Nigenda (Eds.), *Caleidoscopio de la Salud* (pp. 209–225). FUNSalud. México, D.F. United Nations. (2002). General Assembly on Aging, New York, NY 2002.

Wagstaff, A. and Van Doorsaler, E. (1999). Equity in the finance of health care: Some further international comparisons. *Journal of Health Economics, 18,* 263–290.

Wagstaff, A. and Van Doorsaler, E. (1998). Equity in the finance and delivery of health care: An introduction to the Equity Project. In M.L. Barer, T.E. Getzen, and G.L. Stoddart (Eds.), *Health care and health economics: Perspectives on distribution* (pp: 179–207). New York: John Wiley.

Wagstaff, A., and Van Doorsaler, E. (2003). Catastrophe and impoverishment in paying for health care: With applications to Vietnam 1993–98. *Health Economics, 12,* 921–934.

World Health Organization (WHO). (2000). The World Health Report 2000. Health systems: Improving performance. Geneva: WHO.

Xu, K., Evans, D., Kawabata, K., Zeramdini, R., Klavus, J., and Murray, C., (2003). Household catastrophic health expenditure: A multicountry analysis. The *Lancet, 362 (9378),* 111–117.

19
The Health Care Safety
Net for Hispanics

Ronald J. Angel and Jacqueline L. Angel

19.1. Introduction

The research reported in this volume makes it clear that the causes and potential consequences of differential patterns of morbidity and mortality among African-Americans, Hispanics, and Asians are complex, and as yet incompletely understood. Perhaps the only categorical summary statement that one might make concerning the health of older Hispanics is that despite continuing socioeconomic disadvantages, Hispanics, like all other Americans, are living longer and in most cases healthier lives than in previous decades. Yet not all is well, and despite longer life spans, the research reported in this volume reveals that Hispanics and other minority Americans suffer significant morbidity and disability as the result of preventable causes that have their origin in the complex nexus of genetics, lifelong socioeconomic disadvantage, the exposure to environmental and social health risks, and inadequate medical care. Increased longevity in the presence of significant illness and disability means that older Hispanics have elevated levels of need for medical and social services that in many cases are not being addressed.

The papers in this volume examine many population health issues, including the epidemiology of general and specific illnesses, and access to health care in different social and cultural contexts. In this concluding chapter we examine issues related to health care access, and more specifically those related to health care financing as they affect Hispanics and minority Americans generally. In the context of a discussion of the health of older Hispanics the issue of health care financing is particularly salient since there is ample documentation of the fact that individuals with health insurance receive more and better health services and enjoy better health (Institute of Medicine, 2001; Institute of Medicine, 2002). The data also clearly document the negative social and community consequences of inadequate coverage (Institute of Medicine, 2003). Lowered productivity and the costs of treatment for illnesses that might have been avoided with adequate preventive care undermine the welfare of entire groups and represent a major long-term social cost.

Despite the clear individual and collective importance of timely and high-quality preventive and acute care, the United States stands alone among developed nations in not having a universal health care system. Rather, we have developed a system of employment-based group health insurance that covers most middle class Americans, but that leaves individuals and families with tenuous ties to the labor market and those employed in the low-wage service sector without coverage (Quadagno, 2004; Starr, 1982; Weir, Orloff, and Skocpol, 1988). Although Medicaid and the newly introduced State Children's Health Insurance Plan (SCHIP) provide coverage to pregnant women and children in poor families, many other working families with relatively low incomes find themselves without a family health care plan.

Unfortunately, in the United States, employment, and even full-time work, does not guarantee that one will have health coverage. For the working poor, many of whom work full-time, family health plans are unavailable or simply too expensive. In this chapter we briefly review the major holes in the health care safety net for older Hispanics and examine the possibilities for meaningful and realistic health care reform in the United States, especially the ways in which the proposals that have been offered might affect health care coverage and access for Hispanics.

19.2. The Complex Profile of Health Care Coverage Among Hispanics

Before we review the potential remedies to the problem, let us begin by examining the extent of the problem of health care coverage among adult Hispanics. Table 19.1 presents information concerning the sources of coverage of adults aged 18 to 64 and 65 and older by nationality. These data clearly reveal the serious lack of health insurance among working-age Hispanic adults, and especially those of Mexican origin. Although the majority of working-age non-Hispanic white adults have employer-sponsored health insurance, fewer than half of Mexican American or Puerto Rican adults are members of employment-based group plans. Approximately one quarter of black, Cuban American, and Puerto Rican adults report that they have no health insurance. Among those of Mexican origin, nearly half, or 45%, have no health care coverage. It is clear from these data that working-age Mexican-origin individuals are uniquely vulnerable to inadequate health care coverage (Angel and Angel, 1996; National Council of La Raza, 1992; Valdez et al., 1993).

These statistics underscore the differential vulnerability of various groups. Among Hispanics, factors that are as yet poorly understood affect coverage. These may reflect regional concentration and labor market disadvantages, immigration and citizenship status, language difficulties, and other barriers that increase the risk of inadequate coverage (Berk, Schur and Feldman, 2001; Berk, Albers, and Schur, 1996; Berk, Schur, Frankel, and Chavez, 2000). Mexican Americans are concentrated in low-wage service sector jobs and are far less likely than any other group to be employed in managerial or professional occupations (National Council of La Raza, 1992; Schur and Feldman, 2001). The bottom panel of Table 19.1,

TABLE 19.1. Selected type of health insurance coverage for persons under 18 and older: 2001 by race and Hispanic ethnicity[1]

Type of coverage	Non-Hispanic White	Non-Hispanic Black	Mexican American	Cuban American	Puerto Rican
			18–64 years		
Employer	65%	58%	43%	57%	49%
Medicare	—	—	2%	4%	5%
Medicaid	5%	14%	8%	8%	22%
None	13%	24%	45%	25%	23%
Total (in thousands)	122,470	20,648	14,768	795	2,021
			65 years and over		
Employer	36%	29%	17%	15%	17%
Medicare	97%	93%	91%	92%	97%
Medicaid	7%	20%	22%	31%	37%
None	0.03%	2%	5%	3%	0.04%
Total (in thousands)	27,973	2,801	992	311	213

Categories may overlap.
Source: U.S. Census Bureau, *Annual Demographic Supplement*, 2002, and unpublished tabulations for Hispanic subgroups.

which provides information on the coverage of individuals over age 65, clearly demonstrates the equalizing effects of Medicare, which covers over 90% of each group. Even after age 65, though, important group differences in the nature and extent of coverage remain. While approximately 36% of non-Hispanic Whites and 29% of non-Hispanic Blacks report employer-based insurance, only about 15% of any of the three Hispanic groups report employer-based coverage.

Since Medicare does not cover the full cost of health care and only recently began assisting with prescription drugs, individuals without supplemental Medigap plans can accumulate substantial debt or be forced to go without all of the care they need. Table 19.2 presents information from the Hispanic Established Populations for Epidemiologic Studies of the Elderly (H-EPESE), a study that focuses on

TABLE 19.2. Type of health insurance coverage by nativity and age at migration for elderly Mexican Americans: 1993–94 (weighted percent; unweighted sample size (N) in parentheses)

Age at Migration	No Insurance	Medicare Only	Any Medicaid	Private[1]
Late-life	29.8	30.6	34.5	5.2
Middle-age	6.1	46.6	31.6	15.8
Childhood	6.7	40.3	33.0	20.1
Native-born	3.2	38.7	26.5	31.6
Sample size	(137)	(1,205)	(958)	(634)

[1] Includes military health care, i.e., CHAMPUS (Comprehensive Health and Medical Plan for Uniformed Services), CHAMPVA (Civilian Health and Medical Program of the Department of Veteran's Affairs).
Source: H-EPESE

the Mexican-origin population of the Southwest. It reveals that among Mexican-origin foreign-born individuals who immigrated in later life, almost 30% report no insurance and only about 5% report that they have any private Medigap coverage. Native-born elderly Mexican Americans are far more likely to have private medical insurance; 32% report coverage by Blue Cross/Blue Shield or a military health care plan like CHAMPUS.

19.3. The Source of the Problem and Potential Solutions

The source of the problem of differential coverage for different groups can be traced to complex historical, political, and economic forces that have left many poor and minority Americans outside of the post-World War II employment-based health care financing system. If access to health insurance is a benefit tied to employment in high-wage jobs, those workers who are confined to the low-wage sector are almost by definition left out. Although public coverage is available to poor children , the disabled, and the elderly, working-age adults have relatively few options other than charity. Our current health care safety net catches some of those most seriously in need, but it lets many others, including many Hispanics, fall through. Given the structural inability of our current employment-based system to cover all individuals and families, it seems highly unlikely that any reform that does not have as its goal a comprehensive and universal system could ever provide access to basic care for the over 45 million Americans without coverage. Since a disproportionate number of those without coverage are Hispanic, and especially of Mexican origin, for these groups partial reforms that do not address the fundamental problems of access to health care coverage offer little hope.

The question that immediately comes to mind, of course, is whether any significant reform is possible. Although there is reason to imagine that the growing health insurance insecurity of the middle class may provide the impetus for fairly basic reform of the way in which health care is financed in the U.S. in the future, it is difficult to imagine that fundamental changes possibly beneficial to Hispanics will occur in the short term. After the resounding failure of the Clinton health plan in the 1990s, a realistic view might well be that large-scale reform of our employment-based system of health care coverage is simply impossible. After all, the Clinton proposal for universal coverage was only the most recent of a series of attempts to extend health care coverage to all citizens that began with the Progressives, and that system has proven to be highly resistant to fundamental change (Quadagno, 2005; Starr, 1982).

The reasons for the resilience of the current system are not difficult to identify. The employment-based system of group health insurance that flourished after the Second World War has worked well for all of the politically influential players involved. The system has worked well for management, labor, insurers, doctors, hospitals, pharmaceutical companies, and the rest of the health care industry. The health insurance industry has become a major component of the overall economy (Light, 2004). The current arrangement has worked well for employed middle class Americans who correctly perceive that a more inclusive and generous public system would undermine their coverage, restrict their own choices in providers, and result in higher taxes. Medicare and Medicaid, and more recently the State Children Health Insurance Program (SCHIP) have extended coverage to the elderly and children in low-are families, thereby neutralizing two potential sources of serious criticism of the system as a whole. The result is that those segments of the population with the greatest political power have little interest in reform. The fact that 47 million Americans have no health insurance and many more have incomplete or episodic coverage is unlikely to lead to significant health care financing reform. Those 47 million Americans are simply the wrong people to effect reform. They do not have the sort of political or economic power that can prevail against deeply entrenched interests. They are working poor families, unemployed and underemployed adults, children who qualify for Medicaid or SCHIP on the basis of family income criteria but who are not enrolled, and, of course, African Americans and Hispanics.

Yet, as we have suggested, other changes that are undermining the post-war, employment-based health insurance system for middle class Americans may well lead to the need for a more rational and planned system. A recent Kaiser report documents the fact that the cost of health insurance premiums continues to increase at a rate far higher than general inflation and that employers are dropping plans or shifting a greater portion of the cost to their employees (Kaiser Family Foundation, 2005). Increasing health insurance insecurity in the middle class may generate the political impetus for health care financing reform that could benefit Hispanics and other marginalized Americans. On the other hand, if health care financing for the poor continues to rely on federal/state-based funding formulas, financially strapped state legislatures may very well balance their budgets at the expense of the most vulnerable citizens. As the cost of medical care increases, it is difficult to imagine that charitable sources will pick up the slack.

For the remainder of the chapter we focus on the fundamental shortcomings of proposed reforms as they would affect Hispanics and other poor and minority Americans. Our conclusion is that partial reforms that do not lead toward a universal publicly funded system will leave a disproportionate number of minority Americans in the same situation they find themselves in today. Before proceeding, though, we must point out that, as the contributions to this volume show, in addition to having inadequate and incomplete health care coverage, minority Americans are subjected to a wide array of structurally based health risks that represent a major public health crisis. The system fails minority Americans not only in denying them all of the health care they need, but by exposing them to class-based health risks that are the result of an inadequate public health system. A large body of research, in addition to that presented in the previous chapters, shows that the systems for assuring the health of the work force are in serious disarray (Committee on Assuring the Health of the Public in the 21st Century, 2002). Although modern medical care has clearly contributed to both the increased length and higher quality of life, basic public health measures such as insuring clean water and safe food, developing safe workplaces, providing family planning, and especially the delivery of childhood immunization have historically played a major role in increasing life spans (Centers for Disease Control, 1999).

Yet a large fraction of African-American and Hispanic children are incompletely immunized. In 2001 approximately one quarter of African-American and Hispanic children had not received all of their immunizations (Chu, Barker, and Smith, 2004). This fact adds to the already elevated risk profile associated with poverty and threatens the quality of life of millions of young Americans who in the not too distant future will make up a large fraction of the labor force. African-American and Hispanic children suffer from other modifiable health risks such as obesity (Hedley et al., 2004; Hoelscher et al., 2004; Ogden et al., 2002). Clearly, these children are at elevated risk of the early onset of all of the negative consequences of obesity and as a result their health in later life may be seriously compromised.

19.3.1. Are There Realistic Options for Nearly Universal Coverage?

As we mentioned earlier, universal health care coverage has never even come close to passage in the United States. The most recent attempt during the early years of the Clinton Administration clearly revealed the large number of obstacles. Clinton and many others had reason to believe that the nation was ready for major health care reform. After Pennsylvania Senator John Heinz was killed in a plane crash in April 1991, Harris Wofford was appointed interim Senator until a special election could be held that November. Although Wofford began the campaign far behind his opponent, ex-Governor Dick Thornburgh, he defeated his opponent handily in the special election in November, largely as the result of his campaign focus on health insurance. Wofford's victory helped define the Clinton presidential campaign and many observers came to believe that the time for comprehensive coverage had finally arrived. Initially the proposal seemed to have broad support (Skocpol, 1996; Starr, 1994). Unfortunately opposition by segments of the insurance and medical establishments, in combination with serious political blunders by the Clinton Administration itself, resulted in a resounding defeat (Daniels, Light, and Caplan, 1996; Skocpol, 1996). In combination with the previous failures to introduce a system of universal health care coverage the Clinton defeat might well lead one to despair of the possibility of ever seeing universal health care coverage in the United States.

A single-payer system is clearly not possible in the short term, and perhaps even in the long run some other method of insuring universal coverage that is more compatible with the status quo may be necessary. New initiatives based on public/private partnerships that include the insurance industry and other organizational and financing innovations may prove effective in providing health care coverage to at least some fraction of the uninsured (Rosenau, 2000; Sparer, 2000). The growing use of privately managed care plans by state Medicaid administrators is an example of such public/private ventures (Sparer, 2000). If these and similar initiatives result in near universal coverage for basic preventive and acute care, the basic objective of more radical plans will have been achieved.

One of the unique features of U.S. social policy and one of the major roadblocks to reform is the fact that insurance regulation and much social policy for the poor is made at the state level. This fact means that although Washington sets the basic requirements for who and what must be covered we have 50 different systems, some of which are more inclusive than others. Given the fact that Medicaid and other programs for the poor are jointly funded on a federal-state cost-sharing basis, they represent a major drain on state revenues. Increases in health care coverage for poor adults and children are only possible if other programs are cut or if income, sales, and other taxes are raised. Most states face serious opposition to increased taxation and choices among programs are politically difficult. Although Hispanics are fanning out over the entire country, certain states continue to have disproportionately large populations of specific Hispanic groups. In states like California and Texas issues of immigration and the cost of social services to illegal immigrants add an increased acrimony to issues of social service expenditure (Pourat, Lessard, Lulejian, Becerra, and Chakraborty, 2003). Political difficulties aside, though, there are few options other than the safety net. A realistic universal system will probably consist of a publicly subsidized employment-based system accompanied by an expanded safety net.

19.3.2. Proposals for Reform: How Might They Affect Hispanics?

The political and economic challenges faced by would-be health care financing reformers, and especially those who advocate for the most vulnerable Americans, including Hispanics, are formidable. As we have noted, the present employment-based group plan model of health care financing that is unique to the U.S. evolved by incorporating the most politically and economically powerful segments of the population. Given our winner-take-all, two-party system of government and the absence of a Labor party there has never been an organized group furthering the cause of universal health care. The United States is simply not Europe. The result is our unique system of group coverage for those employed in good jobs and no insurance or means-tested coverage for the unemployed or those employed in the low-wage service sector in which health coverage is not offered.

The fundamental problem then lies in the fact that if one does not have access to affordable family coverage through work, there are few options. Individual policies can be purchased, but workers who are employed in jobs that do not offer group coverage are unlikely to be paid enough to afford even a minimally adequate private plan. Even after age 65, when universal coverage through Medicare becomes available, Hispanics, and especially those of Mexican origin, are less likely than non-Hispanic White retirees to have supplemental Medigap plans to cover the premiums and other costs that Medicare will not pay and that can result in burdensome debt for individuals and couples on a low fixed income. Many such plans are part of employment-based retirement packages. Unfortunately, although Hispanics have relatively high employment rates, they are very likely to hold jobs that do not offer coverage (Angel and Angel, 1996). Yet employment often places one above the cutoff for public services. As we have noted, except for pregnant and lactating women and the disabled, the U.S. offers little public coverage other than charitable care for low-income adults (Institute of Medicine, 2001). The only means for assuring greatly increased coverage for Hispanics and other vulnerable Americans, then, is to go beyond the employment-based group plan model.

As many employers, and particularly small businesses in which a disproportionate number of Hispanics are employed, find themselves faced with rapidly rising health insurance costs, many are dropping their plans if they ever had them, and those that continue to offer coverage are passing an increasing proportion of the costs of coverage on to their employees in the forms of higher premiums, co-pays, and deductibles. Since their employers cannot provide the coverage that low-wage working families need, publicly funded care is essential in order to insure and equitable system. Unfortunately the current multi-program public system is incapable of meeting the basic standards of equity and efficiency.

The possibilities for reform of our current health care financing system range from basically no reform, i.e., maintaining the status quo, to moving toward a universal and comprehensive single payer system in which the federal government would assume primary responsibility for the health care needs of all citizens. Many proposals have been offered. Many are based on the expansion of existing programs such as Medicare and Medicaid with incentives and mandates for employers to offer coverage and employees to purchase it. Most of these proposals would probably increase the number of insured Americans, but without universal mandates and adequate funding none of the proposals would cover everyone. Those who would benefit least from proposals that focus heavily on reforms of the existing employment-based system include currently uninsured Hispanics. It is unlikely that any system of tax credits or subsidy, unless they covered basically the entire cost of coverage, would induce employers in low-profit sectors to offer coverage to highly mobile and easily replaced workers.

Since the 1980s Congress has passed significant health care financing reforms. The most important of these have extended coverage for specific groups of the uninsured or underinsured, including the low-income elderly, the disabled, and children in families with incomes below the 200% poverty threshold. Unfortunately, these health care financing reforms have failed to reduce the overall number of uninsured Americans and, in fact, that number has continued to increase along with the size of the population (Institute of Medicine, 2004; Mills and Bhandari, 2003). During the 1980s and 1990s Medicaid and SCHIP eligibility were extended to nearly all children in families with incomes below the 200% poverty threshold (Broaddus and Ku, 2000). Yet fully half of those children who remain uninsured qualify for these programs but are not enrolled (Dubay, Haley, and Kenney, 2002; Kenney, Haley, and Tebay, 2003). In addition to eligible children who are not enrolled these reforms have left many poor elderly individuals, including a large fraction of minority Americans, vulnerable to crushing medical debt. As we noted earlier, even with Medicare the elderly spend a large fraction of their limited incomes on health care and many face impoverishment when they must use long-term care (Angel and Angel, 1997).

For pre-retirement age adults, federal efforts have focused on improving aspects of employment-based insurance though such legislation as the Consolidated Omnibus Budget Reconciliation Act (COBRA) of 1985 and the Health Insurance Portability and Accountability Act (HIPPA) of 1996. These bills are clear benefits to those workers with good jobs since they provide for the portability of insurance from one employer to another and allow individuals to remain in an employer's group plan at his or her own expense for a period after the termination of his or her employment. Unfortunately, Hispanics were again less likely to benefit from these reforms since they were less likely to have had benefits in the first place. Other legislation like the Trade Act (TA) of 2002 provides assistance for individuals who have lost their jobs as the result of international competition (Institute of Medicine, 2004). Again few Hispanics benefit.

Despite the government's piecemeal efforts to retain coverage for the non-welfare unemployed, then, the basic problem remains that if one loses one's job and does not find another one quickly one's group insurance premiums eventually become unaffordable. While portability and continuity of coverage are clear gains for workers in general, if that portability and continuity depend upon the continuity of employment or employment in good jobs, periods of unemployment and employment in low-wage jobs will almost inevitably translate into a chronic lack of coverage and inadequate health care.

19.3.3. Realistic Options for Reform

With few exceptions most proposals for reforming the way health care is provided in the U.S. have been incremental (Maioni, 1998; Marmor and Barer, 1997; Marmor, 1973; Starr, 1982). While most other developed nations introduced some form of universal coverage in the years after World War II in the U.S. private group plans provided by private and public employers became the norm. By the 1950s Blue Cross/Blue Shield plans were the dominant form of coverage for the working-age population (Cunningham III and Cunningham Jr., 1997; Starr, 1982). The focus on employment, though, left many individuals out of the system. Those in jobs in the service or agricultural sectors, the unemployed, the elderly, and a large fraction of children had no regular access to health care. The passage of Medicare and Medicaid addressed some of the most glaring shortcomings of the system and provided coverage to the elderly, a group that wields substantial political clout (Fox, 1986; Hacker, 1997; Schlesinger and Kronebusch, 1990). The remainder of the population with tenuous or no ties to the labor market still lacks adequate coverage, and that group includes a disproportionate number of African-Americans and Hispanic Americans.

Currently, the impetus for the sort of reform that might extend coverage to a large fraction of the un- or underinsured is weak. Even the perceived financial crisis that has accompanied the growth of medical costs has not resulted in serious broad-based support for radical reform (Kahn and Pollack, 2001). In the face of massive resistance to change, the plight of the poor carries little political weight. Yet the exclusion of a large fraction of the population from adequate coverage threatens the very economic fabric of our economy. Our collective welfare depends on the health of the labor force. The situation is particularly clear in the case of Social Security. In 1945, there were approximately 40 workers for each retiree; by the year 2003 there were slightly more than three workers supporting each retired American (Social Security Administration, 2004). In the future the number of workers supporting each retiree will decline even further (Social Security Administration, 2004). Given the higher fertility of minority Americans, and especially Hispanics, the labor force of the future will consist disproportionately of minority group members. At the same time the majority of the retired population will be non-Hispanic and White. If the productivity of those minority workers is undermined by educational and other disadvantages in youth they will not have the resources to, nor will they likely be willing to, shoulder the burden of supporting a disproportionately privileged non-Hispanic White elderly population (Lee, 1997).

Incremental reforms that fail to achieve universality and comprehensiveness are unacceptable because they are inequitable, unjust, and economically short-sighted (Himmelstein and Woolhandler, 2003; Weil, 2001). On the other hand, incremental reforms that eventually arrive at universal and comprehensive coverage are far more logical and should define our longer-term objectives. It would be useful to speculate on how the range of reforms that have been offered might affect minority Americans. Many proposals have been offered and many represent only slight changes in the way health care is financed in the United States. Medical savings accounts, for example, are one version of what are called "high-deductible plans." These are simply traditional insurance policies with high deductibles combined with a plan that allows individuals to save pre-tax dollars to pay them. Like any insurance policy with a high deductible, these plans cost less since they only pay for major medical expenses after one has paid the $1,500 or more deductible from the savings account. These plans have quickly become popular with employers because they help control health care costs. Unfortunately, they do little more than reaffirm the current system (Robinson, 2004). Such plans allow middle class employed workers to set aside pre-tax dollars that can be accumulated from year to year if they are not used. If one is not employed or works for an employer who cannot afford even this sort of plan the reform is of little use. Those who benefit most are individuals in the highest tax brackets for whom pre-tax savings are a real benefit. Poor African-Americans or Hispanics are not often members of that elite group. For poor families with no income or in lower tax brackets such plans are basically irrelevant. Since those at the bottom of the wage distribution must spend all of their income for immediate needs, the concept of savings, even with tax incentives, is not a powerful motivator.

Other reform proposals are more radical and would probably extend coverage to a larger fraction of the uninsured. Again, though, they would bypass many poor and minority Americans. These proposals would extend existing programs such as Medicaid, SCHIP, and Medicare to include a larger fraction of the working poor, the unemployed, and adults younger than 65 who do not have insurance (Davis and Schoen, 2003; Davis, Schoen, and Schoenbaum, 2000; Feder et al., 2001). As with most proposals short of universal coverage they include a federal tax credit that would allow individuals to buy insurance through their employer or some other group or private provider at an affordable cost. Individuals and families with very low incomes who are not required to pay taxes would receive what is essentially a subsidy with which they could purchase coverage (Davis and Schoen, 2003; Feder et al., 2001; Institute of Medicine, 2004; Pauly and Herring, 2001).

Most of these proposals would retain the current two-tiered system in which individuals with employer-sponsored coverage have a different and usually better set of choices than those with publicly-funded coverage. Under such proposals, tax incentives for employers and employees would remain unchanged and individuals could participate or not depending upon whether they wish to avail themselves of the tax credit. One benefit is that these proposals build upon existing administrative structures and experiences. On the other hand, they retain most of the shortcomings of the current system in which families that are eligible simply do not participate. They would not, therefore, result in universal and comprehensive coverage since they reaffirm the two-tiered approach for the middle class and the poor. In all likelihood, a large fraction of African-Americans and Hispanics who are currently uninsured would remain uncovered.

Some proposals go further, but still focus on employment, again limiting their potential impact for minority Americans. Among these are proposals that would require employers to provide coverage to their employees and also require employees to accept that coverage (Institute of Medicine, 2004; Wicks, 2003). Some states, including Massachusetts and Hawaii, have experimented with such systems with great success (Himmelstein and Woolhandler, 2003). Unfortunately, since small employers cannot afford to provide coverage and their employees cannot afford the premiums, large subsidies to small employers would be necessary, and the system would still require a public safety net based on Medicaid and SCHIP or some other program (Institute of Medicine, 2004). These proposals often include the development of large purchaser pools so that small employers and individuals could join together and have the same ability to negotiate with insurers and health care providers as large companies (Curtis, Neuschler, and Forland, 2001). Once again the employment-health insurance connection undermines the utility of these reforms for African-Americans and Hispanics in low-wage jobs. Were employers required to offer coverage, even with tax incentives or subsidies, many might reduce the number of individuals they hire. In the low-wage service sector, employees are easy to hire and easy to fire, and few restrictions are placed on what the employer must offer in terms of benefits. Were such requirements introduced unemployment of the least employable might well increase.

Some proposals do address the employment problem, at least to some degree. Among these are proposals focus on individuals and families directly rather than on employers or employees. These proposals would provide a subsidy directly to individuals and families for the purpose of purchasing health care coverage from a range of options. Since they would eliminate the employment-insurance connection such proposals would clearly increase the number of insured individuals and families. This coverage might be nearly universal if individuals were required to purchase at least a basic plan for themselves and their dependents (Institute of Medicine, 2004). There are many ways in which such plans might work. The subsidy that families would receive would be based on family size and income and could be paid as a tax credit or as a cash subsidy for those with no tax liability. This approach is clearly far more radial than those that would merely modify the current system. It includes much more government involvement in providing health care and it would result in the elimination of the current employer and employee tax credit and the elimination of Medicaid and SCHIP since they would no longer be necessary. The mandatory nature of coverage would require mechanisms, perhaps tied to income tax collection, for assuring that individuals have purchased the required minimal coverage. These proposals clearly come far closer than those that stay closer to the status quo to covering everyone.

The most radical reform, of course, would move us toward a publicly funded system of universal coverage financed through general tax revenues. All developed nations other than the U.S. has some version of universal health care (Himmelstein and Woolhandler, 2003; Institute of Medicine, 2004). Of course, such a universal system would involve the federal government in health care in a much more comprehensive and directive way that Americans are used to. It would require a new federal agency that would set coverage standards and negotiate with and pay doctors, pharmaceutical companies, hospitals, rehabilitation centers, long-term care facilities, and the providers of any other health care including that for dental and vision problems. Today the way in which Medicare is administered comes closest as an example (Institute of Medicine, 2004). Participation might require a small co-payment, although some proposals reject any out-of-pocket payment as a barrier to access (Himmelstein and Woolhandler, 2003). Clearly, such a system would most completely address the needs of the unemployed, the poor, and the uninsured and provide coverage to the working-age African-Americans and Hispanics who today lack complete coverage. Since access to health care would not depend upon employment, the loss of one's job or the lack of employer-sponsored coverage would not represent a barrier to care.

A single payer system would also have the collective benefit of allowing us to better control the national health care budget. Some advocates of universal coverage believe that the potential savings that greater efficiency would bring about would make such a plan almost cost neutral (Himmelstein and Woolhandler, 2003). Such hopes are probably too optimistic since even with substantial savings in terms of efficiency, bringing nearly 50 million uninsured Americans—basically the same number that receive Social Security—into the system will inevitably incur new and greater costs. Many of these uninsured individuals have serious medical care needs that have gone untreated for long periods of time.

Clearly a fully comprehensive and universal health care system would represent a radical departure from our current complex private insurance system, and that fact remains one of the major roadblocks to the adoption of such a system. Many observers believe that because of the opposition from entrenched interests, including the insurance industry, any attempt to introduce universal coverage would precipitate a fully, publicly financed, single-payer system is politically unrealistic (Tooker, 2002). Such pessimism may well be justified. Aside from the opposition of entrenched interest groups, opposition to universal health insurance arises from Americans' distrust of big government. Many people simply do not believe that government can achieve equitable and just ends without distorting the system or introducing a stultifying bureaucracy. These criticisms are not without merit. Yet, the fact that most of the less radical proposals would do little to extend coverage to the large number of uninsured Americans, a population that consists disproportionately of African-Americans and Hispanics, makes the effort to achieve universal coverage essential. However universal coverage is achieved matters less than the fact that it remains our ultimate objective. Such a universal system may not come about quickly and the most realistic option may be a comprehensive system that is based on a mix of mechanisms that functionally achieve universal coverage (Bodenheimer, 2003).

19.4. Conclusion: The Imperative for Universal Coverage

We might end by again returning to the question of where we are at the beginning of the twenty-first century in terms of insuring optimal population health for everyone, but especially for the most economically vulnerable segments of the population, African-Americans and Hispanics. As the chapters in this volume reveal, although during the twentieth century great strides were made in extending life spans and reducing mortality from the most preventable causes for all groups, substantial disparities both in health and access to health care remain. The fact that disparities based on race and ethnicity persist represents nothing short of a major injustice and an indictment of our system of social justice.

A social system that does not guarantee the highest level of health for all segments of the population is not only unjust, but inefficient. As we noted, today a growing elderly population that is predominantly middle class, White, and non-Hispanic is drawing upon the productive potential of a labor force that is increasingly African-American and Hispanic. Today three workers are responsible for the support of each retiree and that number is declining. Those same workers will be called upon to pay the taxes to finance the war on terror, to deal with global warming and our polluted environment, to rebuild of our decaying physical infrastructure, to educate of the next generation, and to pay for everything else that makes up our modern high-tech civilization. If the productivity of the current and future labor force is compromised by low levels of education and poor health, our global dominance and security will be placed in jeopardy.

The utility of this book and of this concluding chapter is to fuel the debate over what might be done to redress the injustice that a system of unequal health care access represents. Health care is only one of the inputs into the production of good health and productivity. A just society assures not only that its citizens receive the basic preventive and acute health care that allows them to thrive and contribute to society, but that they receive the education that will make them productive citizens, that they have adequate housing and adequate diets, and that they live in safe neighborhood in which they can grow and thrive free of fear and chronic stress.

Today those objectives are far from being fully realized and the fact that those who are most socially vulnerable tend to be African-American and Hispanic means that historical injustices are still with us and will likely be passed on to future generations.

In this context the fight for universal health care coverage forms part of the new agenda that goes beyond the traditional agendas of the political right or left. Population health is central to everyone's well-being and prosperity. A healthy labor force is a more productive labor force that brings profits to management and stockholders. In our modern globalized economy workers in all nations of the world compete with laborers in the least developed nations who work for minimal wages. In order to compete effectively, workers in more developed nations must be even more productive than they were in the past. That productivity has its basis in good health and education. Companies in nations with publicly funded universal health care systems are not burdened with the health care costs for their current and retired employees. The fact that they are free of these costs places them at a competitive advantage relative to companies in the United States.

There are many reasons based both in simple equity and justice and in economic efficiency for universal health care coverage. We end with the observation that any reform of our current system that does not eventually result in a tax-based universal system will not adequately address the needs of the most vulnerable groups in our nation, among them African-Americans and Hispanics. Despite the defeat of the Clinton plan, and despite what at times seem like insurmountable barriers to fundamental reform, anything short of universal coverage represents a failure from the perspective of the populations dealt with in this book.

References

Angel, R.J., and Angel, J.L. (1996). The extent of private and public health insurance coverage among adult Hispanics. *The Gerontologist, 36*, 332–340.

Angel, R.J., and Angel, J.L. (1997). *Who will care for us: Aging and long-term care in multicultural America.* New York: New York University Press.

Berk, M.L., Albers, L.A., and Schur, C.L. (1996). The growth in the US uninsured population: Trends in Hispanic subgroups, 1977 to 1992. *American Journal of Public Health, 86*, 572–576.

Berk, M.L., Schur, C., Frankel, M., and Chavez, L. (2000). Health care use and undocumented Latino immigrants.*Health Affairs, 19*, 51–64.

Berk, M.L., and Schur, C. (2001). The effect of fear on access to care among undocumented Latino immigrants. *Journal of Immigrant Health, 3*, 151–156.

Bodenheimer, T. (2003). The movement for universal health insurance: Finding common ground. *American Journal of Public Health, 93*,112–115.

Broaddus, M., and Ku, L. (2000). *Nearly 95 percent of low-income uninsured children now are eligible for Medicad or SCHIP: Measures need to increase enrollment among eligible but uninsured children.* Washington, DC: Center on Budget and Policy Priorities.

Centers for Disease Control. (1999). Ten great public health achievements—United States, 1900–1999. *Morbidity and Mortality Weekly Report, 48*, 241–243.

Chernew, M.E., Jacobson, P.D., Hofer, T.P., Aaronson, K.D., and Fendrick, M.A. (2004). Barriers to constraining health care cost growth. *Health Affairs, 23,* 122–128.

Chu, S.Y., Barker, L.E., and Smith, P.J. (2004). Racial/ethnic disparities in preschool immunizations: United States, 1996–2001. *American Journal of Public Health,* 94,973–977.

Committee on Assuring the Health of the Public in the 21st Century. (2002). *The future of the public's health in the 21st century.* Washington, DC: National Academy of Sciences.

Cunningham III, R., and Cunningham, Jr., R.M. (1997). *The blues: A history of the Blue Cross and Blue Shield system.* De Kalb, IL: Northern Illinois University Press.

Curtis, R.E., Neuschler, E., and Forland, R. (2001). Consumer-choice purchasing pools: Past tense, future perfect? These pools could play and important role in securing coverage for American workers. *Health Affairs, 20,* 164–168.

Daniels, N., Light, D.W., and Caplan, R.C. (1996). *Benchmarks of fairness for health care reform.* Oxford and New York: Oxford University Press.

Davis, K., and Schoen, C. (2003). Creating consensus on coverage choices. *Health Affairs,* hlthaff.w3.199.

Davis, K., Schoen, C., and Schoenbaum, S.C. (2000). A 2020 vision for American health care. *Archives of Internal Medicine, 160,* 3357–3362.

Dubay, L., Haley, J.M., and Kenney, G.M. (2002). Children's eligibility for Medicaid and SCHIP: A view from 2000. In the series *New federalism: National survey of America's families* (Number B-41). Washington, DC: Urban Institute.

Feder, J., Levitt, L., O'Broem, E., and Rowland, D. (2001). Covering the low-income uninsured: The case for expanding public programs. *Heatlh Affairs, 20,* 27–39.

Fox, D.M. (1986). *Health policies, health politics: British and American experience, 1911–1965.* Princeton, NJ: Princeton University Press.

Hacker, J.S. (1997). *The road to nowhere: The genesis of President Clinton's plan for health security.* Princeton, NJ: Princeton University Press.

Hedley, A.A., Ogden, C.L., Johnson, C.L., Carroll, M.D., Curtin, L.R., and Flegal, K.M. (2004). Prevalence of overweight and obesity among U.S. children, adolescents, and adults, 1999–2002. *Journal of the American Medical Association,* 291,2847–2850.

Himmelstein, D.U., and Woolhandler, S. (2003). National health insurance or incremental reform: Aim high, or at our feet? *American Journal of Public Health,* 93,102–105.

Hoelscher, D.M., Day, R.S., Lee, E.S., Frankowski, R.F., Kelder, S.H., Ward, J.L., and Scheurer, M.E. (2004). Measuring the prevalence of overweight in Texas schoolchildren. *American Journal of Public Health, 94,* 1002–1008.

Institute of Medicine. (2001). *Coverage matters: Insurance and health care.* Washington, DC: National Academies Press.

Institute of Medicine. (2002). *Care without coverage: Too little, too late.* Washington, DC: National Academies Press.

Institute of Medicine. (2003). *Hidden costs, value lost: Uninsurance in America.* Washington, DC: National Academies Press.

Institute of Medicine. (2004). *Insuring America's health: Principles and Recommendations.* Washington, DC: National Academies Press.

Kahn III, C.N., and Pollack, R.F. (2001). Building a consensus for expanding health coverage. *Health Affairs, 20,* 40–48.

Kaiser Family Foundation. (2005). Employer health benefits: 2005 summary of findings. Retrieved October 28, 2005, from http://www.kff.org/insurance/7315/sections/upload/7316.pdf

Kenney, G.M., Haley, J.M., and Tebay, A. (2003). Children's insurance coverage and service use improve. Washington, DC: The Urban Institute.

Lee, R. (1997). *Public costs of long life and low fertility: Will the baby boomers break the budget?* Center for Economics and Demography of Aging, Berkeley, CA: University of California. Retrieved December 15, 2004, from University of California at Berkeley Resource Center on Aging Web site: http://socrates.berkeley.edu/~aging /Lee.html.

Light, D.W. (2004). Ironies of success: A new history of the American health care 'system.' *Journal of Health and Social Behavior, 45*(Supplement), 1–24.

Maioni, A. (1998). *Parting at the crossroads: The emergence of health insurance in the United States and Canada.* Princeton, NJ: Princeton University Press.

Marmor, T., and Barer, M.L. (1997). The politics of universal health insurance: Lessons from the 1990s. In T.J. Litman and L.S. Robins (Eds.), *Health politics and policy* (pp. 306–322). Washington, DC: Delmar Publishers.

Marmor, T.R. (1973). *The politics of Medicare.* Chicago, IL: Aldine.

Marshall, T.H. (1950). *Citizenship and social class: And other essays.* Cambridge, England: Cambridge University Press.

Mills, R.J., and Bhandari, S. (2003). Health insurance coverage in the United States: 2002. In *Current Population Reports.* Washington, DC: U.S. Census Bureau.

National Council of La Raza. (1992). Hispanics and health insurance. Washington, DC: Labor Council for Latin American Advancement.

Ogden, C.L., Flegal, K.M., Carroll, M.D., and Johnson, C.L. (2002). Prevalence and trends in overweight among U.S. children and adolescents, 1999–2000. *Journal of the American Medical Association, 288,* 1728–1732.

Pauly, M., and Herring, B. (2001). Expanding coverage via tax credits: Trade-offs and outcomes. *Health Affairs, 20,* 9–26.

Pourat, N., Lessard, G., Lulejian, A., Becerra, L., and Chakraborty, R. (2003). *Demographics, health, and access to care of immigrant children in California: Identifying barriers to staying healthy: Fact sheet.* Los Angeles, CA: The UCLA Center for Health Policy Research, UCLA School of Public Health and the UCLA School of Public Policy and Social Research.

Quadagno, J. (2005). *One nation, uninsured: Why the U.S. has no national health insurance.* New York, Oxford: Oxford University Press.

Quadagno, J. (2004). Why the United States has no national health insurance: Stakeholder mobilization against the welfare state 1945–1996. *Journal of Health and Social Behavior, 45* (Supplement), 25–44.

Robinson, J.C. (2004). Consolidation and the transformation of competition in health insurance. *Health Affairs, 23,* 11–24.

Rosenau, P.V. (Ed.). (2000). *Public-private policy partnerships.* Cambridge, MA: The MIT Press.

Schlesinger, M., and Kronebusch, K. (1990). The future of parental care policy for the poor. *Health Affairs, 9,* 91–111.

Schur, C., and Feldman, J. (2001). Running in place: How job characteristics, immigrant status, and family structure keep Hispanics uninsured. New York: Commonwealth Fund.

Shaffer, E.R. (2003). Universal coverage and public health: New state studies. *American Journal of Public Health, 93,* 109–111.

Skocpol, T. (1996). *Boomerang: Clinton's health security effort and the turn against government in U.S. politics.* New York: W.W. Norton and Co., Inc.

Social Security Administration (2004). *The 2004 Annual Report of the Board of Trustees of the Federal Old-Age and Survivors Insurance and Disability Insurance Trust Fund,* Table IV.B2. Retrieved December 15, 2004, from the Social Security Administration Actuarial Resources Web site: http://www.ssa.gov/OACT/TR/TR04/IV_LRest.html#wp178448.

Sparer, M.S. (2000). Myths and misunderstandings. In P.V. Rosenau (Ed.), *Public-private policy partnerships* (pp. 143–). Cambridge, MA: The MIT Press.

Starr, P. (1982). *The social transformation of American medicine.* New York: Basic Books.

Starr, P. (1994). *The logic of health care reform: Why and how the president's plan will work.* New York: Whittle Books in association with Penguin Books.

Therrien, M., and Ramirez, R. (2000). The Hispanic population in the United States. March 2000: Current Population Reports. P-20-535. Retrieved October 28, 2005, from http://www.census.gov/population/socdemo/hispanic/p20-535/p20-535.pdf

Tooker, J. (2002). Affordable health insurance for all is possible by means of a pragmatic approach. *American Journal of Public Health, 93,* 106–109.

U.S. Bureau of the Census. (2003, January 2). Unpublished tabulations, Robert J. Mills.

Valdez, R., Morgenstern, B.H., Brown, R., Wyn, R., Wang, C., and Cumberland, W. (1993). Insuring Hispanics against the costs of illness. *The Journal of the American Medical Association, 269,* 889–894.

Weil, A. (2001). Increments toward what? Incremental steps taken today affect the options available for future coverage expansions. *Health Affairs, 20,* 68–82.

Weir, M., Orloff, A.S., and Skocpol, T. (1988). *The politics of social policy in the United States.* Princeton, NJ: Princeton University Press.

Wicks, E.K. (2003). Issues in expansion coverage design: Decision points and trade-offs in developing comprehensive health coverage reforms. In *Covering America: Real remedies for the uninsured.* Washington, DC: Economic and Social Research Institute.

Afterword
Globalization and Health: Risks and Opportunities for the Mexico—U.S. Border*

Julio Frenk and Octavio Gómez-Dantés

In this essay we will discuss the phenomenon of globalization as it relates to health. We will focus part of our reflections on one of the most dynamic regions of the planet: the United States and Mexico border, the largest and busiest frontier between the developed and the developing worlds, witness to at least one million crossings every day.

Globalization is evolving at such speed and with such complexity that it challenges our ability to grasp it in its full extent. Obvious as it may be, this dynamism is a good reason to constantly renew the discussion around this phenomenon and its impact on everyday life.

The shift of human affairs from the restricted frame of the nation-state to the vast theater of planet Earth is affecting not only trade, finance, science, the environment, crime, and terrorism, but it is also influencing health (Valaskakis, 2001).

In 1997 an influential report by the U.S. Institute of Medicine stated: "Distinctions between domestic and international health problems are losing their usefulness and are often misleading" (Board on International Health, Institute of Medicine, 1997). This is so because of what Eric Hobsbawm, the great European historian has called the "Perhaps the most dramatic practical consequence of these was a revolution in transport and communications which virtually annihilated time and distance." (Hobsbawm, 1994).

We do not mean to say that intense international contacts are new. From time immemorial the forces of trade, migration, war, and conquest have bound together persons from distant places. After all, the Greek philosopher Diogenes coined the expression "citizen of the world" in the fourth century B.C.

What *is* new is the pace, range, and depth of integration. Like never before, the consequences of actions that are taking place far away show up, literally, at our doorsteps.

The degree of proximity in our world can be illustrated by the fact that the number of international travelers has tripled since 1980, and it now reaches 3 million people

*The general reflections on globalization expressed in this paper are based, with modifications, on parts of the article by J. Frenk and O. Gómez-Dantés. Globalization and the challenges to health systems, in *Health Affairs 2002, 21(3)*, 160–165.

every day. In addition, at the start of the millennium the traffic on international telephone switchboards topped 100 billion for the first time in history (AT Kearny, Inc. and Foreign Policy, 2001). The anti-globalization movement itself went global in 2001 when activists gathered in Porto Alegre, Brazil, in the first annual meeting of the World Social Forum.

We cannot underestimate the implications of these changes for health. In addition to their own domestic problems, all countries must now deal with the international transfer of health risks and opportunities (Frenk, Sepúlveda, Gómez-Dantés, McGuiness, and Knaul, 2001).

The most obvious case of the blurring of health frontiers is the transmission of communicable diseases. Again, this is not a new phenomenon per se. The first documented case of a transnational epidemic was the Athenian plague of 430 B.C. Reputed to have begun in Africa, it spread to the heart of ancient Greece through Persia in grain boats (Porter, 1996). The Black Death of 1347, which killed one third of the European population, was the direct result of international trade. In the sixteenth century, the conquest of the Aztec and Inca empires was an early example of involuntary microbiological warfare through the introduction of smallpox and measles to previously unexposed populations. In this microbial exchange Columbus probably took one dire disease from the Americas to Europe: "great pox" or syphilis (Porter, 2004). In 1829 a cholera pandemic that started in Asia, broke into Egypt and North Africa, entered Russia, and crossed Europe. Three years later it reached the eastern coast of the United States. More recently, the global spread of the influenza pandemic of the early twentieth century accounted for far more casualties than World War I.

What is new, as we said before, is the scale of what has been called "microbial traffic." The explosive increase of world travel produces thousands of potentially infectious contacts daily, and jet planes have made even the longest intercontinental flights briefer than the incubation period of any human infectious disease. Thus, a Peruvian outbreak of cholera that started in January of 1991 turned into a continental epidemic in a matter of weeks. By early 1992 it had reached Mexico's border with the Unites States and affected 400,000 Latin Americans.

Tuberculosis is another re-emerging problem. In 2003 close to 9 million persons worldwide became infected with TB and more than 2 million died from it. Several reasons explain this unexpected comeback. One is the fragility of those who become immune-suppressed. TB is often the first sign that a person harbors HIV. Other reasons include overcrowding, poor nutrition, and inadequate health care, common among the socially marginalized. Migrants are a particularly vulnerable population. Not surprisingly, morbidity and mortality rates for HIV and TB are several times higher among migrants and in the northern border states of Mexico than in the country as a whole. Likewise, more than 50% of TB cases in the United States are reported in the four border states. This is why our two countries have developed the binational TB control card, a successful example of cross-border cooperation to address a common threat.

Another innovative initiative launched by the Mexican government is a special program to protect the health of migrants and their families, which has been called

"Leave Healthy/Return Healthy." The aim of this program is to coordinate the delivery of health education, preventive services, and treatment at the place of origin, during the migratory process, and, at the place of destination. Focalized interventions are already being implemented in those Mexican states with high migration patterns.

The latest additions to the list of global epidemics are severe acute respiratory syndrome, or SARS, and Asian avian flu. The latter is still a regional threat, but specialists believe that a full-scale influenza pandemic may be imminent (Osterholm, 2005).

The radical changes in our environment and lifestyles have led Arno Karlen to speak of a new bio-cultural era (Karlen, 1995). Indeed, to make matters more complex, it is not only people and microbes that travel from one country to another; it is also ideas and lifestyles. Smoking provides a clear example. Whenever a legal or regulatory battle against the tobacco companies is won in the U.S., everyone rejoices for the American public but tremble for the consequences in other countries, because those same victories give those same companies the incentive to look for new markets with less stringent regulations. Already about 5 million people are dying of smoking-related causes every year. By 2020 that number will grow to 10 million, making tobacco the leading killer worldwide.

This shows why effective national policies must be coupled with global action, like the Framework Convention on Tobacco Control, aimed at curbing tobacco-related deaths and diseases, which was unanimously adopted by the World Health Assembly two years ago. We feel proud that Mexico became the first country in the Americas to ratify the Framework Convention.

Furthermore, the globalization of health goes beyond diseases and risk factors to include also health products. To mention but one example, careful regulations on access to prescription drugs in one country may be subverted when its neighbor allows the unrestricted purchase of antibiotics, thereby stimulating the appearance of resistant microbes that show up in the first country. This is a particularly relevant issue for our common border. A recent study estimated that 5% of the roughly 350 million annual crossings at the U.S.-Mexico border may be health-related (Sekri, Gómez-Dantés, and Macdonald, 1999). Seventy-five percent of these health-related crossings are from the U.S. into Mexico, most often by persons wanting to purchase pharmaceuticals without prescription, including antibiotics.

The growing commerce of health care services also illustrates the blurring of political frontiers. Mexican nationals, for instance, are regular users of health care in the border states of the United States. To mention but two examples, the Methodist Hospital in Dallas and the MD Anderson Hospital in Houston are receiving an ever increasing number of patients from Mexico, both for outpatient and inpatient care (Warner and Reed, 1993).

But health care consumers are also moving in the opposite direction. Residents in border regions of the United States go to Mexico on a regular basis in search of less expensive health care and dental services. This has been documented in surveys developed since the 1970s in places as diverse as Nogales, Brownsville, and Cameron County (Sekri et al., 1999).

Insurance plans have also been developed since the early 1980s to provide health care in Mexico to migrant farm workers. One of these plans chose a group of Mexicans physicians and a 40-bed hospital in San Luis Rio Colorado, across from Yuma, Arizona, as regular providers. More recently, an Aetna-Meximed joint venture started providing HMO coverage for workers who commute across the border.

The transborder movement of providers is also increasing. The most dramatic example is the migration of nurses. Intermittent nurse shortages are a common phenomenon in rich countries, but are particularly acute in the United States. In the last 15 years the shortages in the U.S. have been met through the importation of nurses mostly from the Philippines but also from Jamaica, Nigeria, and India, sometimes leaving a weakened health care system in source countries. This trend is increasing given the enormous predicted shortfall in the U.S. for the next decade and is already showing its effects on the Mexican nurse community (Aitken, Buchan, Sochalski, Nichols, and Powell, 2004). The government of Zacatecas in Mexico recently implemented a pilot program aimed at helping Mexican nurses get jobs in hospitals in Texas, Arizona, Florida, New York, and Washington.

The risks and opportunities involved in this process are considerable and require careful examination. The international debate around the responsibilities of all actors has produced an interesting range of proposals, which include ethical recruitment guidelines and financial compensation for exporting countries (Brush, Sochalski, and Berger, 2004).

Allow us here, now that we are discussing the transborder movement of workers, to make a brief comment on migration from Mexico to the United States. According to the U.S.-Mexico Binational Council, around 400,000 Mexicans have been crossing the U.S.-Mexico border annually since the late 1990s (U.S.-Mexico Binational Council, 2004). Recent surveys indicate that a large proportion of this migration is permanent. This large migration of the Mexican working force to the U.S is having a differential impact on these two economies. In a sense, one could say that Mexico is exporting part of its demographic bonus to the United States. The influx of young workers is helping to mitigate the financial impact of the aging of the American population. In contrast, Mexico, although benefiting from the remittances of its young migrant workers, is partially losing the benefits of its current demographic bonus, which would otherwise promote local savings and investment to eventually face its own aging process.

The spirit of collaboration imbedded in NAFTA should allow for the bilateral design and implementation of economic and migration policies that can, on the one hand, provide the numbers and types of jobs to encourage most Mexican workers to stay in Mexico and, on the other, offer those willing to migrate to the U.S. safe conditions to do so. Cooperation should always be the motto of interdependence.

In the field of health, in fact, interdependence has opened novel avenues for international collective action. For instance, initial efforts in the 1990s to secure cheaper drugs for AIDS victims in poor countries yielded only modest results. A few years ago, however, strong international mobilization persuaded several major multinational drug companies to establish agreements with developing countries to sell AIDS drugs at heavily discounted prices. Mexico benefited from these agreements and thanks to them universal access to anti-retrovirals has become a reality in Mexico since 2003.

Forces related to globalization also prompted the organization of the U.N. General Assembly Special Session on HIV/AIDS on June 2001, which approved a historical Declaration of Commitment. This was the first time in U.N. history that a session of the General Assembly was devoted to a health topic, thus underscoring the growing link between pandemics such as AIDS, economic development, and global security.

Social mobilization has also played a key role along the U.S.-Mexico border in strengthening cooperative public health interventions. A good example of its results is the Binational Health Week, organized for the first time in 2001 by the Mexican government and the California health authorities, along with several social, academic, and philanthropic organizations. This innovative program included health education and awareness activities, as well as specific prevention campaigns, such as immunization of migrant children. In light of its success, the Binational Health Week has expanded every year, so that in 2004 it covered 15 American states with large populations of Mexican origin.

This sort of initiatives illustrates the growing complexity of health systems. Such complexity has made international comparisons more valuable than ever. Given the enormous economic and social impact of policy decisions, countries can benefit from a process of shared learning. This was the significance of the effort carried out by WHO five years ago to assess the performance of all 191 health systems of the world (World Health Organization, 2000). The identification of relatively good and bad performances will, hopefully, increase the interest around successful experiences and promote the international dissemination of good practice. Perfectible as it is, this exercise has nourished an intense and fruitful debate, which builds on previous efforts by many academic and intergovernmental organizations. This kind of comparative analysis has the virtue of turning information into a global public good (Kaul, Grumberg, and Stein, 1999).

Allow us to present an example of how international methods coupled with excellent country-specific data generated evidence that catalyzed a major health care reform in our own country. The process began with the development of a novel analytical tool: national health accounts. Thanks to several initiatives by academic and international organizations, the Mexican Health Foundation was able to develop in the mid-1990s a system of national health accounts, which showed that more than half of total health expenditure in Mexico was out-of-pocket. This proved to be a direct result of the fact that approximately half of the country's population lacks health insurance.

These findings were unexpected, as it was generally believed that the Mexican health system was based on public funding. The realization that households had been paying impoverishing out-of-pocket sums to private providers generated a different perspective on the operation of the health system in Mexico. Policymakers extended their focus to include financial issues that proved to have a great impact on the provision of health care and levels of poverty among Mexican households.

As a direct result of its high levels of out-of-pocket spending, Mexico was seen to perform very poorly on fair financing in the international comparative analysis developed by WHO. Instead of generating a defensive reaction, this poor result spurred detailed country-level analysis in 2001 that showed that catastrophic and impoverishing health expenditures were concentrated among poor and uninsured households. More specifically, uninsured households with senior citizens or women that had recently delivered a baby showed higher risks (Sesma-Vázquez, Pérez-Rico, Sosa-Manzano, and Gómez-Dantés, 2005). The risk of incurring in this kind of expenditure was 2.5 times higher in uninsured households with senior citizens than in uninsured households that had no elderly members. The analysis was undertaken jointly by the Ministry of Health of Mexico, WHO, and the Mexican Health Foundation, an example of how national governments, international organizations, and non-governmental institutions can join forces.

The country-level analysis was based on data from the National Health Income and Expenditure Surveys for Mexico, yet another effort made for the global public knowledge good. These surveys are produced by many countries in the world and provide homogenous datasets that are key for cross-national comparisons.

The careful interplay between national and international analyses generated the advocacy tools to promote a major legislative reform establishing a system of social protection in health, which was approved in 2003 by a large majority of the Congress. This system will reorganize and increase public funding over seven years in order to provide universal health insurance, including the 45 million Mexicans (3 million of which are over 65), most of them poor, who have been excluded until now from formal social insurance schemes for workers in the formal sector of the economy.

The standardized methodology developed by WHO to assess fairness of finance is also being applied at the state level in Mexico to benchmark health system performance and mark progress toward implementation of the reform.

This is a clear example of how the process of globalization can turn knowledge into an international public good that can then be brought to the center of the domestic policy agenda in order to address a local problem. Such application, in turn, feeds back into the global pool of experience, thus generating a process of shared learning among countries.

The performance of local health systems can also be enhanced by one of the most potent motors of globalization: the telecommunications revolution. Telemedicine is opening vast perspectives for improving access to care by underserved populations and points the way to a future when physical distance may no longer be a significant barrier to health care.

The challenge, of course, will be to make sure that the distance divide is not merely replaced by the digital divide, and that the new technologies do not generate new forms of social exclusion. The magnitude of this challenge becomes clear when we realize that the 80% of the human population living in developing countries represents less than 10% of internet users (United Nations Development Programme, 1999). Canada, the United States, and Sweden rank among the most wired nations, with 40% of their population regularly connected to cyberspace

(Prescott-Allen, 2001). In contrast, most African and Southern Asian countries count less than 10 internet users per 10,000 population.

The new forms of social exclusion feed on the old scourges of poverty and inequality. The 1.3 billion people who survive on one dollar per day are a reminder to all of the enormous gaps that must still be overcome within and between countries. These gaps have major consequences for health as expressed by the existing and, in some regions, increasing inequalities in health conditions and access to health care.

Exclusion and inequality are one dark side of globalization. Insensitivity to local cultures is another. Together they may explain a painful paradox of our days: Precisely when technology has brought human beings closer to each other than ever before, we are witnessing the reappearance of intolerance in its ugly guises of xenophobia, ethnic cleansing, and oppression. And with intolerance, as a Siamese twin comes terrorism, traditionally the instrument of offended fanatic minorities that resist believing in persuasion. At its essence terrorism is the worst form of dehumanization, as it turns innocent people into mere targets. The arsenal of terrorism has expanded to include chemical and biological arms. According to intelligence agencies in recent years several militant groups across the world started developing or tried to purchase biological weapons for terrorist use.

There is a lot of discussion around the viability and possible magnitude of these kinds of attacks. What seems clear, though, in the face of the New York, Madrid, and London events and the rapidly growing power of biotechnology, is the urgent need to strengthen our surveillance capabilities through actions such as building international networks of public health laboratories, sharing information among national surveillance authorities, and developing training programs for specialized personnel.

The American and Mexican health authorities are already moving in this direction. We are convinced that, whether a bioterrorist attack materializes or not, these measures will in themselves improve the daily functioning of our public health systems for the general good (Henderson, 2001). In the process of strengthening those systems, the recent "Healthy Border Initiative" can play a key role to promote a healthy environment, anticipate risks, and foster timely intervention. This initiative was developed by the Mexico-U.S. Border Health Commission, a powerful instrument we now have at our disposal, which brings together the top health authorities of the four American and six Mexican border states under the joint chairmanship of the two federal secretaries of health.

In the long run, the challenge we have before us is to build a world order characterized by peace in the midst of diversity. Instead of asserting one's identity by rejecting or destroying what is different, we must try to soften collisions, balance claims, and reach compromises (Berlin, 1992).In this way, we may try living according to what Vaclav Havel, former President of the Czech Republic, has called a "basic code of mutual coexistence" (Havel, 1997). Not an easy task when sharing a diverse and extended neighborhood. Border regions tend to be places of intense contact, cultural clashes, and competing interests; disagreements are common and resentments, longstanding.

But we have no choice. The U.S.-Mexico border is already an environmental and epidemiological unit, and the demographic, cultural, and political influence of the Mexican population in the region is increasing at a very fast pace. There is no doubt: we have a common destiny, and we should find new ways of making our interdependence a force for peace and prosperity.

Health may contribute to this pursuit because it involves those domains that unite all human beings. It is there, in birth, in sickness, in recovery, and ultimately in death that we can all find our common humanity. In our turbulent world, health remains one of the few truly universal aspirations. We can make health a powerful force for diplomacy because it offers a concrete opportunity to reconcile national self-interest with international mutual interest. More today than ever, health is a bridge to peace, a common ground, a source of shared security.

But for this ideal to happen, we must renew international cooperation for health. We suggest three key elements for such renewal—the three e's of exchange, evidence, and empathy.

Health systems around the world are facing similar challenges; many of them, as we have just discussed, are related to globalization. Developed countries are witnessing problems of cost explosion, irrational use of technologies, and consumer satisfaction. Developing nations are dealing with problems of access to health care, quality of services, and lack of financial protection for millions. The communication revolution provides the opportunity to *exchange* information about the challenges facing national health systems and about the initiatives to deal with them.

To be informative, such exchange should be based on sound *evidence* about alternatives so that we may build a solid knowledge base of what really works, one that may be transferred across countries when its culturally, politically, and financially reasonable.

But there is another value. The British philosopher Isaiah Berlin has proposed the comparative study of other cultures as an antidote against intolerance, stereotypes, and the dangerous delusion by individuals, tribes, states, ideologies, or religions that each, in its view, is the sole possessor of truth (Berlin, 2001). And this leads us to the third element, *empathy*, that human characteristic which allows us to emotionally participate in a foreign reality, understand it, relate to it and, in the end, value the core elements that make us all members of the human race.

As we engage in the process of renewal, we would do well to remember the words of a great American, Dr. Martin Luther King, Jr., who wrote in 1968:

It really boils down to this: that all life is interrelated. We are all caught in an inescapable network of mutuality, tied into a single garment of destiny. Whatever affects one directly, affects all indirectly (King, p. 94).

Let us continue to weave together the destiny of better health for all the citizens of our common border and, indeed, for all the citizens of the world.

References

Aitken, L., Buchan, J., Sochalski, J., Nichols, B., and Powell, M. (2004). Trends in international nurse migration. *Health Affairs, 23*, 69–77.

AT Kearny, Inc., and Foreign Policy. (2001). Midiendo la globalización. *Este País, May*, 2–9.

Berlin, I. (2001). Nacionalismo: Notas para una conferencia futura. *Letras Libres, 3*, 105–106.

Berlin, I. (1992). The pursuit of the ideal. In I. Berlin (Ed.), *The crooked timber of humanity* (pp. 1–19). New York: Vintage Books.

Board on International Health, Institute of Medicine. (1997). America's vital interest in global health: Protecting our people, enhancing our economy, and advancing our international interests. Washington, DC: National Academies Press.

Brush, B., Sochalski, J., and Berger, A. (2004). Imported care: Recruiting foreign nurses to U.S. health care facilities. *Health Affairs, 23*, 78–87.

Frenk, J., Sepúlveda, J., Gómez-Dantés, O., McGuiness, M., and Knaul, F. (1997). The new world order and international health. *British Medical Journal, 314*, 1404–1407.

Havel, V. (1997). Harvard University June 1995. In V Havel, The art of the impossible (pp. 216–230). New York: Alfred A. Knopf.

Henderson, D.A. (2001, September 5). Testimony before the Foreign Relations Committee of the U.S. Senate.

Hobsbawm, E. (1994). The age of extremes: A history of the world, 1914–1991. New York: Pantheon Books.

Karlen, A. (1995). Man and microbes: Disease and plagues in history and modern times. New York: Simon & Schuster.

Kaul, I., Grumberg, Y., and Stein, M. A. (Eds.). (1999). Global public goods: International cooperation in the 21st century. New York: Oxford University Press for the United Nations Development Programme.

King Jr., M.L. (1968). *The trumpet of conscience.* Scottking, C. (Ed.) 1993. The Martin Luther King Jr. Companion. New York: St. Martin Press, Harper.

Osterholm, M. (2005). Preparing for the next pandemic. *Foreign Affairs, 84*, 24–37.

Porter, R. (Ed.). (1996). Illustrated history of medicine. Cambridge, UK: Cambridge University Press.

Porter, R. (2004). Blood and guts: A short history of medicine. New York and London: WW Norton & Company.

Prescott-Allen, R. (2001). The well being of nations: A country-by-country index of quality of life and the environment. Washington, DC: Island Press, International Development Research Center.

Sekri, N., Gómez-Dantés, O., and Macdonald, T. (1999). Cross-border health insurance: An overview. Oakland, CA: California HealthCare Foundation.

Sesma-Vázquez, S., Pérez-Rico, T., Lino Sosa-Manzano, C., and Gómez-Dantés, O. (2005). Gastos catastróficos por motivos de salud en México: Magnitud, distribución y determinantes. *Salud Pública de México, 47*, S37–S46.

U.S.-Mexico Binational Council. (2004). Managing Mexican migration to the United States: Recommendations for policymakers. Washington, DC: Center for Strategic and International Studies.

United Nations Development Programme. Human Development Report (1999). New York: Oxford University Press for the United Nations Development Programme.

Valaskakis, K. (2001). Westfalia II: Por un nuevo orden mundial. *Este País, 126*, 4–13.

Warner, D., and Reed, K. (1993). Healthcare across the border. Austin, Texas: LBJ School of Public Affairs, The University of Texas at Austin.

World Health Organization (WHO). (2000). World Health Report. Health systems: Improving performance. Geneva: WHO.

Index

Acculturation Rating Scale for Mexican
    Americans II, 52
activities of daily living (ADL)
    disabilities in, 27, 29
    of Mexican population, 104
    risk ratio limitations for, 45–46
African Americans, 1
Al-Snih, Soham, 40–47
American Diabetes Association (ADA), 211
Angel, Jacqueline L., 1–12, 263–276
Angel, Ronald J., 263–276
Annual Social and Economic Supplement
    (ASEC), 224
Arreola-Ornelas, Héctor, 237–260
Asian American men, 31
Assets and Health Dynamics Survey (AHEAD),
    88

Barker-type conjectures, 23
Bastida, Elena, 222
Binational Health Week, 284
Black Americans, 72–73, 85
    men, 31
Black Death of 1347, 281
Blue Cross/Blue Shield plans, 271
Border Epidemiologic Study on Aging (BESA),
    223–224
    health and healthcare utilization, 229
    health insurance coverage and doctor visits,
        230–231
    socio-demographic characteristics, 228
Brown, H. Shelton, 222–232
Brown, Sharon A., 211–218

Census's *five percent public-use micro data
    sample* (PUMS) file, 28
Center for Medicarr and Medicaid services
    (CMS), 208

Central American women, 33
cholesterol medication, 198
cognitive functioning, of elder U.S. Mexican
    Americans
    assessing instruments, 56–59
    and dementia, 59–61
    nativity and cultural orientation as predictors,
        58
    and socioeconomic status, 59, 61
cognitive impairment not meeting dementia
    (CIND) criteria, 51
colonias, 150–158
Consolidated Omnibus Budget Reconciliation
    Act (COBRA) of 1985, 271
Coustasse, Alberto, 165–178
Crimmins, Eileen M., 85–93
Current Population Survey (CPS), 224

de Snyder, V. Nelly Salgado, 121–131
diabetes, 2
Diabetes Control and Complications Trial
    (DCCT), 218
Diabetes Prevention Program (DPP), 218
Dominican women, 33

Elderly Mexican immigrants, 2
elder U.S. Mexican Americans
    disability and life expectancy of
        analysis, 43–44
        data, 42
        measures, 42–43
        result analysis, 44–47
    incidence of dementia and cognitive
        impairment in
        assessing instruments, 56–59
        discussion, 61–63
        measurements, 52
        methods, 51

elder U.S. Mexican Americans (*cont.*)
    mortality rates, 56
    result analysis, 53–56, 59–61
    statistical methods, 53
employment disability, 29
environmental health hazards and Hispanics, 2
epidemiologic paradox, for Hispanic health,
        67–70
Eschbach, Karl, 26–38, 40–47
Espino, David V., 195–200
Extra Territorial Jurisdiction (ETJ), 156

fasting blood glucose (FBG), 212
fatalism, 214–215
foreign-born Mexican Americans
    disability and life expectancy of
        analysis, 43–44
        data, 42
        measures, 42–43
        result analysis, 44–47
Framework Convention on Tobacco Control,
        282
Frenk, Julio, 280–287
functional impairments, 30

General Agreement on Trade and Services
        (GATS), 2
going outside the home disability, 29
Gómez-Dantés, Octavio, 280–287
Gonzalez, Hector M., 50–63
Goodwin, James S., 40–47

Ham-Chande, Roberto, 134–140
Hayward, Mark D., 85–93
Health and Retirement Survey (HRS), 8, 88
health care consumers, 282
health care utilization
    in U.S.-Mexico border region
        data and methodology, 225–227
        policy research, 224–225
        result analysis, 227–231
health insurance coverage
    of Latino elderly
        methods, 184–186
        results, 186–190
    in U.S.-Mexico border region
        data and methodology, 225–227
        policy research, 224–225
        result analysis, 227–231
Health Insurance Portability and Accountability
        Act (HIPPA) of 1996, 271
Higgins, Monica, 99–119
Hinton, Ladson, 50–63

Hispanic Established Populations for
        Epidemiologic Studies of the Elderly
        (H-EPESE), 10, 19, 21, 27, 182, 184, 265
    study of life expectancy rates, 42–47
Hispanic men, 33
Hispanic paradox, 17, 26, 86–87, 91
Hispanic population
    access to health services, 6–7
    and age factor, 19
    and environmental health hazards, 2
    aging and immigrant health, 4–6
    child health status, 23–24
    disabilities in activities of daily living (ADL),
        27
    elders, 1
        epidemiological paradox, 18–22
        high disability rates, 27–29
        immigrants, 2, 22–23, 86–87
        living in U.S., 22–24
    epidemiological profile of, 4
    foreign-born, 6, 19
    health care coverage among, 264–266
    and health care financing system, 266–275
    health disparities associated with, 2
    influence of western culture, 23
    instrumental activities of daily living (IADL),
        27
    mortality profiles in different countries, 26–27
    mortality risks
        background, 66–70
        data set, 70
        measurement, 70–71
        methodology, 71
        results, 71–80
    mothers, 19–20
    social stratification system of immigrants, 22
    women, 3
human capital, 22
Hummer, Robert A., 65–83

infant mortality, in San Antonio, 18
informal homestead subdivision (IFHS),
        150–158
Institute of Medicine (IOM) report, 222–223
Instituto Mexicano de Seguro Social (IMSS), 207
instrumental activities of daily living (IADL),
        27, 29
    of Mexican population, 104
intergenerational transfers, in developing
        countries, *see* Mexican Health and Aging
        Study
Interpolation of Markov Chains (IMACH)
        method, 43

Katz ADL index, 42
Knaul, Felicia Marie, 237–260

language acculturation, 43
Latin American countries, 1
Latino elderly in U. S., insurance coverage of
    methods, 184–186
    results, 186–190
Leave Healthy/Return Healthy program, 282
life expectancy, 3
    educational gradient by race/ethnicity in, 92
    of foreign-born Mexican Americans, 42–47
    by gender and immigration place of birth,
        46
    hazard factors in, 90
    in Mexico, 143
    of older U.S. Mexican Americans, 42–47
Lopez, Mary N. HaanVivian Colon, 50–63

maquila nursing homes, 143
Markides, Kyriakos S., 26–38, 40–47
McKinnon, Sarah A., 65–83
Medicare Modernization Act of 2003, 206
Medicare Prescription Drug, Improvement, and
    Modernization Act, 196
Medicare program, 6
Mehta, Kala, 50–63
Méndez-Carniado, Oscar, 237–260
mental disability, 29
Mexican American natives, health status of
    acculturation and dietary practices, 215–216
    data and measures, 88–89
    fatalism, 214–215
    methodology, 89–90
    multistate life style models, 91–93
    recruitment and retention in research studies,
        213–214
    result analysis, 90–91
    traditional gender roles, 215
    trends, 216–217
    and type 2 diabetes, 214, 216
    weight loss, 216
Mexican elder population, in U.S.
    barriers to acute and long-term health care for
        access barriers, 169–173
        methodology, 167–169
        primary barriers, 174–175
        secondary barriers, 175–176
        tertiary barriers, 176–177
    cross border health insurance for, 202–206
    elder care, 146–149
    financial access issues, 196–197
    financial protection in health, 239–240

catastrophic and impoverishing health
    expenditure, 249–254
    indicators, methodology and data,
        243–249
    logit estimations, 255–256
    tobit regression analysis, 257–258
growth of colonias and informal homestead
    subdivisions for, 150–158
health financing for, 241–243
irregular settlement and informal housing,
    143–146
issues with quality of care, 199–200
kinesthetics and communication styles,
    197–198
migratory experiences of
    materials and methods, 123–124
    perception of U.S., 126–127
    perceptions about Mexico, 127
    and poverty, 121–122, 124–125
    social support, 129–130
population living in poverty, 149–150
Mexican Health and Aging Study (MHAS),
    146
Mexican Health Foundation, 284
Mexican/Latin-born participants, 53–56
Mexican panel rider, 206
Mexican population, health and aging study of
    data, 101–102
    demographic statistics of, 104
    distribution of economic aids, 106–109
    dynamics of transfers and economic factors,
        115–116
    dynamics of transfers and gender, marital
        status and number of children, 114–115
    dynamics of transfers with age, 114
    economic transfers, 110–114
    health and illness, 128–129
    and health shocks, 116
    methodology, 102–103
    principal health conditions, 105–106
    result analysis, 103–116
Mini-Mental State Exam-Modified, 52, 58
mobile home communities, 156
Moore, Kari M., 50–63
mortality rates for adults
    for Hispanics and non-Hispanics, 73, 75, 77,
        79
    by race/ethnicity, 72
Multiple Cause of Death File (MCD), 66

National Center for Health Statistics, 70
National Death Index (NDI), 8, 21, 26, 43, 70
National Health Disparity Report, 173

National Health Interview Survey (NHIS), 8, 66,
    70
National Longitudinal Mortality Survey, 67
National Mortality Followback Survey, 66
Native American men, 31
non-Hispanic whites, 212
    men, 33, 35
    women, 31, 33
        foreign born, 35
North American Free Trade Agreement
    (NAFTA), 2
NUMIDENT file, 26

Oakes, Liliana, 195–200
old age disability
    age-standardized census, 31–32, 34–35
    for Asian Americans, 36–38
    ethnic patterns in, 28–29
    gender differences, 37
    methodology, 29–31
    result analysis, 31–36
    types of, 29
    of U.S. Mexican Americans, 42–47
Old Age Survivors, Disability, and Health
    Insurance program (OASDIHI), 2
Organization for Economic Cooperation &
    Development (OECD), 208

Pagán, José A., 222–232
Palloni, Alberto, 17–24
Peek, M. Kristen, 26–38
Performance-Oriented Mobility Assessment, 43
Personal Responsibility andWork Opportunity
    Reconciliation Act (PRWORA), 3
physical disability, 29
Puerto Ricans, 27, 148
    males and females, 32–33, 69, 72–75, 79–82

Ray, Laura A., 26–38
retirement colonias, 155
Rubio, Mercedes, 181–192

Sacramento Area Latino Study on Aging
    (SALSA), 8, 51–63
Saenz, Rogelio, 181–192
self-care disability, 29
sensory disability, 29
Small Area Health Insurance Estimates (SAHIE)
    program, 224
Social Security Administration's Master
    Beneficiary record, 26
Social Security program, 2
Social Security systems, 3

South Americans, 69
    men, 32–33
    women, 33
Spaniard men, 32–33
Spanish English Verbal Learning Test (SVELT),
    52, 59
Spanish women, 33
Starr County intervention studies, 213–217
State Children Health Insurance Program
    (SCHIP), 7, 264, 267
Sullivan Commission Report, 172
Supplemental Security Income, 2

Texas immigrant children, 7
Texas-Mexico border community, 211
therapeutic processes, 197–198
Tijuana population, 139–140
Torres, Ana Cristina, 237–260
Trade Act (TA) of 2002, 271
trailer parks, 156
Treviño, Fernando, 165–178
tuberculosis, 281

UK Prospective Diabetes Study (UKPDS), 218
U.S.-born Asian Americans, 37
U.S.-born Asian women, 33
U.S.-born Mexican Americans men, 36
U.S.-born Mexican women, 36
U.S.-born women, 33
U.S. Census 1990 Public Use Microdata
    Sample, 2
U.S. Death abroad, 21
U.S.-Mexico border region
    demographic characteristics of, 134, 137–138
    development of population sizes, 134–135
    globalization and health across, 280–287
    health insurance coverage and health care
        utilization
        data and methodology, 225–227
        policy research, 224–225
        result analysis, 227–231
    health outcomes, 138
    Mexicanization of, 136–137
    perspective from the population in Tijuana,
        139–140
U.S.–Mexico borderlands, 2

Ward, Peter M., 141–160
Warner, David C., 202–209
Warner, David F., 85–93
Whitfield, Keith E., 1–12
Wong, Rebeca, 99–119
World Health Organization (WHO), 121, 284

About the Authors

Jacqueline L. Angel received her Ph.D. from Rutgers University, and is currently a Professor of Public Affairs and Sociology and a Faculty Affiliate at the Population Research Center at The University of Texas at Austin. In 1990–92, she was a NIA Postdoctoral Fellow in the Demography of Aging Training Program at The Pennsylvania State University. She has contributed to the scholarly literature that addresses the relationships linking family structures, inequality, and health across the life course with a special emphasis on populations of Hispanic origin and specifically those persons of Mexican ancestry. Her research along these lines examines observable factors implicated in the health and economic security of Hispanic elders, life cycle demographic events, and the constraints imposed by the policy environment. Since 1992, she has been a co-investigator with colleagues from the UT medical schools in Galveston and San Antonio on the Hispanic Established Population for Epidemiologic Studies of the Elderly (H-EPESE), a benchmark study of elderly Mexican Americans' health in the Southwestern United States.

She has published 60 peer-review articles and chapters related to Hispanic health and social welfare policy as well as three books, *Health and Living Arrangements of the Elderly* (Garland Publishing, 1991), and jointly with Ronald J. Angel, *Painful Inheritance: Health and the New Generation of Fatherless Families* (University of Wisconsin Press, 1993), and *Who Will Care for Us? Aging and Long-Term Care in Multicultural America* (New York University Press, 1997). She is a Fellow of The Gerontological Society of America and Chair of the Behavior and Social Science of Aging and Review Committee, National Institute on Aging and the Section on Aging and the Life Course of the American Sociological Association.

Dr. Keith Eric Whitfield is a Research Professor in the Department of Psychology and Neuroscience at Duke University. He earned a Bachelors degree in Psychology from the College of Santa Fe, Santa Fe NM and a MA and Ph.D. in Life-Span Developmental Psychology from Texas Tech University in Lubbock Texas. He also received post doctoral training in quantitative genetics from the University of Colorado, Boulder, CO. His research on individual differences in minority aging employs a two prong approach that includes studying individual people as well as members of twin pairs. Dr. Whitfield's research examines the individual variation in health and individual differences in cognition due to health conditions. Dr. Whitfield has worked with researchers from Sweden, Russia, and the United States to examine how social, psychological, and cultural factors of cognition and healthy aging. He has completed a study that involves examining health and psycho-social factors related to health among adult African American twins. His current research project is a longitudinal study of cognition and health among older African Americans.

He is the member of several professional associations including the American Psychological Association, the Gerontological Society of America, the Society for Behavioral Medicine, and the Society for Multivariate Experimental Psychology. In the Gerontological Society of America, he serves as a chair of the Fellows Committee and is past chair of the Task Force on Minority Issues which published *Closing the Gap: Improving the health of Minority Elders in the New Millennium.* He was a member of the National Research Council/National Academy of Sciences "Aging Mind" committee and "Research Agenda for the Social Psychology of Aging" committee and the Institute of Medicine committee's recent report on "Assessing Interactions Among Social, Behavioral, and Genetic Factors on Health." He is a member of the National Advisory Board for the Center for Urban African American Aging Research at the University of Michigan, the Health and Adherence in Rural Practice (HARP) Data Safety Monitoring Board for the University of Alabama at Birmingham, the Advisory Board for Institute on Aging at Wayne State University, the Advisory Committee for The Export Center to Reduce Health Disparities in Rural South Carolina at Clemson He was also recently named as a member to the Board of Scientific Counselors for the National Institute on Aging.

Printed in the United States
110289LV00002B/177/A

9 780387 472065